Novel Approaches of Physical Therapy-Based Rehabilitation

Novel Approaches of Physical Therapy-Based Rehabilitation

Guest Editors

Andreia Noites
Andreia S. P. Sousa

MDPI
Basel • Beijing • Wuhan • Barcelona • Belgrade • Novi Sad • Cluj • Manchester

Guest Editors

Andreia Noites
Department of Physiotherapy
CIR, ESS, Polytechnic of Porto
Porto
Portugal

Andreia S. P. Sousa
Department of Physiotherapy
CIR, ESS, Polytechnic of Porto
Porto
Portugal

Editorial Office
MDPI AG
Grosspeteranlage 5
4052 Basel, Switzerland

This is a reprint of the Special Issue, published open access by the journal *Applied Sciences* (ISSN 2076-3417), freely accessible at: https://www.mdpi.com/journal/applsci/special_issues/V9Z2MRLF51.

For citation purposes, cite each article independently as indicated on the article page online and as indicated below:

Lastname, A.A.; Lastname, B.B. Article Title. *Journal Name* **Year**, *Volume Number*, Page Range.

ISBN 978-3-7258-7965-6 (Hbk)
ISBN 978-3-7258-7966-3 (PDF)
https://doi.org/10.3390/books978-3-7258-7966-3

Contents

About the Editors

Andreia Noites

Andreia Noites graduated in Physiotherapy from the School of Health, Polytechnic of Porto (ESS | P.Porto) in 2001. She holds a master's degree in Sports Sciences—Physical Activity for the Elderly and a Ph.D. in Health Sciences and Technologies. She also holds the professional title of Expert in Therapy and Rehabilitation.

Currently, she is a Full Professor of Physiotherapy at the School of Health, Polytechnic of Porto. Her main academic activities include teaching in both graduate and postgraduate programs, course coordination in the master's program in Dermatofunctional (Integumentary and Metabolic) Physiotherapy, and supervising research projects, master's and doctoral theses, and clinical internships. She is also involved in consultancy, community intervention, and health promotion initiatives.

Her research is conducted at the Center for Human Activity and Movement Studies (CEMAH) of the Polytechnic of Porto. Andreia Noites is the author of several scientific papers, book chapters, and books, and has presented her work at numerous national and international conferences. She has participated in national and international funded research projects as an investigator and serves as a Guest Editor for the *Applied Sciences* journal's Special Issue "Novel Approaches of Physical Therapy-Based Rehabilitation".

Andreia S. P. Sousa

Andreia S. P. Sousa has a degree in Physiotherapy from the School of Health of the Polytechnic Institute of Porto, Portugal (2006), and a master's (2009) and PhD (2013) in Biomedical Engineering, in the area of human movement and postural control, from the University of Porto. In 2024, she obtained the Aggregation in Physiotherapy from the Faculty of Sport of the University of Porto. She is an Adjunct Professor with Aggregation in the Scientific Area of Physiotherapy at the School of Health of the Polytechnic Institute of Porto, a member of the Board of Directors of the Center for Rehabilitation Research, an integrated researcher, and coordinator of the human movement research since the Center's creation in 2017. With 16 years of research experience dedicated to optimizing human movement in rehabilitation sciences, she has participated in over 13 funded R&D projects involving applied research, including 6 with competitive funds (TECH (P2020), SMARTHEALTH (P2020), BIP (Erasmus), SAFHE (P2020), FCT R&D Units, and FES-HAND (Fundácion General CSIC)), 2 funded by companies, and 5 funded through PhD fellowships. She is the co-author of 121 articles: 7 book chapters, 70 articles in indexed journals with an impact factor, 7 in other journals, 14 in international conferences, 9 in national conferences, and 14 conference abstracts. For her contributions to knowledge generation, she has received 14 scientific recognition awards. She has (co-)supervised 108 bachelor's theses, 43 master's theses, and 6 doctoral theses, and is currently (co-)supervising 3 doctoral projects. She was also involved in awarding 20 research grants and 3 research contracts for PhD researchers as well as supervising 10 research grants or internship programs. Her main research areas include biomechanics and motor control in rehabilitation sciences.

applied sciences

MDPI

Editorial

Novel Approaches of Physical Therapy-Based Rehabilitation

Andreia Noites *, Ana Figueiredo and Andreia S. P. Sousa

Center for Rehabilitation Research (CIR), ESS, Polytechnic of Porto, Rua Dr. António Bernardino de Almeida, 400, 4200-072 Porto, Portugal; amrf@ess.ipp.pt (A.F.); asp@ess.ipp.pt (A.S.P.S.)
* Correspondence: arn@ess.ipp.pt

1. Introduction

The field of physical therapy and rehabilitation sciences is currently undergoing a profound transformation. This Special Issue, titled "Novel Approaches in Physical Therapy and Rehabilitation", was conceived to bring together cutting-edge research and evidence-based practices that are redefining the discipline, promoting knowledge exchange across research, clinical, and academic communities.

The 13 published papers in this edition offer diverse perspectives on how technology and novel clinical strategies can improve functional outcomes across a wide range of populations.

2. Overview of Published Articles

The contributions to this Special Issue provide valuable insights into several key pillars of modern rehabilitation. These can be broadly categorized into three main areas: telerehabilitation, advanced technologies in rehabilitation, and holistic clinical interventions.

2.1. Telerehabilitation

Traditional models of rehabilitation usually require regular face-to-face interactions between patients and healthcare professionals for assessment, monitoring, and intervention. Although this approach offers clear benefits, the conventional rehabilitation method is not accessible or feasible for everyone. Worldwide, millions of people encounter either geographical, socioeconomic, or physical barriers that limit their access to adequate rehabilitation [1,2].

In this context, digital health has emerged as a promising alternative to expand healthcare access. Commonly defined as the use of information and communication technologies in healthcare, digital health can improve healthcare delivery and service coverage, and reinforce health promotion [3]. The WHO Global Health Strategy for Digital Health (2020–2025) recognizes the pivotal role of digital health in strengthening healthcare systems, addressing global health issues, and improving healthcare outcomes.

Telerehabilitation is one of the digital health strategies that deliver remote rehabilitation services through digital tools (smartphones, wearable devices, or virtual reality systems), allowing clinicians to conduct virtual consultations, monitor patient progress, and guide rehabilitation at a distance [4]. Despite the challenges associated with implementing telerehabilitation [5,6], it can represent an effective and even preferable treatment option for certain populations and particularly challenging contexts, such as during COVID-19 [7]. Two of the papers included in this Special Issue reinforce this approach by demonstrating the positive effects of telerehabilitation programs based on therapeutic exercise among home-office workers (Contribution 2) and of gamification for children with neurodevelopmental disabilities in war-affected Ukraine (Contribution 1). The positive effects were

 https://doi.org/10.3390/app16062954

expressed in improvements in both musculoskeletal and mental health domains in home-office workers and in mood, balance, and functional mobility in children with neurodevelopmental disabilities in war-affected Ukraine. Another paper included in this special study demonstrated the feasibility of gamification for postoperative hand mobilization (Contribution 3). In fact, similar benefits have been observed when comparing gamified interventions to conventional exercise programs across different post-surgical conditions, such as anterior cruciate ligament reconstruction (ACL) [8,9] and shoulder arthroscopy [10], suggesting the potential of this tool to extend rehabilitation beyond healthcare settings after surgery. By incorporating elements of reward and challenge, gamification can stimulate dopaminergic pathways in the brain, improving synaptic connections and facilitating learning through long-term potentiation [11]. This shows that, beyond demonstrating feasibility, therapy gamification may also increase patient motivation and engagement.

2.2. Advanced Technologies in Rehabilitation

Precision in diagnosis and therapy is increasingly supported by high-tech solutions, and Artificial Intelligence (AI) is gradually transforming the field of rehabilitation by enabling more data-driven, personalized, and effective approaches to patient care. Several key applications have already been identified, including the development of personalized treatment plans, the use of predictive analytics to support clinical decision-making, the implementation of remote monitoring systems, and advances in rehabilitation robotics [12].

Two of the studies included in this Special Issue demonstrate the advantages of these technologies in neurological rehabilitation. Rojek et al. (Contribution 4) demonstrated the suitability and superiority of Convolutional Neural Networks (CNNs) for capturing complex gait patterns and their potential for supporting clinical decision-making and planning personalized stroke rehabilitation. In a complementary area, Casanova et al. (Contribution 5) presented a gamified robotic framework for upper-limb rehabilitation that integrates kinematic and electromyography (EMG) metrics of different upper-limb muscles to monitor real-time performance and progress, with potential for exploration in patients with acquired brain injury (ABI). Both studies highlight the potential and shortcomings of these advanced technologies in rehabilitation.

Beyond neurological rehabilitation, technological advances are also expanding our understanding of neuromuscular control in musculoskeletal conditions. Vélez-Gutiérrez et al. (Contribution 6) conducted a scoping review examining functional and neuroplastic adaptations during ACL rehabilitation. Their findings suggest that unilateral training of the uninjured limb induces bilateral neuroplastic adaptations, representing a cost-effective strategy during early rehabilitation stages. Neuroplasticity refers to the brain's capacity to reorganize its structure and function in response to experiences, learning processes, or recovery from injury or disease [13]. Although it was once assumed that significant neuroplasticity occurred primarily during early development, research now shows that the brain retains this adaptive capacity throughout the lifespan [13]. These adaptive changes are fundamental for skill acquisition, rehabilitation, and healthy aging [14]. For these reasons, a deeper understanding of neuroplastic mechanisms is critical for designing targeted interventions that promote beneficial neuroplasticity while minimizing potentially harmful neural reorganization [15]. Supporting these findings, Mateus et al. (Contribution 7) used advanced three-dimensional motion capture systems to identify persistent interlimb asymmetries in kinetic and muscle-level variables up to 9 months after ACL reconstruction during drop-jump tasks, revealing long-lasting neuromuscular adaptations. In the context of injury prevention, Tedeschi et al. (Contribution 8) emphasized the role of high-speed sprint training in enhancing hamstring tissue resilience. They advocated integrating it with eccentric strengthening protocols to reduce injury risk.

https://doi.org/10.3390/app16062954

Overall, the integration of AI, robotics, and advanced biomechanical analysis has the potential to improve the delivery of care by offering flexible, tailored, and highly effective treatment interventions [4,16]. These technologies enable clinicians not only to monitor performance with greater precision but also to better understand the neurophysiological mechanisms underlying recovery. By combining data-driven approaches with insights into neuroplasticity and neuromuscular control, rehabilitation can evolve toward more adaptive, individualized, and mechanism-based interventions that improve patient engagement, satisfaction, and long-term clinical outcomes [12].

2.3. Clinical Interventions and Holistic Care

Holistic health incorporates multiple dimensions of well-being, including physical, mental, social, emotional, and spiritual health [17]. Within this framework, rehabilitation should be understood not only as a process of restoring physical function but also as an opportunity to promote broader well-being. Physiotherapy, in particular, can positively influence physical, mental, and social health outcomes among individuals with mental health conditions [18].

Despite the evidence, rehabilitation programs have traditionally focused primarily on physical recovery, often overlooking the psychological and social dimensions of health, posing a significant risk for long-term mental health issues and overall rehabilitation outcomes [19,20]. While restoring physical function remains crucial, rehabilitation must consider well-being as a central component, addressing emotional and social aspects as well.

A meta-analysis of rehabilitation programs found that programs incorporating psychological and social support strategies reported significantly better long-term mental health outcomes (GBD 2021 Diseases and Injuries Collaborators [21]). These findings reinforce the need for rehabilitation models to shift from a physical restoration to a more comprehensive, integrated model in which well-being is viewed as an integral part of recovery [22].

Several studies focused on specific populations and the psychosocial dimensions of health. Morais et al. (Contribution 9) highlighted the effectiveness of a structured group exercise and relaxation plan, which fosters body awareness and exercise physiology, and its role in promoting well-being, directly encouraging self-care and thus improving the health literacy intervention program in improving the health literacy of informal caregivers of children with special needs. Carcelén-Fraile et al. (Contribution 12) performed a meta-analysis confirming the positive impact of aerobic training on the mental health of children and adolescents, particularly in reducing stress and enhancing self-esteem. Finally, Golec et al. (Contribution 13) investigated diaphragmatic therapy, revealing its potential to reduce resting pelvic floor muscle activity and alleviate emotional stress in women. In oncology, de Hoop et al. (Contribution 10) explored physiotherapists' perspectives on exercise immunology, emphasizing the need for tailored training to support the immune function of cancer patients. Oncology physiotherapists recognized the potential value of incorporating knowledge of exercise immunology to better tailor training and inform patients about the health benefits of exercise. They were eager to learn more about the topic. For post-surgical recovery, Guerero et al. (Contribution 11) evaluated adhesive elastic taping after saphenectomy, noting significant benefits in ecchymosis control, skin elasticity, and functional recovery against conventional post-surgical treatment.

3. Conclusions

The research presented in this Special Issue confirms that the future of physical therapy lies in the synergy between technological innovation and personalized clinical reasoning. From the use of AI for diagnostics to health literacy to support the vital role of informal caregivers and patients, these findings underscore that modern rehabilitation must be

https://doi.org/10.3390/app16062954

both interdisciplinary and adaptable. We hope this collection serves as a benchmark for clinicians and researchers, inspiring further innovations that enhance the quality of life and functional independence of diverse patient populations.

Author Contributions: Conceptualization, A.N., A.S.P.S. and A.F.; investigation, A.N., A.S.P.S., A.F.; writing—original draft preparation, A.N., A.S.P.S. and A.F.; writing—review and editing, A.N., A.S.P.S. and A.F. All authors have read and agreed to the published version of the manuscript.

Funding: This research received no external funding.

Acknowledgments: The Guest Editors would like to thank all the authors for their innovative contributions and the reviewers for their rigorous work and constructive feedback, which ensured the high scientific quality of this first edition.

Conflicts of Interest: The authors declare no conflicts of interest.

List of Contributions:

1. Kushnir, A.; Kachmar, O.; Bonnechère, B. Feasibility and Perceptions of Telerehabilitation Using Serious Games for Children with Disabilities in War-Affected Ukraine. *Appl. Sci.* **2025**, *15*, 8526.
2. Szilágyi, T.; Veres-Balajti, I.; Lukács, B.; Király, E.; Laczkó, A.M.; Jenei, Z. Pilot Study of an Online Exercise Therapy Programme for Home Office Workers in Terms of Musculoskeletal and Mental Health. *Appl. Sci.* **2025**, *15*, 10963.
3. Abdallah, H.B.; Tennler, J.; Seelmann, C.; Haverkamp, V.; Schmitz, C.; Homann, H.H.; Ohmann, T. Does a Novel Gamification Approach Improve Hand Function in Patients Undergoing Rehabilitation? A Mono-Centric Randomized Controlled Trial. *Appl. Sci.* **2026**, *16*, 987.
4. Rojek, I.; Mikołajewska, E.; Małolepsza, O.; Kozielski, M.; Mikołajewski, D. Comparative Analysis of Different AI Approaches to Stroke Patients' Gait Analysis. *Appl. Sci.* **2025**, *15*, 10896.
5. Casanova, A.; Sempere, N.; Romero, C.; Porcel, K.; Ubeda, A.; Jara, C.A. A Robotic Gamified Framework for Upper-Limb Rehabilitation. *Appl. Sci.* **2025**, *15*, 11007.
6. Vélez-Gutiérrez, J.M.; Rojas-Jaramillo, A.; Ascuntar-Viteri, J.D.; Quintero, J.D.; García-Muro San José, F.; Bazuelo-Ruiz, B.; Cannataro, R.; Bonilla, D.A. Functional and Neuroplastic Effects of Cross-Education in Anterior Cruciate Ligament Rehabilitation: A Scoping Review with Bibliometric Analysis. *Appl. Sci.* **2025**, *15*, 8641.
7. Mateus, R.B.; Cabral, S.; Richter, C.; Veloso, A.P. Between-Limb Asymmetry Assessment During a Double-Leg Drop Jump Task After Anterior Cruciate Ligament Reconstruction—A Musculoskeletal Modelling Approach. *Appl. Sci.* **2025**, *15*, 12347.
8. Tedeschi, R.; Giorgi, F.; Donati, D. Sprint Training for Hamstring Injury Prevention: A Scoping Review. *Appl. Sci.* **2025**, *15*, 9003.
9. Morais, S.; Pimenta, R.E.; Morais, C.; Macedo, R.; Ribeiro, I.; Ferreira, P.L. Effectiveness of an Intervention Program for Informal Carers of Children Admitted to a Rehabilitation Centre. *Appl. Sci.* **2025**, *15*, 8544.
10. De Hoop, A.M.S.; Jäger, K.; Dronkers, J.J.; Veenhof, C.; Ruurda, J.P.; Krul, C.A.M.; Pieters, R.H.H.; Valkenet, K. Perspectives of Physiotherapists on Immune Functioning in Oncological Rehabilitation in the Netherlands: Insights from a Qualitative Study. *Appl. Sci.* **2025**, *15*, 8673.
11. Guerero, R.M.; Zordão, C.C.; Zampieri, E.H.S.; Noites, A.; Guirro, E.C.d.O. Effect of Taping on Postoperative Recovery Following Saphenectomy. *Appl. Sci.* **2025**, *15*, 9227.
12. Carcelén-Fraile, M.d.C.; Aibar-Almazán, A.; Hita-Contreras, F. Aerobic Training on Mental Health in Children and Adolescents: A Systematic Review with Meta-Analysis. *Appl. Sci.* **2025**, *15*, 9572.
13. Golec, J.; Gamrot, S.; Michalik, M.; Sulowska-Daszyk, I.; Nowak, M.; Balicka-Bom, J. Effects of Diaphragmatic Therapy on Pelvic Floor Muscle Activity, Stress Levels, and Sexual Life Satisfaction in Polish Women. *Appl. Sci.* **2025**, *15*, 12055.

References

1. Andrade, R.M.; Santana, B.G.; Schmidt, A.V.; Barsotti, C.E.; Baroni, M.P.; Saragiotto, B.T.; Ribeiro, A.P. Effect of traditional rehabilitation programme versus telerehabilitation in adolescents with idiopathic scoliosis during the COVID-19 pandemic: A cohort study. *J. Rehabil. Med.* **2024**, *56*, jrm5343. [CrossRef]

https://doi.org/10.3390/app16062954

2. McMaughan, D.J.; Oloruntoba, O.; Smith, M.L. Socioeconomic Status and Access to Healthcare: Interrelated Drivers for Healthy Aging. *Front. Public Health* **2020**, *8*, 231. [CrossRef]
3. Ahmed, M.M.; Okesanya, O.J.; Olaleke, N.O.; Adigun, O.A.; Adebayo, U.O.; Oso, T.A.; Eshun, G.; Lucero-Prisno, D.E., 3rd. Integrating Digital Health Innovations to Achieve Universal Health Coverage: Promoting Health Outcomes and Quality Through Global Public Health Equity. *Healthcare* **2025**, *13*, 1060. [CrossRef] [PubMed]
4. Calabrò, R.S.; Mojdehdehbaher, S. AI-Driven Telerehabilitation: Benefits and Challenges of a Transformative Healthcare Approach. *AI* **2025**, *6*, 62. [CrossRef]
5. Leochico, C.F.D.; Espiritu, A.I.; Ignacio, S.D.; Mojica, J.A.P. Challenges to the Emergence of Telerehabilitation in a Developing Country: A Systematic Review. *Front. Neurol.* **2020**, *11*, 1007. [CrossRef] [PubMed]
6. Rabanifar, N.; Abdi, K. Barriers and Challenges of Implementing Telerehabilitation: A Systematic Review. *Iran. Rehabil. J.* **2021**, *19*, 121–128. [CrossRef]
7. Matsumoto, M.E.; Wilske, G.C.; Tapia, R. Innovative approaches to delivering telehealth. *Phys. Med. Rehabil. Clin. N. Am.* **2021**, *32*, 451–465. [CrossRef]
8. Clausen, J.D.; Nahen, N.; Horstmann, H.; Lasch, F.; Krutsch, W.; Krettek, C.; Weber-Spickschen, T.S. Improving maximal strength in the initial postoperative phase after anterior cruciate ligament reconstruction surgery: Randomized controlled trial of an app-based serious gaming approach. *JMIR Serious Games* **2020**, *8*, e14282. [CrossRef]
9. McClincy, M.; Seabol, L.G.; Riffitts, M.; Ruh, E.; Novak, N.E.; Wasilko, R.; Hamm, M.E.; Bell, K.M. Perspectives on the gamification of an inter-active health technology for postoperative rehabilitation of pediatric anterior cruciate ligament reconstruction: User-centered design approach. *JMIR Serious Games* **2021**, *9*, e27195. [CrossRef]
10. Marley, W.D.; Barratt, A.; Pigott, T.; Granat, M.; Wilson, J.D.; Roy, B. A multicenter randomized controlled trial comparing gamification with remote monitoring against standard rehabilitation for patients after arthroscopic shoulder surgery. *J. Shoulder Elb. Surg.* **2022**, *31*, 8–16. [CrossRef]
11. Janssen, J.; Verschuren, O.; Renger, W.J.; Ermers, J.; Ketelaar, M.; van Ee, R. Gamification in Physical Therapy: More Than Using Games. *Pediatr. Phys. Ther.* **2017**, *29*, 95–99. [CrossRef] [PubMed]
12. World Health Organization. Global Strategy on Digital Health 2020–2025. Licence: CC BY-NC-SA 3.0 IGO. 2021. Available online: https://www.who.int/docs/default-source/documents/gs4dhdaa2a9f352b0445bafbc79ca799dce4d.pdf (accessed on 2 March 2026).
13. Gazerani, P. The neuroplastic brain: Current breakthroughs and emerging frontiers. *Brain Res.* **2025**, *1858*, 149643. [CrossRef] [PubMed]
14. Marzola, P.; Melzer, T.; Pavesi, E.; Gil-Mohapel, J.; Brocardo, P.S. Exploring the role of neuroplasticity in development, aging, and neurodegeneration. *Brain Sci.* **2023**, *13*, 1610. [CrossRef] [PubMed]
15. Hu, X.; Xu, W.; Ren, Y.; Wang, Z.; He, X.; Huang, R.; Ma, B.; Zhao, J.; Zhu, R.; Cheng, L. Spinal cord injury: Molecular mechanisms and therapeutic interventions. *Signal Transduct. Target. Ther.* **2023**, *8*, 245. [CrossRef]
16. Rahman, S.; Sarker, S.; Haque, A.K.M.N.; Uttsha, M.M.; Islam, M.F.; Deb, S. AI-driven stroke rehabilitation systems and assessment: A systematic review. *IEEE Trans. Neural Syst. Rehabil. Eng.* **2023**, *31*, 192–207. [CrossRef]
17. Mikkelsen, K.; Stojanovska, L.; Polenakovic, M.; Bosevski, M.; Apostolopoulos, V. Exercise and mental health. *Maturitas* **2017**, *106*, 48–56. [CrossRef]
18. Vancampfort, D.; Probst, M.; Helvik Skjaerven, L.; Catalán-Matamoros, D.; Lundvik-Gyllensten, A.; Gómez-Conesa, A.; Ijntema, R.; De Hert, M. Systematic review of the benefits of physical therapy within a multidisciplinary care approach for people with schizophrenia. *Phys. Ther.* **2012**, *92*, 11–23. [CrossRef]
19. Kirkbride, J.B.; Anglin, D.M.; Colman, I.; Dykxhoorn, J.; Jones, P.B.; Patalay, P.; Pitman, A.; Soneson, E.; Steare, T.; Wright, T.; et al. The social determinants of mental health and disorder: Evidence, prevention and recommendations. *World Psychiatry* **2024**, *23*, 58–90. [CrossRef]
20. Richardson, C.R.; Franklin, B.; Moy, M.L.; Jackson, E.A. Advances in rehabilitation for chronic diseases: Improving health outcomes and function. *BMJ* **2019**, *365*, l2191. [CrossRef]
21. GBD 2021 Diseases and Injuries Collaborators. Global incidence, prevalence, years lived with disability (YLDs), disability-adjusted life-years (DALYs), and healthy life expectancy (HALE) for 371 diseases and injuries in 204 countries and territories and 811 subnational locations, 1990–2021: A systematic analysis for the Global Burden of Disease Study 2021. *Lancet* **2024**, *403*, 2133–2161. [CrossRef]
22. Marques, L.M.; Battistella, L.R. Well-being as a core element of rehabilitation: Preventing mental health challenges in a global context. *Glob. Health J.* **2025**, *9*, 1–5. [CrossRef]

 https://doi.org/10.3390/app16062954

applied sciences

MDPI

Article

Does a Novel Gamification Approach Improve Hand Function in Patients Undergoing Rehabilitation? A Mono-Centric Randomized Controlled Trial

Heithem Ben Abdallah [1,*], Janina Tennler [2], Christine Seelmann [1], Viola Haverkamp [3], Christian Schmitz [4], Heinz-Herbert Homann [3] and Tobias Ohmann [1]

1 Centre for Clinical Research, BG Klinikum Duisburg, 47249 Duisburg, Germany; tobias.ohmann@bg-klinikum-duisburg.de (T.O.)
2 Department for Arthroscopic Surgery, Sports Traumatology & Sports Medicine, BG Klinikum Duisburg, 47249 Duisburg, Germany
3 Department of Hand and Plastic Surgery & Severe Burn Injuries, BG Klinikum Duisburg, 47249 Duisburg, Germany
4 Department of Rehabilitation, Conservative and Technical Orthopedics, BG Klinikum Duisburg, 47249 Duisburg, Germany
* Correspondence: heithem.ben.abdallah@bg-klinikum-duisburg.de

Abstract

Background: Gamification is a promising approach to support rehabilitation, but implementing new devices after hand surgery (HS) faces challenges, especially in patient motivation and compliance. Technology-based, personalized solutions may encourage patients to perform gamified exercises consistently. This study investigated the impact of a tablet-based gamified hand-mobilization system on functional outcomes after HS compared to standard finger-expander training. **Methods**: Forty inpatients at BG Klinikum Duisburg were randomly assigned to a gamified or control group. Both groups underwent four weeks of stationary rehabilitation. Outcomes were assessed using the Disabilities of Arm, Shoulder, and Hand questionnaire (DASH), SF-36 health-related quality of life, and a visual analog scale (VAS) at pre- (T0) and post-training (T1). Tablet use frequency and duration were recorded. **Results**: DASH scores improved significantly over time ($F(1.55, 58.85) = 16.36$, $p < 0.001$, partial $\eta^2 = 0.137$) with no Group or Time $\times$ Group effects ($p > 0.40$), exceeding the MCID in both groups. SF-36 MCS, PCS, and VAS pain also improved over time (all $p < 0.05$) with no between-group differences ($p > 0.40$). Exercise duration differed ($p = 0.007$), but training frequency did not. **Conclusions**: Both gamified and conventional rehabilitation programs led to clinically meaningful improvements in hand function after hand surgery. No significant differences were observed between groups, indicating that the tablet-based system was feasible and well-accepted but not superior to standard training.

Keywords: rehabilitation; gamification; DASH; SF36; tablet

1. Introduction

Hand function is essential for independence in activities of daily living, work participation, and quality of life. In a setting such as a German statutory accident insurance hospital (BG Klinikum), the overarching goal after severe occupational injuries is not only symptom reduction but the most complete functional recovery possible to enable timely and safe return to work. Traumatic hand and wrist injuries frequently require surgical treatment, including fracture fixation and soft-tissue procedures for tendon and nerve lesions [1,2]. Despite adequate operative care, postoperative impairment can be substantial—particularly after severe injuries—making structured rehabilitation a key determinant of functional recovery. Structured, goal-oriented hand rehabilitation is critical not only for restoring functional capacity but also for achieving independence in daily activities and facilitating a sustainable return to work. It directly targets strength, dexterity, coordination, and sensorimotor control, bridging the gap between surgical repair and real-world functional performance. In clinical practice, rehabilitation typically combines therapist-guided manual techniques and progressive active training, including mobilization, proprioceptive exercises, strengthening, nerve gliding, and task-oriented functional practice, often supported by simple exercise devices (e.g., finger-expanders) [3,4].

However, repetitive exercises may be perceived as monotonous or uncomfortable, which can reduce engagement and adherence factors closely linked to functional outcome, especially when a high practice dose over several weeks is required [5,6]. In working-age patients, an insufficient training dose and delayed functional gains may ultimately prolong work disability and delay return to work. Recent systematic reviews and randomized trials indicate that gamified or VR-based rehabilitation, when combined with standard therapy, can enhance upper-limb motor recovery, improve hand dexterity and strength, and increase patient engagement compared with conventional methods [5–7]. In particular, recent work on bionic-constrained, piano-based interactive training for anthropomorphic hands highlights the potential of advanced technology-assisted, goal-oriented exercises to further improve motor learning and hand function [7].

Patients were recruited irrespective of the specific hand surgery performed (e.g., carpal tunnel release, ganglion excision, tendon repair, fracture fixation). Study eligibility was determined by postoperative functional limitation of the hand, not the type of surgical procedure. Only patients with thumb/finger amputations or major neurological/psychiatric conditions were excluded. The tablet-based program is intended as an adjunct to standard postoperative care, not a replacement for surgical procedures. Home-based, patient-oriented rehabilitation has been shown to accelerate functional recovery and reduce pain duration [8,9].

Importantly, the time point at which patients can begin postoperative rehabilitation varies considerably. Depending on the severity and type of surgery, as well as early postoperative complications, some patients may initiate active mobilization as early as one week after surgery, whereas others require several weeks before safe mobilization is possible. This variability highlights the need for flexible, patient-adapted rehabilitation approaches. The integration of gamified systems in hand rehabilitation may offer several advantages. By transforming conventional exercises into interactive and enjoyable tasks, patients may be more inclined to adhere to their therapy regimens. Furthermore, digital platforms can provide real-time feedback, monitor progress, and allow for the personalization of exercises, potentially improving clinical outcomes [10]. Within this context, any digital intervention should be conceptualized as a supplement ("plus one") rather than a replacement of standard care.

Tablet applications as part of gamification have been proposed as a technique to stimulate cortical reorganization through goal-oriented feedback-guided tasks [11]. Wrist, hand, and finger injuries can be treated through smartphone touchscreens, as proposed by Algar et al. [12]. This affects the proprioceptive and neuromuscular control levels. Tablet-based exercises were applied to patients with wrist, hand, and finger injuries, including post-surgical soft-tissue and fracture cases. Although the exercises were performed on a 2D tablet surface, they are designed to support the early-phase recovery of foundational motor functions such as range of motion, coordination, precision, and movement initiation,

 https://doi.org/10.3390/app16020987

which are essential prerequisites for later 3D, task-specific activities in daily life. Therefore, tablet-based gamification serves as a complementary tool that enhances early rehabilitation rather than replacing conventional 3D functional training [13].

Tablet-based hand rehabilitation with gamification elements might have the potential for a shorter recovery period than traditional therapy methods, with reduced healthcare dependence (requiring a smaller number of required healthcare appointments beyond a home activity program prescribed on paper). This could enhance clinical recovery and improve the short-term recovery of hand function.

Therefore, the research question for this randomized trial was how to evaluate the clinical effectiveness of a tablet-based gamified application as an add-on to standard inpatient multimodal rehabilitation compared with standard rehabilitation plus a conventional exercise device (finger-expander). We hypothesized that the gamified approach would enhance patient adherence and yield comparable or superior improvements in hand function, pain, and health-related quality of life, thereby supporting a rehabilitation process aligned with vocational reintegration goals.

2. Materials and Methods

A randomized controlled trial enrolled patients with wrist, hand and/or finger injury undergoing their four-week rehabilitation in the BG Klinikum Duisburg, which is a trauma center belonging to the BG Kliniken Hospital Group ("BG = Berufsgenossenschaftliche") of the German Federal Statutory Accident Insurance ("Deutsche Gesetzliche Unfallversicherung") in Germany. Randomization was performed using simple allocation by chance. An independent administrative staff member prepared identical, opaque, folded papers labeled either 'Tablet group' or 'Finger-expander group' and placed them in a sealed box. Participants in the tablet group received a tablet-based gamified application (Routine Health GmbH, Düsseldorf, Germany), while participants in the finger-expander group used a conventional exercise device (manufacturer information not available). Each participant drew one paper at the start of rehabilitation, ensuring allocation concealment. The staff member preparing the papers was not involved in patient recruitment, while the study coordinator supervised the draw but remained blinded to the upcoming assignments. Participants were randomly allocated to one of two groups (tablet group vs. finger-expander group) in a 1:1 allocation. Patients provided written informed consent for the study. The study was approved by the local Ethics Committee (Ethik-Antrag-Nr: 2019119) and conducted in accordance with the Declaration of Helsinki.

A total of 127 patients were assessed for eligibility: those with thumb or finger amputation, arthrodesis of finger joints, severe neurological comorbidities, or visual impairments interfering with tablet use were excluded. Forty participants met the inclusion criteria and were randomized, with twenty patients in each group (Figure 1). Outcome parameters were assessed three times, at the beginning of the rehabilitation (Pre), after the four-week rehabilitation (Post), and twenty-four weeks after the beginning (follow-up, FU, via telephone to collect additional outcome information).

2.1. Inclusion and Exclusion Criteria

Participants were eligible if they were at least 18 years old and possessed sufficient cognitive and communicative abilities to understand instructions and complete study questionnaires, as assessed by the recruiting clinician. Exclusion criteria included thumb or finger amputations, severe neurological comorbidities, manifest psychiatric disorders (e.g., post-traumatic stress disorder, anxiety disorders), and visual impairments that could interfere with tablet use, based on clinical judgment.

 https://doi.org/10.3390/app16020987

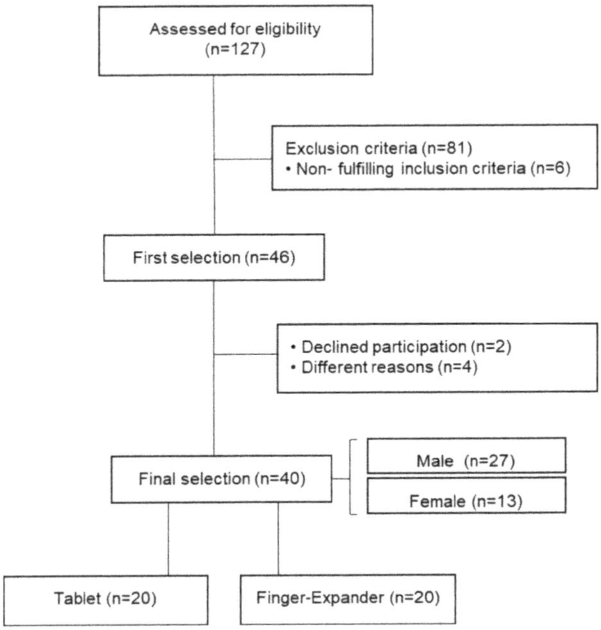

Figure 1. Flow chart of the study.

Tablet Group: Participants in the intervention group used a commercially available iPad application developed by Routine Health GmbH (Düsseldorf, Germany). The app includes three game-based exercises:

1. Colored Circle: Players rotate their fingers to align a spaceship with target holes to maintain balance in space. This game promotes rotational finger movements similar to actions like bottle-opening, and enhances eye–hand coordination.
2. Gravity Crush: Players use two fingers to collect candy and prevent theft by an antagonist. This game develops fine motor skills such as pincer grip and fast, reactive finger movements.
3. Karate Chop: Players swipe green balls away by tapping with two fingers and flicking one finger. This game improves finger mobility and promotes finger extension.

Performance data, including Critical Repetition Frequency (CRF) and training duration (in minutes), were automatically recorded by the app.

The finger-expander group received an exercise device for the hand (Figure 2). The device strengthens both the flexor and extensor muscles of the hand and the forearm. The finger-expander is designed for use across various age groups. Participants could carry and use the device at any time. Exercise frequency and duration were recorded manually.

Our outcome measures are summarized in Table 1 and included the following: (1) The Disability of Arm, Shoulder, and Hand (DASH) questionnaire, which was designed to measure total upper-extremity functional disability [14]. The total score ranges from 0 to 100, with 0 being equivalent to no disability and 100 equivalent to maximum disability. (2) Additionally, the visual analog scale (VAS) was recorded to measure pain. (3) Determination of health-related quality of life was investigated using the German version of the Short-Form 36 (SF-36) [15]. The SF-36 has been frequently used in hand rehabilitation studies and was recommended in the literature [16–18]. The result of the SF-36 is reported in two subscores: the Physical Composite Score (PCS) and the Mental Composite Score (MCS) [19]. Each subscore ranges from 0 to 100. A lower score indicates a higher limitation

 https://doi.org/10.3390/app16020987

of health. In our study, the SF-36 was collected at baseline and at the end of the 4-week rehabilitation period; however, no SF-36 follow-up data were obtained because a follow-up assessment for this questionnaire was not included in the original study protocol. (4) Clinical variables of the fingers and hand were assessed. Key handle was measured using a pinch gauge placed between the thumb pad and the lateral side of the index finger. A key handle is an ergonomic aid designed to attach to a standard key, making key-turning tasks easier for individuals with arthritis, neuromuscular conditions, limited grip strength, or reduced thumb and finger function [20].

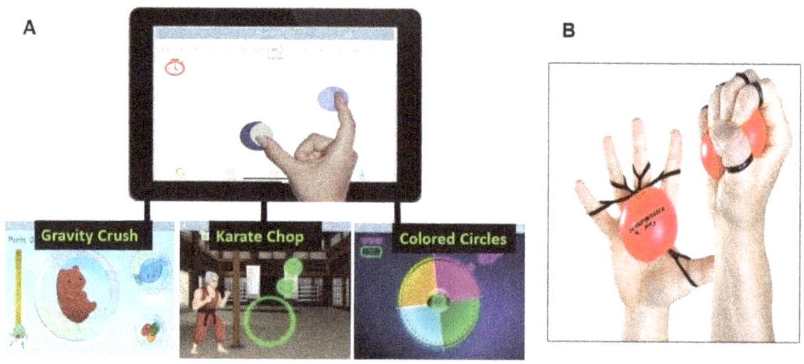

Figure 2. (**A**) Game-like tasks of the tablet app. (**B**) Example of performance of an exercise in the Control group using the finger-expander.

Table 1. Exercise group characteristics: type, frequency, duration, supervision, and recording method.

Group	Exercise Type	Frequency	Duration	Supervision	Recording Method
Tablet	3 game-based exercises: mobility, coordination, grip	Daily	Self-paced	Intro by physiotherapist; Independent	Automatic via app
Finger Expander	Hand/forearm strengthening: flexors and extensors	Daily	Self-paced	Intro by physiotherapist; Independent	Manual log

To assess functional hand performance, we included the following clinical variables. Case handle: An adapted or auxiliary handle added to tools, instruments, or devices to improve grasp, leverage, and comfort for patients with limited hand or wrist function. Case handles support power grips and engage larger muscle groups, providing a functional complement to precision grips. Pointed handle: A slim, precision-oriented handle that enables controlled pressure and tactile feedback during exercises. Pointed handles minimize fatigue, maintain functional wrist posture, and facilitate goal-oriented finger movements (e.g., abduction/adduction). Key handle: A handle designed to simulate real-life tasks such as turning keys, providing an assessment of grip precision and functional coordination.

These measures were included to capture aspects of hand performance using objective functional outcomes, focusing on strength, mobility, and dexterity along with patient-reported outcome measures (e.g., DASH, SF-36). While not yet validated in large psychometric studies, they were selected based on clinical relevance and prior use in goal-oriented hand rehabilitation exercises. All data were collected and managed in accordance with clinical data protection and handling procedures.

https://doi.org/10.3390/app16020987

2.2. Intervention Protocols

Participants in both groups received standardized instructions from a physiotherapist on the correct use of their assigned device. They were encouraged to perform additional independent exercises during evenings, on weekends, or after completing therapy sessions. Exercise progression, frequency, and duration were recorded automatically by the app in the tablet group and manually in a logbook for the finger-expander group.

A summary of the exercise components, dose, progression, and supervision is provided in Table 2.

Table 2. Overview of outcome measures and evaluation time point.

	Pre	Post	FU
DASH	x	x	x
Pain (VAS)	x	x	x
SF-36	x	x	
Clinical variables	x	x	

2.3. Statistics

Statistical analysis was performed using IBM® SPSS® Statistics Version 27.0 (IBM Corporation, Armonk, NY, USA). The normality assumption was verified for each variable using the Shapiro–Wilk test. Despite the small group sizes (n = 20), all parameters met the assumption of normality ($p > 0.05$), allowing for the use of parametric analyses [19]. A significance level of $p < 0.05$ was applied for all analyses. Repeated measures ANOVA was used for variables measured at three time points (DASH and VAS: pre, post and follow-up). In cases where follow-up data were missing, linear mixed-effects models were applied. Sphericity was assessed using Mauchly's test, and Greenhouse–Geisser correction was applied if the assumption of sphericity was violated. A post hoc test was performed to identify exactly which groups differed from each other. Independent samples t-tests were used to compare gamification parameters (e.g., usage duration and frequency). Mean values and standard deviation (STD) were calculated for all measured parameters. Single stars (*) means that p-value is ≤ 0.05, two stars (**) means that p-value $p \leq 0.01$, and three stars (***) means $p \leq 0.001$.

3. Results

3.1. Sample Demographic Characteristics

The final analysis included 40 participants (13 females and 27 males), with a mean age of 51 $\pm$ 9.5 years. Results showed that our patients BMI was 26.19 $\pm$ 5.9. The detailed demographic characteristics of participants are shown in Table 3.

Table 3. Demographic characteristics of participants.

Characteristic	N	%
Gender		
Female	13	32
Male	27	68
Age (min–max)	20–71	
BMI (mean)	26.19	

https://doi.org/10.3390/app16020987

Table 3. *Cont.*

Characteristic		N	%
Nicotine consumption:	Yes No	16 24	38 62
Alcohol consumption:	Yes No	12 28	28 72
Affected side:	Right Left	17 23	43 57
Handedness:	Right Left	37 3	95 5

3.2. Clinical Outcome Measures

3.2.1. DASH Score

The reference values from a non-clinical population were 17.49 and those from the clinical population were 40.36 [21]. Our data are therefore consistent with the outcome of the clinical population. In the finger-expander group, the DASH score decreased from pre- (57 ± 21) to post-training (46.8 ± 17) and to follow-up (FU) (34.9 ± 23), showing a difference of 10.2 (pre–post) and 21 points (pre–FU). In the tablet group, the results between pre- (58 ± 19) and post-training (45.5 ± 22) and FU (37 ± 28) show a difference of 12.5 (pre–post) and 21 points (pre–FU). Mauchly's test indicated that the assumption of sphericity was violated (W = 0.709, $\chi^2(2) = 12.747$, $p = 0.002$); therefore, Greenhouse–Geisser correction ($\varepsilon = 0.774$) was applied for the repeated-measures analysis.

Clinical relevance: Changes from pre-training to FU in both groups exceed the Minimum Clinically Important Difference (MCID) of 15 points [22,23]. The differences between pre- and post-training are less than the MCID in both the finger-expander (11.5 points) and tablet groups (10.2 points), respectively.

Statistical significance: Repeated measures ANOVA showed a significant main effect of Time, ($F(1.55, 58.85) = 16.36$, $p < 0.001$, partial $\eta^2 = 0.137$). Tukey-adjusted post hoc analysis revealed a significant ($p < 0.001$) difference between pre- and post-training, as well as pre-training and FU. However, no significant group or interaction effects were detected (Figure 3).

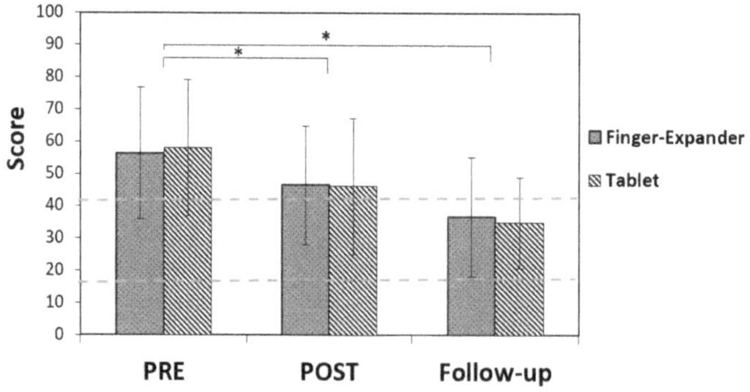

Figure 3. DASH score of the finger-expander (n = 20) and tablet group (n = 20) displayed at the three time points (pre, post, and FU). A single asterisk (*) indicates a statistically significant difference ($p < 0.05$).

https://doi.org/10.3390/app16020987

3.2.2. SF36

According to Ogura et al. [24], a five-point change in SF-36 is clinically relevant. In the Mental Component Summary (MCS), the tablet group improved by 10.5 points from pre- (30.5 ± 14) to post-training (41.0 ± 13.5), exceeding the clinical threshold, while the finger-expander group increased by 5.1 points (pre: 28.5 ± 14; post: 33.6 ± 14.2). Linear mixed-effects analysis indicated no significant effect of time or group on MCS (intercept = 0.660, SE = 0.264, t = 2.500, $p = 0.017$; pre = 0.00078, SE = 0.006, t = 0.130, $p = 0.897$; post = −0.005, SE = 0.006, t = −0.822, $p = 0.416$). These results suggest that although both groups' mental health improved during rehabilitation, there was no statistically significant advantage of tablet-based gamification over conventional exercises (Figure 4). In the PCS, a non-clinically relevant difference of 1.42 points was observed in the finger-expander group between pre- (40.6 ± 10) and post-training (42.04 ± 8.6), and a difference of 0.42 points was observed in the tablet group between pre- (42.7 ± 5) and post-training (42.6 ± 14.2). Linear mixed-effects analysis, with group (tablet vs. finger-expander) and time (pre, post) as fixed effects and patient ID as a random intercept, revealed no significant main effects or interactions. Specifically, the model coefficients were as follows: intercept = 0.157, SE = 0.558, t = 0.282, $p = 0.779$; pre = 0.008, SE = 0.010, t = 0.768, $p = 0.447$; post = 0.00068, SE = 0.010, t = 0.065, $p = 0.949$. These results indicate that neither intervention produced significant improvements in physical health scores over the rehabilitation period (Figure 4).

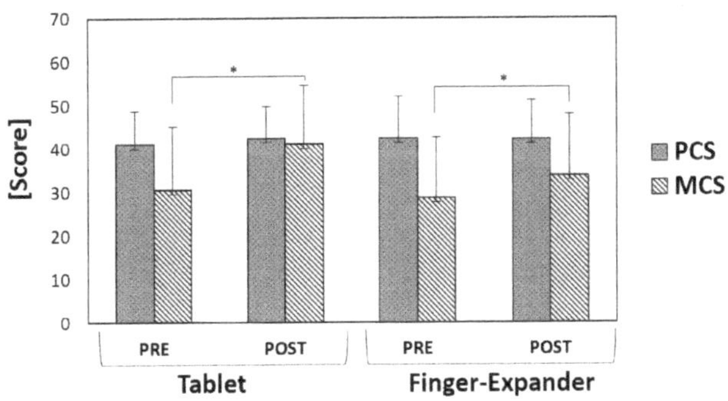

Figure 4. PCS and MCS subgroups of the SF36 of the finger-expander (n = 20) and tablet group (n = 20) displayed at two time points (pre- and post-training). A single asterisk (*) indicates a statistically significant difference ($p < 0.05$).

3.2.3. Visual Analog Scale

A change of three points on the VAS is considered a clinically significant difference in pain intensity according to Lee et al. (2003) [25]. Both groups showed similar improvements: in the finger-expander group, the results between pre-training (4.63 ± 1.9), post-training (3.41 ± 1.9) and FU (3.20 ± 1.4) showed differences of 1.22 (pre–post) and 1.43 points (pre–FU). In the tablet group, the results between pre-training (4.63 ± 2), post-training (3.85 ± 1), and FU (3.22 ± 2.18) showed differences of 0.78 (pre–post) and 1.44 points (pre–FU). Mauchly's test indicated that the assumption of sphericity was not violated (W = 0.838, $\chi^2(2) = 4.779$, $p = 0.092$); therefore, sphericity was assumed (Greenhouse–Geisser $\varepsilon = 0.860$). There was a significant main effect over time: Greenhouse–Geisser $F(2, 16.3) = 7.16$; $p = 0.002$; partial $\eta^2 = 0.096$. Tukey-adjusted post hoc analysis revealed significant differences between pre- and post-training, as well as pre-training and FU (Figure 5). We could not find any clinically relevant differences between the two groups or interaction effect.

https://doi.org/10.3390/app16020987

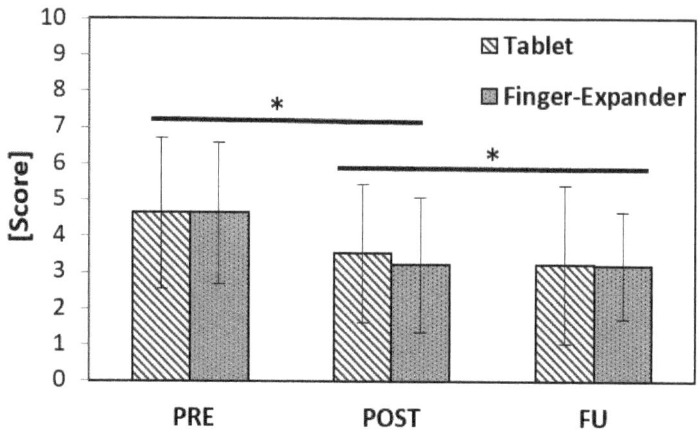

Figure 5. Visual analog scale of the finger-expander (n = 20) and tablet group (n = 20) displayed at the three time points (pre, post, and FU). Values of $p < 0.05$ were considered statistically significant (*).

3.2.4. Exercise Duration and Frequency

There is a significant difference in the duration of the exercise between groups: F(1, 38) = 8.13; $p = 0.007$; partial $\eta^2 = 0.0176$. The results show that the finger-expander group has a significantly longer exercise time (325 ± 295 min) than the tablet group (107 ± 172 min). No significant difference was observed in training frequency between the tablet and finger-expander groups: F(1, 38) = 2.15; $p = 0.15$; partial $\eta^2 = 0.054$ (Figure 6).

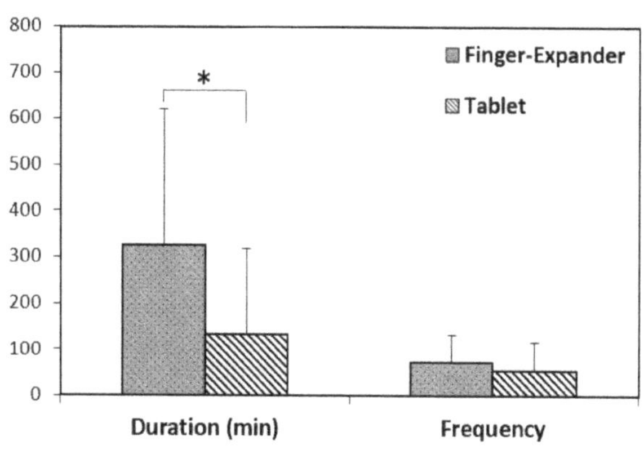

Figure 6. Comparison between tablet and finger-expander groups after 4 weeks of training. n = 40; A single asterisk (*) indicates a statistically significant difference ($p < 0.05$).

3.2.5. HAND Measurements

For case handle measurements, a significant improvement was observed over time—(1, 66) = 34.08; $p = 0.001$; partial $\eta^2 = 0.0056$—but there was no significant difference between the groups: (1, 26) = 0.13; $p = 0.135$; partial $\eta^2 = 0.00022$. Pointed and key handle measurement results showed no significant difference between the two groups or between the two-time points, with $p > 0.05$ (Supplementary Material Figure S1).

4. Discussion

Previous studies have reported that tablet or smartphone-based hand rehabilitation programs may offer broader benefits compared with conventional paper-based exercise

https://doi.org/10.3390/app16020987

instructions. In particular, these digital interventions have been associated with shorter recovery times, earlier return to work, and reduced healthcare utilization in patients with wrist, hand, or finger injuries [11,13,26].

4.1. Functional Outcomes

In our study, we wanted to show whether the additional integration of tablet-based gamification in a four-week inpatient hand rehabilitation can support improvements in hand function and enhance patients' motivation compared to the additional usage of a simple training device (hand exerciser). The limited additional effect of gamification in the structured inpatient setting is likely due to the high number of exercises performed daily, which can lead to fatigue or pain. While patients received consistent guidance and support, these intensive training demands may have reduced the measurable benefit of the gamified intervention. Overall, these findings suggest that the structured inpatient context may have minimized differences between interventions, highlighting the importance of context in interpreting intervention effects. Despite statistically significant improvements in hand function, these effects did not always reach the threshold for clinical relevance, which may relate to the severity of injuries in our trauma center cohort and the relatively low baseline symptom burden. Similar work published in 2020 also reported significant improvements in hand function [13]. No group effect was observed between the tablet group and the control group with regard to self-reported hand function based on the DASH score. DASH values remained within the range of reference values for a clinical population in both groups at post, and FU values approached normal population values but still indicated significant functional limitations. Improvements over time were significant in both groups, with only the long-term pre–FU changes exceeding the MCID.

4.2. Quality of Life

As part of the SF-36, we found significant differences in the MCS between pre- and post-rehabilitation, but not between the two groups. The differences between pre- and post-rehabilitation were, in both groups, above the MCID, with an obviously higher difference in the tablet group than in the control group. This is in line with a recent review (2024), which highlighted that digital health technologies, including gamification and virtual reality, can enhance motivation, adherence, and mental well-being in rehabilitation settings. No significant group differences were observed [27]. These results indicate that improvements in mental well-being may not necessarily translate into measurable gains in physical function in the short term. However, improvements in mental health did not translate into better PCS or DASH values, indicating physical impairment post-rehabilitation [28,29]. This pattern reflects the general observation that generic instruments like the SF-36 are less sensitive to changes in upper-limb function during early recovery phases compared to injury-specific tools like DASH [30]. Mahabier et al. found a strong correlation between shoulder function and the SF-12 PCS in patients recovering from humeral shaft fractures, whereas the correlation with the MCS was considerably weaker [31]. The absence of FU data for SF-36 in our study limits our ability to assess long-term trends in mental versus physical health outcomes.

4.3. Pain (VAS)

The differences in pain measured between pre-, post, and follow-up (FU) by the visual analog scale (VAS) were statistically significant, but did not reach the MCID of three points and are therefore not clinically relevant in either group. Neither exercise method showed superiority in terms of impact on pain level. This emphasizes that, while statistically detectable, changes in pain may not be clinically meaningful, and intervention differences are minimal in a well-supported inpatient program. These results are in

https://doi.org/10.3390/app16020987

line with other studies that report modest but significant short-term pain improvements following hand rehabilitation [32,33]. In our study, we wanted to examine whether the additional integration of tablet-based gamification in a four-week inpatient hand rehabilitation program could accelerate improvements in hand function and increase patient motivation compared to the additional use of a simple training device (hand exerciser). Similar work illustrates significant improvements in hand function with gamified rehabilitation approaches [34]. Moreover, previous research has found that physical health outcomes correlate more strongly with functional recovery than mental health outcomes. For instance, in patients with humeral shaft fractures, Mahabier et al. observed a strong correlation between shoulder function and the PCS of the SF-12, whereas the MCS was less strongly associated with function [31].

4.4. Exercise Adherence and Engagement

Interestingly, the hand exercise group reported significantly longer exercise durations than the tablet group. This discrepancy may reflect differences in measurement methods: exercise time in the tablet group was automatically recorded, while the control group self-reported their durations, potentially overestimating them. There was no significant difference in practice frequency between groups. This finding contradicts expectations that gamification would increase adherence through greater engagement. Taken together, these results indicate that adherence may be influenced more by context and supervision than by gamification alone.

4.5. Objective Hand Function

Objective measures of grip and pinch strength revealed no significant advantage of gamification. Despite encouraging evidence from prior research—such as "Effects of virtual reality based rehabilitation on distal upper extremity function and health related quality of life: a single blinded, randomized controlled trial" [35], which found improvements in upper extremity function and strength following VR based training, and studies like "Feedback guided exercises performed on a tablet touchscreen improve return to work, function, strength and healthcare usage more than an exercise program prescribed on paper for people with wrist, hand or finger injuries: a randomized trial" [13], which demonstrated that tablet-based motor training can outperform paper-based programs in hand/wrist injury rehabilitation—our inpatient context may have diluted these benefits. The handle-based clinical measures (case, pointed, key handles) are exploratory and inadequately validated; results from these outcomes should be interpreted cautiously and complement the primary outcome findings.

4.6. Interpretation and Context

Contrary to our hypothesis, the gamification group did not outperform the control group in either subjective or objective outcomes. One plausible explanation is the structured inpatient environment: participants received frequent therapist support and real-time feedback, potentially reducing the benefit of in-app feedback and motivation mechanisms. In contrast, gamification may prove more beneficial in home-based settings where external guidance is minimal. Moreover, the existing literature suggests that gamified exercises tend to encourage midline, limited range movements, potentially restricting range-of-motion benefits. These findings highlight that intervention effects are context-dependent, and structured support may level differences between novel and conventional approaches. For example, Shin et al. (2016) reported significant improvements in distal upper-limb function following VR-based rehabilitation, although the effects on grip strength were not explicitly highlighted [35].

 https://doi.org/10.3390/app16020987

Rehabilitation technologies such as soft wearable gloves equipped with motion sensors, as well as haptic devices targeting pinch and grip tasks, have been proposed to encourage specific movement patterns and functional goals [36]. Tablet-based game exercises may promote repetitive, constrained movements, underscoring the importance of designing gamified interventions that encourage full functional ranges and task-specific training [13].

Study Limitations: A key limitation of this study is that the duration and intensity of exercises varied considerably between patients, as they could perform tablet- or finger expander-based exercises independently in the evenings or on weekends. However, the intensity of exercises during supervised physiotherapy sessions was adapted individually for each patient, not only in the ongoing rehabilitation process but also as part of the overall therapy planning. In addition, exercise duration was measured differently between groups (self-reported for the finger-expander group vs. automated recordings for the tablet group), which may introduce systematic measurement bias and limits direct comparability. This variability may have affected the comparison between digital and non-digital exercises. Missing follow-up SF-36 and clinical hand measurements limit the assessment of long-term physical and mental health outcomes. Finally, the handle-based clinical variables (case, pointed, and key handles) represent non-standard, exploratory measures with limited psychometric validation, which we acknowledge as an additional study limitation.

5. Conclusions

Gamified hand rehabilitation via a tablet application is feasible and well accepted by patients, as indicated by recorded exercise duration and frequency. However, it did not outperform conventional hand exercises in terms of functional outcomes or quality of life during a four-week inpatient program. Mental health improvements in both groups suggest a general benefit of rehabilitation rather than a gamification-specific effect. Engagement was inferred from usage patterns and patient feedback, as no validated scales were applied. Future research should explore gamification in home-based rehabilitation, incorporate motion-sensitive devices for precise tracking, and consider psychological factors earlier in the care pathway. With further refinement, gamified rehabilitation tools could become a cost-effective and engaging component of modern therapy.

Supplementary Materials: The following supporting information can be downloaded at: https://www.mdpi.com/article/10.3390/app16020987/s1, Figure S1: Case handles (a), Pointed and key handle measurement. (b) Comparison between Tablet and Finger-expander group. Data is mean ± SD. * Significant difference at $p < 0.05$.

Author Contributions: T.O., H.-H.H., C.S. (Christian Schmitz) and H.B.A. contributed to the conception and design of the study. T.O., H.B.A., C.S. (Christine Seelmann), V.H. and J.T. conducted the data analysis and data interpretation as well as the manuscript preparation and writing. T.O. contributed to the revision. All authors have read and agreed to the published version of the manuscript.

Funding: The study is funded by the DGUV (German Social Accident Insurance).

Institutional Review Board Statement: The study was conducted in accordance with the Declaration of Helsinki, and approved by the Ethik-Kommission der Ärztekammer Nordrhein (ÄkNo-EK), protocol code 2019119 and date of approval is 21 October 2019.

Informed Consent Statement: All participants provided written informed consent prior to enrollment in the study, in accordance with the Declaration of Helsinki and institutional ethical guidelines.

Data Availability Statement: The original contributions presented in this study are included in the article/Supplementary Materials. Further inquiries can be directed to the corresponding author.

https://doi.org/10.3390/app16020987

Acknowledgments: The authors gratefully acknowledge all colleagues and reviewers for their critical feedback, which significantly improved the quality of this manuscript. All aspects of the study including design, data collection, analysis, figures, tables, and photographic materials were independently developed and executed by the authors without the use of artificial intelligence tools. The authors take full responsibility for the content and integrity of the work and confirm that all reviewers thoroughly evaluated the manuscript prior to submission.

Conflicts of Interest: The authors declare no conflicts of interest.

Abbreviations

The following abbreviations are used in this manuscript:

FU	Follow-up
STD	Standard deviation
DASH	Disabilities of Arm, Shoulder, and Hand questionnaire
HS	Hand Surgery
MCID	Minimal Clinically Important Difference
MCS	Mental component summary
PCS	Physical component summary

References

1. Angermann, P.; Lohmann, M. Injuries to the hand and wrist. A study of 50,272 injuries. *J. Hand Surg.* **1993**, *18*, 642–644. [CrossRef]
2. Sorock, G.S.; Lombardi, D.A.; Courtney, T.K.; Cotnam, J.; Mittleman, M.A. Epidemiology of occupational acute traumatic hand injuries: A literature review. *Saf. Sci.* **2001**, *38*, 241–256. [CrossRef]
3. Schaub, T.A.; Chung, K.C. Systems of provision and delivery of hand care, and its impact on the community. *Injury* **2006**, *37*, 1066–1070. [CrossRef]
4. Alimanova, M.; Kozhamzharova, D.K.; Zholdygarayev, A.; Meraliyev, M. Adaptation of Gaming Process to improve Hand Rehabilitation. In *International Scientific-Practical Conference "Innovations in Education and Science" Dedicated to the 25th Anniversary of Independence of the Republic of Kazakhstan and 20th Anniversary of Suleyman Demirel Univeristy, 2 June 2016*; SDU Scientific Library Office: Kaskelen, Kazakhstan, 2016; pp. 34–37. Available online: https://repository.sdu.edu.kz/items/557ed584-9c09-4fcb-9f0f-4b4903eabdf1 (accessed on 15 December 2025).
5. Polygerinos, P.; Lyne, S.; Wang, Z.; Nicolini, L.; Mosadegh, B.; Whitesides, G.; Walsh, C. Towards a soft pneumatic glove for hand rehabilitation. In *2013 IEEE/RSJ International Conference on Intelligent Robots and Systems (IROS)*; IEEE: New York, NY, USA, 2013; pp. 1512–1517.
6. Alimanova, M.; Borambayeva, S.; Kozhamzharova, D.; Kurmangaiyeva, N.; Ospanova, D.; Tyulepberdinova, G.; Gaziz, G.; Kassenkhan, A. Gamification of hand rehabilitation process using virtual reality tools: Using leap motion for hand rehabilitation. In *2017 First IEEE International Conference on Robotic Computing (IRC)*; IEEE: New York, NY, USA, 2017; pp. 336–339.
7. Yang, Y.; Wang, Z.; Xing, D.; Wang, P. Learning playing piano with bionic-constrained diffusion policy for anthropomorphic hand. *Cyborg Bionic Syst.* **2024**, *5*, 0104. [CrossRef]
8. Hoffman, H.G.; Boe, D.A.; Rombokas, E.; Khadra, C.; LeMay, S.; Meyer, W.J.; Patterson, S.; Ballesteros, A.; Pitt, S.W. Virtual reality hand therapy: A new tool for nonopioid analgesia for acute procedural pain, hand rehabilitation, and VR embodiment therapy for phantom limb pain. *J. Hand Ther.* **2020**, *33*, 254–262. [CrossRef] [PubMed]
9. Kay, S.; McMahon, M.; Stiller, K. An advice and exercise program has some benefits over natural recovery after distal radius fracture: A randomised trial. *Aust. J. Physiother.* **2008**, *54*, 253–259. [CrossRef] [PubMed]
10. Mubin, O.; Alnajjar, F.; Jishtu, N.; Alsinglawi, B.; Al Mahmud, A. Exoskeletons with virtual reality, augmented reality, and gamification for stroke patients' rehabilitation: Systematic review. *JMIR Rehabil. Assist. Technol.* **2019**, *6*, e12010. [CrossRef]
11. Blanquero, J.; Cortés-Vega, M.D.; García-Frasquet, M.Á.; Sánchez-Laulhé, P.R.; Suero-Pineda, A. Exercises using a touchscreen tablet application improved functional ability more than an exercise program prescribed on paper in people after surgical carpal tunnel release: A randomised trial. *J. Physiother.* **2019**, *65*, 81–87. [CrossRef]
12. Algar, L.; Valdes, K. Using smartphone applications as hand therapy interventions. *J. Hand Ther.* **2014**, *27*, 254–257. [CrossRef]
13. Blanquero, J.; Cortés-Vega, M.-D.; Rodríguez-Sánchez-Laulhé, P.; Corrales-Serra, B.-P.; Gómez-Patricio, E.; Díaz-Matas, N.; Suero-Pineda, A. Feedback-guided exercises performed on a tablet touchscreen improve return to work, function, strength and healthcare usage more than an exercise program prescribed on paper for people with wrist, hand or finger injuries: A randomised trial. *J. Physiother.* **2020**, *66*, 236–242. [CrossRef]

14. Germann, G.; Harth, A.; Wind, G.; Demir, E. Standardisierung und Validierung der deutschen Version 2.0 des "Disability of Arm, Shoulder, Hand" (DASH)-Fragebogens zur Outcome-Messung an der oberen Extremität. *Unfallchirurg* **2003**, *106*, 13–19. [CrossRef] [PubMed]
15. Bullinger, M.; Kirchberger, I.; Ware, J. Der deutsche SF-36 Health Survey Übersetzung und psychometrische Testung eines krankheitsübergreifenden Instruments zur Erfassung der gesundheitsbezogenen Lebensqualität. *J. Public Health* **1995**, *3*, 21–36. [CrossRef]
16. MacDermid, J.C. Outcome evaluation in patients with elbow pathology: Issues in instrument development and evaluation. *J. Hand Ther.* **2001**, *14*, 105–114. [CrossRef]
17. Angst, F.; John, M.; Goldhahn, J.; Herren, D.B.; Pap, G.; Aeschlimann, A.; Schwyzer, H.K.; Simmen, B.R. Comprehensive assessment of clinical outcome and quality of life after resection interposition arthroplasty of the thumb saddle joint. *Arthritis Care Res. Off. J. Am. Coll. Rheumatol.* **2005**, *53*, 205–213. [CrossRef]
18. MacDermid, J.C.; Richards, R.S.; Donner, A.; Bellamy, N.; Roth, J.H. Responsiveness of the short form-36, disability of the arm, shoulder, and hand questionnaire, patient-rated wrist evaluation, and physical impairment measurements in evaluating recovery after a distal radius fracture. *J. Hand Surg.* **2000**, *25*, 330–340. [CrossRef] [PubMed]
19. Ware, J.E. *SF-36 Physical and Mental Health Summary Scales: A User's Manual*; The Health Institute, New England Medical Center Hospitals: Boston, MA, USA, 1994.
20. Mohammadian, M. A review on grip and pinch strength tests. *Ergon. Int. J.* **2023**, *7*, 2–8. [CrossRef]
21. Jester, A.; Harth, A.; Rauch, J.; Germann, G. DASH-Daten nicht-klinischer versus klinischer Personengruppen–eine Vergleichsstudie mit T-Normen für die klinische Praxis. *Handchir. Mikrochir. Plast. Chir.* **2010**, *42*, 55–64. [CrossRef] [PubMed]
22. Beaton, D.E.; Bombardier, C.; Katz, J.N.; Wright, J.G.; Wells, G.; Boers, M.; Strand, V.; Shea, B. Looking for important change/differences in studies of responsiveness. OMERACT MCID Working Group. Outcome Measures in Rheumatology. Minimal Clinically Important Difference. *J. Rheumatol.* **2001**, *28*, 400–405.
23. Beaton, D.E.; Katz, J.N.; Fossel, A.H.; Wright, J.G.; Tarasuk, V.; Bombardier, C. Measuring the whole or the parts?: Validity, reliability, and responsiveness of the Disabilities of the Arm, Shoulder and Hand outcome measure in different regions of the upper extremity. *J. Hand Ther.* **2001**, *14*, 128–142. [CrossRef]
24. Ogura, K.; Yakoub, M.A.; Christ, A.B.; Fujiwara, T.; Nikolic, Z.; Boland, P.J.; Healey, J.H. What are the minimum clinically important differences in SF-36 scores in patients with orthopaedic oncologic conditions? *Clin. Orthop. Relat. Res.* **2020**, *478*, 2148. [CrossRef]
25. Lee, J.S.; Hobden, E.; Stiell, I.G.; Wells, G.A. Clinically important change in the visual analog scale after adequate pain control. *Acad. Emerg. Med.* **2003**, *10*, 1128–1130. [CrossRef]
26. Suero-Pineda, A.; Oliva-Pascual-Vaca, Á.; Durán, M.R.-P.; Sánchez-Laulhé, P.R.; García-Frasquet, M.Á.; Blanquero, J. Effectiveness of a Telerehabilitation Evidence-based Tablet App for Rehabilitation in Traumatic Bone and Soft-Tissue Injuries of the Hand, Wrist and Fingers. *Arch. Phys. Med. Rehabil.* **2023**, *104*, 932–941. [CrossRef]
27. Prahm, C.; Bressler, M.; Heinzel, J.; Lauer, H.; Ritter, J.; Daigeler, A.; Kolbenschlag, J. Digital technologies and strategies in amputation medicine. *Unfallchirurgie* **2024**, *127*, 637–643. [CrossRef]
28. Røe, C.; Dalen, H.; Lein, M.; Bautz-Holter, E. Comprehensive rehabilitation at Beitostølen Healthsports Centre: Influence on mental and physical functioning. *J. Rehabil. Med.* **2008**, *40*, 410–417. [CrossRef]
29. Berget, A.M.; Moen, V.P.; Hustoft, M.; Eide, G.E.; Skouen, J.S.; Strand, L.I.; Hetlevik, Ø. Long-term change and predictors of change in physical and mental function after rehabilitation: A multi-centre study. *J. Rehabil. Med.* **2023**, *55*, 2809. [CrossRef] [PubMed]
30. Cheema, A.S.; O'Brien, P.J.; Broekhuyse, H.M.; Lefaivre, K.A. Measuring outcomes over time in distal radius fractures: A comparison of generic, upper extremity-specific and wrist-specific outcome measures. *J. Hand Surg. Glob. Online* **2020**, *2*, 272–276. [CrossRef]
31. Shields, E.; Sundem, L.; Childs, S.; Maceroli, M.; Humphrey, C.; Ketz, J.; Gorczyca, J.T. Factors predicting patient-reported functional outcome scores after humeral shaft fractures. *Injury* **2015**, *46*, 693–698. [CrossRef]
32. Ye, L.; Kalichman, L.; Spittle, A.; Dobson, F.; Bennell, K. Effects of rehabilitative interventions on pain, function and physical impairments in people with hand osteoarthritis: A systematic review. *Arthritis Res. Ther.* **2011**, *13*, R28. [CrossRef]
33. Harth, A.; Germann, G.; Jester, A. Evaluating the effectiveness of a patient-oriented hand rehabilitation programme. *J. Hand Surg. (Eur. Vol.)* **2008**, *33*, 771–778. [CrossRef] [PubMed]
34. Naqvi, W.M.; Nimbulkar, G.; Umate, L. Gamification for distal radius fracture rehabilitation: A randomized controlled pilot study. *Cureus* **2022**, *14*, e29333. [CrossRef] [PubMed]

 https://doi.org/10.3390/app16020987

35. Shin, J.-H.; Kim, M.-Y.; Lee, J.-Y.; Jeon, Y.-J.; Kim, S.; Lee, S.; Seo, B.; Choi, Y. Effects of virtual reality-based rehabilitation on distal upper extremity function and health-related quality of life: A single-blinded, randomized controlled trial. *J. Neuroeng. Rehabil.* **2016**, *13*, 17. [CrossRef] [PubMed]
36. Chen, X.; Gong, L.; Wei, L.; Yeh, S.-C.; Da Xu, L.; Zheng, L.; Zou, Z. A wearable hand rehabilitation system with soft gloves. *IEEE Trans. Ind. Inform.* **2020**, *17*, 943–952. [CrossRef]

 https://doi.org/10.3390/app16020987

applied sciences

MDPI

Article

Between-Limb Asymmetry Assessment During a Double-Leg Drop Jump Task After Anterior Cruciate Ligament Reconstruction—A Musculoskeletal Modelling Approach

Rodrigo B. Mateus [1], Sílvia Cabral [1], Chris Richter [2,†] and António P. Veloso [1,*]

[1] CIPER, Faculty of Human Kinetics, University of Lisbon, Cruz Quebrada-Dafundo, 1495-751 Oeiras, Portugal; rodrigomateus@fmh.ulisboa.pt (R.B.M.); scabral@fmh.ulisboa.pt (S.C.)
[2] UPMC Sports Surgery Clinic, Santry Demesne, D09 C523 Dublin, Ireland; chris.richter@byte.bayern
* Correspondence: apveloso@fmh.ulisboa.pt
† Current address: byte-Bavarian Digital Agency, 81669 Munich, Germany.

Abstract: Approximately two-thirds of athletes who are submitted to Anterior Cruciate Ligament Reconstruction (ACLR) never return to their preinjury level of performance, potentially due to muscle strength deficiencies or altered loading patterns during landing or jumping tasks. This study aimed to estimate individual muscle forces during a double-leg drop jump task, and assess sagittal plane between-limb asymmetries in muscle forces and ground reaction forces using a musculoskeletal modelling approach, in athletes who underwent ACLR. Thirty male field-sport athletes (age: 18–35 years; mass: 84.3 ± 12.3 kg; height: 180.2 ± 8.4 cm) post-ACLR (39.8 ± 3.9 weeks) using patellar or quadriceps tendon grafts were tested. Scaled musculoskeletal models were implemented in OpenSim, and muscle forces were estimated using the Computed Muscle Control optimization method. The contralateral limb exhibited greater vertical ground reaction forces across most of the rebound phase (d = 2.01). Compared with the contralateral limb, the ACLR limb showed reduced quadriceps (d = 1.72), soleus (d = 0.95), and gluteus maximus (d = 0.83) forces, indicating deficits in knee extensor, plantarflexor, and hip extensor neuromuscular function. Smaller asymmetries were found for the gluteus medius (d = 0.60) and hamstrings (d = 0.72), while other muscles showed symmetrical activation patterns. These results reveal persistent between-limb asymmetries in muscle recruitment and loading up to nine months post-ACLR, emphasizing the importance of targeted rehabilitation to restore symmetrical neuromuscular control during explosive movements.

Keywords: knee injury; limb symmetry; musculoskeletal modelling; sports injury

1. Introduction

In field sports, such as football, knee injuries account for approximately one-fifth of all injuries, with the majority being non-contact injuries. Anterior cruciate ligament (ACL) injury is one of the most common of all knee injuries, with its injury mechanisms relating to landing and change in direction tasks. The most frequent management of ACL injury is ACL reconstruction (ACLR), intending to regain stability and prevent subsequent injuries at the knee joint, as well as ensuring an athlete's level of function has been appropriately restored before returning to play (RTP) [1].

Following ACLR, previous reports state that between 49% and 92% of athletes reach a level sufficient to return to play and return to a competitive level of sports performance after a 12-month rehabilitation period [2,3]. However, between 35% and 60% of athletes who undergo this procedure do not reach their self-declared knee function level prior to injury [4,5]. The low rate of success regarding regaining a similar level of preinjury performance may be due to muscle strength deficiencies [6], and modified loading patterns whilst performing landing [7–10], jumping, and squatting [11] tasks. Moreover, the rate of reinjury following ACLR is high, where approximately 22% of the athletes suffer an ACL injury in the contralateral lower limb after returning to sport [12]. Therefore, it is important to define objective criteria to evaluate RTP.

Biomechanical analyses of landing tasks have been used to study the effectiveness of injury prevention protocols and the impact of rehabilitation after ACLR. Previous studies reported differences in kinematic and kinetic variables along all three anatomical planes at the hip, knee, and ankle joints between the ACLR lower limb and the healthy contralateral lower limb during jumping and cutting tasks, in male athletes who underwent ACLR [13,14]. Research also demonstrated that ground reaction forces (GRF) reported on the reconstructed lower limb were lower than those reported on the healthy contralateral side, nine months after surgery [15]. In addition to this, muscle activations are shown to change in patients following ACLR, so muscle dynamics should be taken into account [15,16].

Musculoskeletal modelling provides insights into muscle function and internal loads through computational simulations using a musculoskeletal model. When combined with motion analysis biomechanical data, it may offer more clarity and is even able to estimate muscle forces in high-risk actions. For instance, research on landing maneuvers revealed that the GRF, quadriceps and hamstring muscle forces, and tibiofemoral force were the main causes of ACL loading [17]. Additionally, muscles that do not span the knee joint, like the soleus, were also found to contribute ACL loading [18,19]. Much of the growing body of research for landing tasks using musculoskeletal modelling focused on the estimation of force produced by the quadriceps and hamstrings [14], estimation of muscle contributions to tibial anterior shear forces during a drop vertical jump tasks [20], and muscle contributions to the ACL loading during a stop-jump task [19]. More recently, computational modelling approaches have been used to predict optimal neuromuscular control strategies that minimize ACL forces during landing tasks, demonstrating how muscle recruitment patterns influence ACL loading [21], highlighting the value of computational modelling in understanding movement strategies that either increase or reduce ACL loading. However, while these studies provide theoretical insights, few have compared actual muscle force patterns between ACL-reconstructed and healthy limbs during high-risk movements, such as the drop jump.

Furthermore, inter-limb asymmetries in athletes create constraints that may limit how they perform certain movements, especially physical constraints, which can put the athletes in an injury-prone position that hinders musculoskeletal health. Evaluating limb symmetry has been a useful tool in the assessment of injury risk, of player readiness to RTP [15], as well as predicting the possibility of reinjury [13]. Common movements employed in testing batteries consisted mainly of unilateral and bilateral jumping and landing tasks [13,15], such as drop jumps and countermovement jumps, running [22], and change in direction tasks [15]. More specifically, landing tasks like drop jumping have been intimately associated with an ACL injury, and thoroughly implemented in biomechanical analyses following ACLR [16]. Studying between-limb asymmetry is an important tool for evaluating deficits in neuromuscular control during jumping and landing tasks [16]. In a recent study, ACL, and peak tibiofemoral contact forces, at the time of RTP, were estimated using an electromyography (EMG)-informed musculoskeletal modelling approach in male athletes, and with this workflow, it was possible to conclude that between-limb asymmetry concerning both ACL and tibiofemoral loading was present even after RTP clearance criteria were met [1]. However, the description of both musculotendon forces and GRF, in terms of timing and magnitude would result in a better understanding of the biomechanical differences between the ACL reconstructed and contralateral limb during a double-legged drop jump task [23], and no prior studies have provided a detailed description of the strategy used by the major lower limb muscles to perform double-leg drop jump. Bearing this in mind, investigating inter-joint coordination and assessing inter-limb asymmetry during jumping and landing tasks like the drop jump may be a useful approach to monitor the risk of lower-limb injuries in athletes and their performance.

Therefore, this study aimed to use musculoskeletal modelling to quantify individual lower limb muscle forces and vertical GRF during a double-leg drop jump in athletes following ACLR and to compare these variables between the ACLR and contralateral limbs. Based on prior evidence of reduced knee extensor and plantarflexor forces, altered muscle activation, and persistent between-limb asymmetry, we hypothesized that the ACLR limb would show lower knee extensor and plantarflexor forces and reduced vertical GRF, and peak muscle activations would occur in the knee extensors, hip extensors, and ankle plantarflexors, reflecting their primary role during the rebound phase of the movement.

2. Materials and Methods

A total of 30 eligible male participants (mass: 84.3 ± 12.34 kg; height: 180.2 ± 8.4 cm; time post-op: 39.8 ± 3.9 weeks) were recruited between November 2019 and January 2020. Participants were recruited from the caseload of two Orthopedic surgeons in the Sports Surgery Clinic, Santry, Dublin, and were part of a longer-term research project consisting of physical testing at 6 and 9 months post ACL reconstruction (NCT02771548). All subjects suffered an ACL injury and an ACLR was completed using either a patellar or a quadriceps tendon graft technique. Subjects were male, between 18 and 35 years old, participated in a field-based sport involving cutting/change in direction (Gaelic Football, Hurling, Soccer, Basketball, etc.), underwent primary ACL injury reconstruction, and had the intention to return to the same level of physical activity following rehabilitation. All surgeries were performed at a single site, but rehabilitation was completed locally under the supervision of referring physiotherapists. Participants followed an accelerated program progressing from early weightbearing to strength, power, and plyometric training, then on-field running and change in direction drills. Orthopedic reviews occurred at 2 weeks, 3 months, and 6–9 months post-surgery, with final physical testing at 9 months. Participants who underwent multiple concurrent ligament reconstructions, and meniscal repair, suffered previous ACL or chondral injuries, and had no intention to return to the same level of preinjury activity were excluded. Data collection took place in a biomechanics laboratory where a ten-camera motion analysis system (200 Hz; Bonita-B10, Vicon, Oxford, UK), synchronized (Vicon Nexus 2.3) with two force platforms (1000 Hz BP400600, AMTI, Watertown, MA, USA) was used to record the three-dimensional positions of 42 reflective markers (14 mm diameter). The motion capture and force plate systems used exhibit high instrument accuracy (errors < 1 mm and <1 N) [24,25]. The employed marker set was a combination between the Plug-in Gait model [26] and the 6 Degrees of Freedom [25] model, where the reference markers were positioned on the lower limb according to the former and the cluster markers inserted into the thigh and shanks of both lower limbs following the latter (Figure 1). Calibration markers were placed only during the static trial to estimate joint widths. Markers located at anatomical landmarks present in both models were also placed. In addition to this, six other markers were incorporated in the marker set to be employed as a quality measure for data processing of the Plug-in Gait model. These marker sets have shown excellent intra-rater and test–retest reliability for lower-limb kinematics, with high values of the coefficient of multiple correlation and the intraclass correlation coefficient (greater than 0.8) and low measurement error (less than 5°) for key sagittal-plane variables [27].

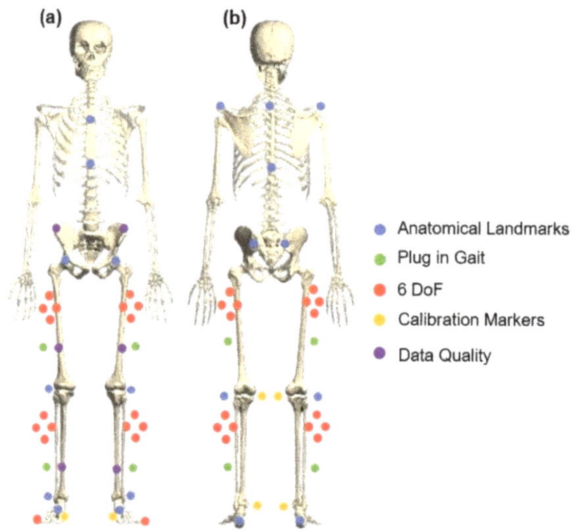

Figure 1. Representation of the marker set employed in the experimental data acquisition. Anterior (**a**) and posterior (**b**) views are given.

All subjects gave their written informed consent after an explanation of the experimental protocol and before the beginning of the study, which was approved by the institution's research ethics committee (25-AFM-010). Before data collection, all participants completed a standardized warm-up consisting of jogging, squats, and submaximal countermovement jumps. A static trial was collected after which the calibration markers were removed. As described by King et al., 2018 [13], participants then completed a testing battery consisting of several bilateral jumping and landing tasks, as well as planned and unplanned change in direction tasks. The double leg drop jump (DLDJ) was carried out from a 30 cm box. Participants started with their feet approximately hip-width apart and landed with one foot on each of the laboratory force plates. Participants were instructed to drop off the box with both feet simultaneously, and upon landing to jump vertically as high as possible while spending as little time as possible on the force platforms. Participants completed three trials with a 30 s rest period allowed between each trial. The trial with the highest reactive strength index was selected for analysis to ensure optimal performance consistency and data quality. Drop jump is a plyometric exercise involving an eccentric muscle action followed by an explosive concentric muscle contraction—rebound phase—a flight phase, and a landing phase. During the rebound phase of this task, the muscles work eccentrically (eccentric stage) to prepare for a rapid concentric contraction (concentric stage). Therefore, to perform this movement efficiently, forces need to be dampened, stabilized, and accelerated. To identify each phase, the vertical centre of pressure data was used. Hence, this study focuses on the rebound phase, with results presented separately for the eccentric and concentric stages to provide a detailed understanding of both performance and movement strategy. Custom software (Version R2019a; Mathworks Inc.; Natick, MA, USA) was used for filling gaps in marker trajectories and screening data for errors. A filtering step was performed on the marker trajectory and force data using a fourth-order zero-lag lowpass Butterworth filter (cut-off frequency 15 Hz), to remove high-frequency noise while preserving movement characteristics relevant for a DLDJ. The position and orientation of the local coordinate systems were computed using a global optimization algorithm [28], using the Visual 3D software (Version 2021.06.02; C-Motion, Inc.; Germantown, MD, USA). Segment masses were determined per Dempster [29], and the remaining inertial parameters were estimated based on Hanavan [30]. For this work, the generic musculoskeletal model, called "Gait 2392", comprising 12 rigid bodies, 29 degrees of freedom (DoF), and 92 musculotendon actuators, representing both lower limbs and a head-arms-torso segment, was inserted in the simulation using OpenSim (Version 4.2; Stanford University, Stanford, CA, USA) [31], an open-source musculoskeletal simulation software. This model was selected due to its well-established representation of lower-limb musculoskeletal anatomy and the inclusion of the major muscles involved in a DLDJ. The hip and back joints were modelled as ball-and-socket joints, allowing 3 DoF, the ankle as a revolute joint (1 DoF), and the knee joints were modelled as a custom joint allowing the motion of flexion/extension [32]. This single-axis representation was adopted to reflect the predominant sagittal-plane motion of the knee and ankle during drop-jump tasks within the Gait2392 modelling framework. The implemented Hill-type muscle model is a modification from the Thelen model [33], to avoid numerical singularities. Scale factors, and inverse kinematics outputs used in the musculoskeletal modelling portion of this work were exported from Visual3D.

Firstly, a scaling step with scale factors was performed, so that the scaled model would match the subject's joint frame locations, mass center location, force application points, and muscle attachment points. A residual reduction algorithm (RRA) step was implemented to reduce errors due to modelling simplifications, and kinetic and kinematic inconsistencies arising from the experimental procedure. Muscle forces were estimated with the computed muscle control (CMC) optimization method [33]. CMC uses a proportional-derivative controller to track the experimental kinematics while solving an optimization problem that minimizes the sum of squared actuator controls and the sum of squared acceleration tracking errors, consequently obtaining the desired set of muscle excitations. Muscle force-length–velocity properties were taken into account and used in this step. The estimated excitations are then converted into individual muscle forces through the implemented muscle model, accounting for the aforementioned muscle properties and activation dynamics. The musculoskeletal simulation steps followed the current guidelines for best practices in musculoskeletal modelling simulations, to ensure reproducible modelling outcomes [34].

To assess differences in muscle force variables between lower limbs, a paired t-test was implemented into parametric and nonparametric analyses using the Statistical Parametric Mapping (SPM) software (SPM spm1d version M.0.4.8 (2021-09-09); https://spm1d.org) [35]. Normality was assessed qualitatively by comparing the results between the parametric and nonparametric analyses (Supplementary Table S1). Cohen's d was calculated on a frame-by-frame basis across the entire task to quantify effect sizes (small: $d < 0.2$; medium: $0.2 \leq d < 0.8$; large: $d \geq 0.8$). In this context, small effects were interpreted as minor asymmetries between limbs, medium effects indicated moderate, easily noticeable and often clinically meaningful differences in muscle loading, and large effects represented substantial asymmetries likely reflecting altered neuromuscular strategies following ACLR. For statistically significant clusters, suprathreshold regions of the SPM trajectory indicated clusters where ACLR lower limb forces were significantly lower than those of the healthy limb (ACLR < Healthy, $p < 0.05$). Conversely, subthreshold clusters were identified where ACLR lower limb forces exceeded those of the healthy limb (ACLR > Healthy). The maximum effect size was calculated to estimate the magnitude of differences for each statistically significant cluster, and the coefficient of variation (CV) was computed on the cluster-wise differences between limbs to quantify between-subject variability. These CV values were reported for each significant cluster, providing a relative measure of dispersion and relative reliability of the asymmetry across participants.

3. Results

To begin with, validation and verification were performed to give confidence to the results. RRA and CMC results confirmed acceptable model performance. Translational and rotational kinematic errors remained within recommended limits (Supplementary Table S2), with normalized root mean square errors (nRMSE) between RRA-derived and inverse dynamics joint moments of 19.2 ± 1.19%. Mean peak reductions were observed for rotational (37.8 ± 20.0%) and translational (66.1 ± 15.4%) residuals at the pelvis joint. CMC kinematic errors also remained below the maximum accepted threshold (Supplementary Table S2).

3.1. Time-Series Comparisons (SPM Analysis)

3.1.1. Vertical Ground Reaction Force

Mean vertical GRF were highly similar between limbs during the initial moments of the rebound phase. However, one suprathreshold cluster spanning 8.82–95.53% of the rebound phase exceeded the critical threshold, indicating significantly lower forces in the ACLR limb compared to the healthy limb. The probability of this occurring by random chance was extremely low ($p < 0.00001$), leading to rejection of the null hypothesis. This suggests a consistent unloading strategy of the ACLR limb throughout most of the rebound phase (Table 1, Figure 2).

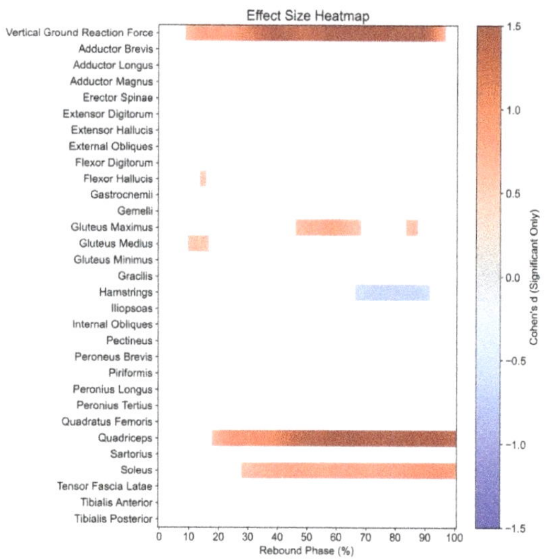

Figure 2. Heatmap of statistically significant clusters illustrating between-limb differences (Healthy vs. ACLR) across the task. Each row represents a muscle, and the horizontal axis corresponds to the percentage of the rebound phase (0–100%). Color intensity reflects the magnitude of the effect size (Cohen's *d*), with warmer colors indicating greater absolute differences between lower limbs. Negative values (blue) indicate higher values in the ACLR lower limb relative to the Healthy one, while positive values (red) indicate higher values in the Healthy lower limb.

Table 1. Summary of significant clusters detected by the parametric paired SPM t-tests comparing the ACLR limb to the contralateral healthy limb during the rebound phase. Cluster ranges are reported as percentages of the rebound phase, with corresponding p-values, maximum effect sizes, and coefficients of variation (CV(%)). Cluster Type indicates whether the ACLR limb exhibited higher (Subthreshold) or lower (Suprathreshold) values compared to the healthy limb, as indicated in the final column.

	Critical Threshold	Cluster Type	Cluster Range (% Rebound Phase)	p-Value	Max Effect Size	CV (%)	ACLR vs. Healthy
			Ground Reaction Forces				
Vertical GRF	3.268	Suprathreshold	8.82–95.53	<0.00001	2.012	97.53	ACLR < Healthy
			Muscles				
Flexor Hallucis	3.143	Suprathreshold	13.81–15.44	0.0485	0.59	174.33	ACLR < Healthy
Gluteus Maximus	3.239	Suprathreshold	45.85–67.63	<0.00001	0.834	139.67	ACLR < Healthy
		Suprathreshold	82.50–86.59	0.0358	0.635	162.47	ACLR < Healthy
Gluteus Medius	2.953	Suprathreshold	9.43–16.28	0.0413	0.599	176.52	ACLR < Healthy
Hamstrings	3.128	Subthreshold	65.57–90.14	0.00007	0.717	153.57	ACLR > Healthy
Quadriceps	3.047	Suprathreshold	17.33–100	<0.000001	1.716	77.97	ACLR < Healthy
Soleus	3.044	Suprathreshold	27.93–100	<0.000001	0.950	126.41	ACLR < Healthy

Table 1 summarizes the key characteristics of all significant clusters identified by the paired SPM t-tests. For each variable, the table reports the critical threshold for statistical inference, the cluster type (suprathreshold = ACLR < Healthy; subthreshold = ACLR > Healthy), and the percentage of the rebound phase in which the between-limb difference occurred. The corresponding p-values reflect the likelihood of these clusters arising by chance, while the maximum effect size (Cohen's d) indicates the magnitude of asymmetry (small < 0.2; medium 0.2–0.8; large $\geq$ 0.8). The coefficient of variation (CV(%)) reflects how consistently each cluster appeared across participants. Collectively, these metrics outline when asymmetries emerged, how large they were, and how reliably they were expressed between individuals.

3.1.2. Muscle Forces

Paired SPM t-tests were conducted to compare muscle force profiles between the ACLR and healthy limbs during the double-leg drop jump. For most muscles, forces were similar between limbs during large portions of the rebound phase. However, several suprathreshold clusters exceeded the critical threshold, indicating time-dependent asymmetries in force production for specific muscles (Table 1, Figure 2).

A large suprathreshold cluster was observed for the quadriceps, spanning 17.33–100% of the rebound phase ($p < 0.000001$). During this cluster, the ACLR limb generated consistently lower quadriceps forces than the healthy limb. Random curves of identical smoothness would produce this result with a probability of less than 0.001, leading to rejection of the null hypothesis.

Similarly, the soleus displayed a suprathreshold cluster from 27.93 to 100%, with lower forces in the ACLR limb throughout the rebound phase ($p < 0.000001$). This indicates systematic deficits in ankle plantarflexor function during the latter part of the rebound phase.

The gluteus maximus exhibited two distinct suprathreshold clusters: One from 45.85 to 67.63% ($p < 0.00001$), and another from 82.50 to 86.59% ($p = 0.0358$). In both clusters, the ACLR limb produced lower forces than the healthy limb, reflecting deficits in hip extensor engagement during mid- and late-rebound phases.

A brief early suprathreshold cluster was detected for the gluteus medius between 9.43 and 16.28% ($p = 0.0413$), indicating lower ACLR limb forces during the initial part of the rebound phase.

The hamstrings showed a subthreshold cluster from 65.57 to 90.14% ($p = 0.00007$). Unlike other muscles, the ACLR limb generated greater hamstring forces than the healthy limb during this period, possibly reflecting a compensatory strategy to stabilize the knee during late rebound.

The flexor hallucis exhibited a small suprathreshold cluster between 13.81 and 15.44% ($p = 0.0485$), where ACLR limb forces were lower than those of the healthy limb. This brief asymmetry likely reflects individual differences in forefoot loading at initial contact.

No significant clusters were detected for other muscles, indicating symmetrical force production throughout the rebound phase.

3.2. Cluster-Level Characteristics and Effect Sizes

3.2.1. Vertical Ground Reaction Force

The vertical ground reaction force exhibited a single, large suprathreshold cluster spanning 8.82–95.53% of the rebound phase ($p < 0.00001$). Within this cluster, the ACLR limb generated consistently lower vertical forces than the healthy limb. The maximum effect size reached 2.01, indicating a very large difference between limbs. The CV for cluster extent was 97.53%, suggesting moderate inter-individual variability in the timing of this asymmetry (Table 1, Figure 3). This result indicates a robust and systematic unloading of the ACLR limb during rebound phase of this movement.

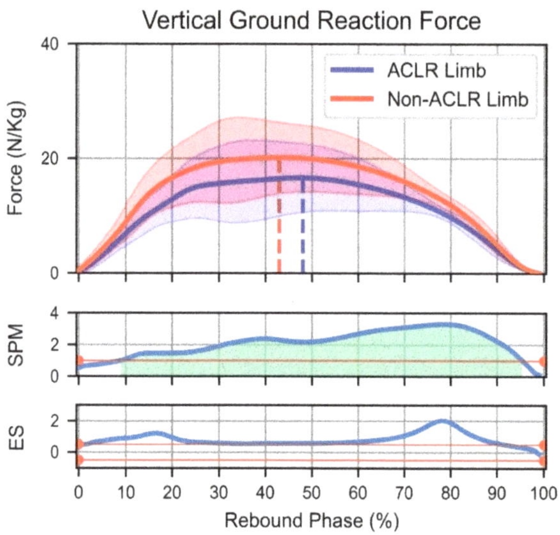

Figure 3. Between-limb differences in vertical ground reaction forces during the rebound phase of a double-leg drop jump. The top panel illustrates the mean ± SD for the ACLR and Healthy contralateral lower limbs of 30 subjects. The vertical lines represent the timings of peak vertical ground reaction forces on the ACLR (blue) and healthy (red) lower limbs. The middle panel illustrates the normalized SPM{t}—the t-statistic as a function of time describing the difference between the two lower limbs. The red line represents the critical threshold and the shaded portion of the curve indicates $p < 0.05$. The bottom panel illustrates the effect size over the range of the movement, and two red horizontal lines are placed at 0.5 and −0.5.

3.2.2. Muscle Forces

Among individual muscles, the quadriceps showed the largest and most consistent asymmetry, with a large suprathreshold cluster extending from 17.33 to 100% of the rebound phase ($p < 0.000001$). The maximum effect size within this cluster was 1.72, indicating a very large difference, with lower forces consistently produced by the ACLR limb. The CV for cluster extent was 77.97%, reflecting moderate variability in the timing of this asymmetry across participants, showing a deficit in quadriceps force generation during the entire rebound phase (Table 1, Figure 4).

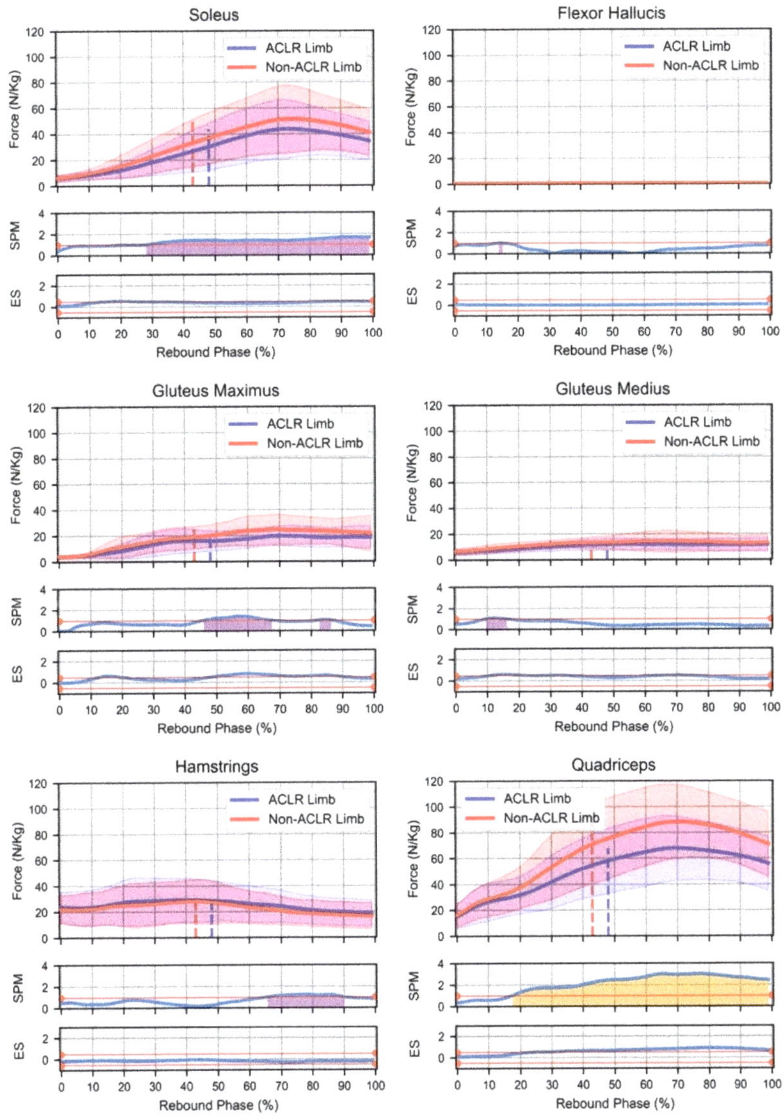

Figure 4. Between-limb differences in muscle forces during the rebound phase of a double-leg drop jump. The top panel illustrates the mean ± SD for the ACLR and Healthy contralateral lower limbs of 30 subjects. The vertical lines represent the timings of peak vertical ground reaction forces on the ACLR (blue) and healthy (red) lower limbs. The middle panel illustrates the normalized SPM{t}—the t-statistic as a function of time describing the difference between the two lower limbs. The red line represents the critical threshold and the shaded portion of the curve indicates $p < 0.05$. The bottom panel illustrates the effect size over the range of the movement, and two red horizontal lines are placed at 0.5 and −0.5.

The soleus also demonstrated a large and sustained suprathreshold cluster from 27.93 to 100% ($p < 0.000001$), with a maximum effect size of 0.95. This represents a large magnitude of asymmetry, where the ACLR limb exhibited consistently lower plantarflexor forces. The cluster's CV was 126.41%, indicating greater inter-individual variability compared to the quadriceps, particularly regarding when the asymmetry emerged within the rebound phase (Table 1, Figure 4).

The gluteus maximus displayed two distinct suprathreshold clusters: the first occurring between 45.85 and 67.63% ($p < 0.00001$) and the second between 82.50 and 86.59% ($p = 0.0358$).

The first cluster had a maximum effect size of 0.83, while the second reached 0.64, both indicating moderate to large differences. Cluster variability was relatively high, with CVs of 139.67% and 162.47%, suggesting that the timing and duration of hip extensor asymmetry varied greatly between participants (Table 1, Figure 4).

The gluteus medius showed a brief suprathreshold cluster early in the rebound phase (9.43–16.28%, $p = 0.0413$), with a maximum effect size of 0.60. The CV for this cluster was 176.52%, indicating considerable variability in when and how strongly this asymmetry appeared across individuals (Table 1, Figure 4).

The hamstrings were unique in displaying a subthreshold cluster between 65.57 and 90.14% ($p = 0.00007$). Here, the ACLR limb produced greater hamstring forces than the healthy limb, with a maximum effect size of 0.72. The CV was 153.57%, reflecting a wide range of participant-specific compensatory strategies in knee flexor activation (Table 1, Figure 4).

Finally, the flexor hallucis showed a small suprathreshold cluster between 13.81 and 15.44% ($p = 0.0485$), with a maximum effect size of 0.59 and a CV of 174.33%. This brief cluster suggests highly variable asymmetries at initial ground contact (Table 1, Figure 4).

3.3. *Peak Muscle Activations*

Mean peak muscle activations were compared between the ACLR and healthy limbs (Figure 5). The quadriceps showed the highest peak activation across all muscles, with lower values in the ACLR limb, confirming the strong, sustained asymmetry observed in the SPM analysis.

The gluteus maximus and soleus also exhibited reduced peak activations in the ACLR lower limb, consistent with their significant clusters, indicating deficits in hip extension and plantarflexor function during the rebound phase.

For the hamstrings, peak activations were similar between limbs despite the SPM analysis showing greater ACLR limb activity late in the rebound phase.

Smaller differences were seen in the gluteus medius and flexor hallucis, with slightly lower peaks in the ACLR limb.

No notable asymmetries were found in other muscles, indicating symmetrical peak activation patterns.

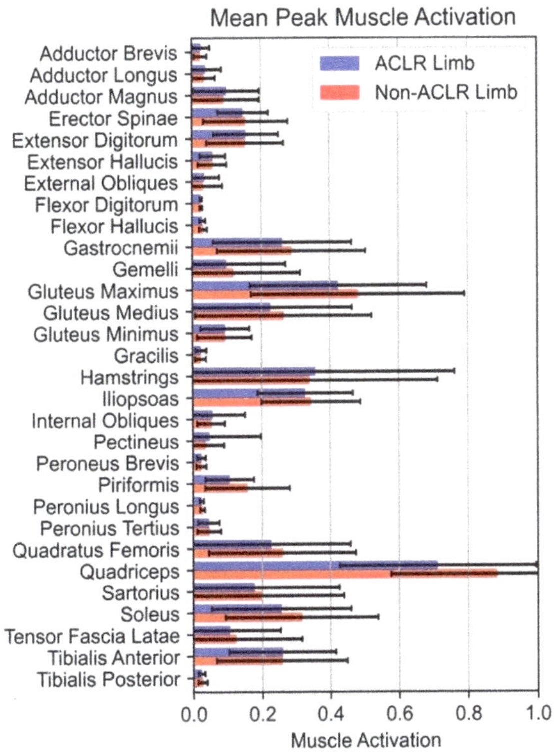

Figure 5. Mean peak muscle activations for both lower limbs during the rebound phase of a double-leg drop jump. Each bar illustrates the mean ± SD for the ACLR and Healthy contralateral lower limbs of 30 subjects.

4. Discussion

This work aimed to estimate muscle forces and investigate inter-limb asymmetries during a DLDJ task, in athletes after ACLR. Using musculoskeletal modelling and SPM, we identified persistent and systematic deficits in the ACLR limb. These deficits occurred mainly in muscles responsible for knee extension and ankle plantarflexion, despite athletes performing the task symmetrically. This highlights that bilateral tasks can still reveal meaningful internal asymmetries and supports their continued use in post-ACLR assessments [9].

Peak muscle activation analysis supported SPM results, confirming that muscles with large asymmetries, such as the quadriceps, soleus, and gluteus maximus, also showed reduced peak activations in the ACLR limb. Conversely, muscles with smaller or variable clusters, such as the hamstrings and gluteus medius, showed inconsistent or minimal peak differences, suggesting that timing and coordination, rather than peak magnitude, drive these asymmetries.

The main knee extensors, ankle plantar flexors, and hip extensors showed significantly higher muscle activations throughout the entire movement, which aligns with a study of individual muscle contributions to ground reaction forces during a drop jump task in elite basketball players [36]. These results also agree with findings from other tasks, where quadriceps isometric strength was reported to be lower in the ACLR limb during gait [37]. Moreover, Salem et al. observed a reduced knee extensor moment in the ACLR limb during squatting [38]. They also found a higher hip extensor moment in the ACLR limb, which may explain the higher mean peak hamstrings activations in the ACLR lower limb found in the present study. These estimates and statistically significant between-limb differences observed in this work suggest this muscle group may act as a dynamic protector of the ACL, by pulling the tibia posteriorly and increasing knee joint compression [39], likely reflecting a compensatory mechanism to stabilize the knee and compensate for quadriceps deficits. While such compensation may help maintain joint stability, excessive reliance on hamstrings could increase the risk of secondary injuries or hamstring strain. The systematic unloading of the quadriceps and soleus observed in the ACLR limb may reflect a protective strategy to reduce ACL loading. This aligns with model-based predictions showing that altered neuromuscular control can minimize ACL forces during landing tasks [21]. However, while such strategies may reduce ACL loading, they may also lead to persistent asymmetries and impaired performance if not addressed during rehabilitation.

It is expected that muscles spanning the knee joint, such as the quadriceps and hamstrings, are important for the loading of the ACL [19]. However, significant differences were also found in non-knee spanning muscles, such as the soleus, gluteus maximus, and gluteus medius, which supports to the concept of dynamic coupling. Furthermore, their relation to ACL loading has been studied during other tasks. The soleus has been shown to have an important role in preventing anterior tibial translation, which is correlated to ACL loading and subsequent injury [18]. The soleus showed large, continuous clusters indicating systematic unloading throughout the rebound phase. These findings align with previous reports of persistent deficits in plantarflexor function after ACLR, which may impair landing mechanics and explosive take-off performance. Persistent unloading may also contribute to muscle weakness and delayed return to sport if not addressed during rehabilitation. Regarding the gluteus maximus, its role in preventing the femur from internally rotating relative to the tibia might play an important role in ACL loading [40]. The gluteus maximus exhibited two moderate clusters, while the gluteus medius showed brief, variable asymmetries early in the rebound phase. The high variability in cluster extent suggests that individuals adopt different hip strategies to compensate for knee extensor deficits. Some athletes may rely heavily on hip extensors to stabilize and generate force, whereas others do not. This heterogeneity highlights the need for individualized rehabilitation programs targeting hip strength and control.

Concerning the GRF, the results showed increased peak vertical GRF on the healthy contralateral lower limb compared to the ACLR lower limb. Between-limb differences (asymmetries) were also evident during the rebound phase. Research has shown that vertical GRF in the ACLR limb was lower than in the healthy contralateral side, nine months following ACLR [15]. This asymmetry might reflect continued offloading of the ACLR limb possibly due to insufficient rehabilitation following ACLR. Moreover, this asymmetry has been identified as a risk factor not only for primary ACL injuries but also for ACL injuries to the contralateral limb after ACLR [9]. The vertical ground reaction force mirrored the muscle-level findings, with the ACLR limb consistently producing lower forces throughout most of the rebound phase. This reduction reflects the combined effect of diminished quadriceps and soleus contributions, leading to overall unloading of the ACLR lower limb and have been linked to an increased risk of both re-injury and contralateral ACL injury. Previous studies demonstrated that muscle strength deficits in the ACLR limb, notably in the quadriceps muscle group, can induce between-limb discrepancies in vertical GRF, as well as hip and knee joint moments along the sagittal plane [15]. However, further research is needed on the muscle contributions to GRF during this task, to clarify how muscle strength deficits alter between-limb differences in vertical GRF.

Albeit the majority of the calculated effect size fell within he medium to large range, the low effect size observed for the hamstrings highlights the substantial inter-subject variability in the dataset. This was further supported by the high CV across all muscles, indicating substantial variation in muscle force estimates, which likely reflects the heterogeneous nature of ACLR recovery among athletes. Given that rehabilitation protocols were consistent across participants, neuromuscular compensation strategies and graft type may have contributed to the variability in muscle force production observed here. Overall, the effect size represents the meaningfulness of the difference between the variables. These findings underscore the importance of individualized assessment and rehabilitation following ACLR, as group-level averages may obscure clinically relevant deviations at the individual level.

This work provides novel insights into the neuromuscular adaptations that persist in athletes following ACLR during an explosive, sport-specific task. Using a musculoskeletal modelling approach in OpenSim, individual muscle forces were estimated, and significant between-limb asymmetries were identified in key muscles—including the quadriceps, soleus, gluteus maximus, gluteus medius, and hamstrings. The results demonstrate that muscle recruitment and ground reaction force loading patterns remain altered in the ACLR limb, even during symmetrical bilateral tasks. These internal biomechanical asymmetries may contribute to impaired performance or an increased risk of reinjury, and may remain undetected through conventional clinical assessments.

This study has several limitations that should be considered. First, a generic musculoskeletal model scaled to each participant was used, with a planar knee model allowing only flexion/extension and locked subtalar and metatarsophalangeal joints to maintain neutral foot positioning. While this simplification is appropriate for sagittal-plane analysis of the DLDJ, it does not capture multi-planar knee or foot mechanics. In addition to this, the model parameters related to the musculotendon actuator architecture, like tendon slack length were obtained from different studies that did not relate to the population studied in this study [41]. Although previous work suggests that scaled generic and subject-specific models produce similar results for gait [42], subject-specific parameters may improve accuracy for high-demand tasks. Even though this gives confidence to the results of this study, future studies must maintain this line of thought in choosing the appropriate model for the implementation of a musculoskeletal modelling approach in scientific research work. Also, this study did not acquire electromyographic data, which limits direct validation of estimated muscle activations. However, throughout the simulation, positional errors and reserve actuators were all kept within the acceptable range of values according to Hicks et al., 2015 [34], so this gives confidence to the results obtained and their interpretations. Future research should consider using subject-specific models, multi-degree-of-freedom joints, and EMG acquisition to refine muscle force estimates and better capture neuromuscular adaptations after ACLR.

5. Conclusions

This study implemented a musculoskeletal modelling framework to evaluate changes in muscle recruitment after ACLR during a DLDJ. The ACLR limb produced consistently lower vertical GRF throughout the rebound phase and reduced quadriceps, gluteus maximus, and soleus forces. Peak activations occurred in the quadriceps, gluteus maximus, and soleus, reflecting their main role during the rebound phase of a DLDJ, while increased hamstring forces late in the movement suggested a compensatory strategy to achieve knee stability. Together, these kinetic and muscle-level asymmetries reveal persistent neuromuscular adaptations following ACL reconstruction and highlight the importance of rehabilitation and RTP assessments that restore symmetrical neuromuscular function and improve landing mechanics.

Supplementary Materials: The following supporting information can be downloaded at: https://www.mdpi.com/article/10.3390/app152212347/s1, Table S1: Critical *t*-values and significant cluster ranges identified by parametric and non-parametric paired SPM *t*-tests comparing the ACLR and contralateral healthy limbs across muscles and variables during the rebound phase. Cluster ranges are expressed as percentages of the rebound phase, with corresponding peak *t*-values. Positive peak t-values indicate higher values in the ACLR limb, while negative values indicate higher values in the contralateral limb. "-----" indicates no significant clusters were detected; Table S2: Positional errors. Translational errors (Pelvis_tx, Pelvis_ty, Pelvis_tz) are given in cm and rotational errors (Pelvis Tilt, Pelvis List, Pelvis Rotation, Hip Flexion, Hip Adduction, Hip Rotation, Knee Flexion, Ankle Plantarflexion, Lumbar Extension, Lumbar Bending, Lumbar Rotation) are given in degrees.

Author Contributions: R.B.M. drafted the manuscript; R.B.M., C.R. and S.C. performed analysis; R.B.M., S.C., C.R. and A.P.V. conceived the study, and carried out critical analysis of the paper. All authors have read and agreed to the published version of the manuscript.

Funding: This work was supported by Fundação para a Ciência e a Tecnologia (FCT): Grant numbers DOI: 10.54499/2020.07356.BD (PhD Grant), and DOI: 10.54499/UIDB/00447/2020, attributed to CIPER—Centro Interdisciplinar de Estudo da Performance Humana (unit 447).

Institutional Review Board Statement: The study was conducted in accordance with the Declaration of Helsinki, and approved by the Ethics Committee of UPMC Sports Surgery Clinic (25-AFM-010; Approval Date: 29 May 2014). The study was registered at ClinicalTrials.gov (NCT02771548).

Informed Consent Statement: Informed consent was obtained from all subjects involved in the study.

Data Availability Statement: The data presented in this study are available on request from the corresponding author. The datasets presented in this article are not readily available due to patient privacy and ethical restrictions, as they contain sensitive clinical information.

Acknowledgments: The authors would like to thank and acknowledge the staff at the UPMC Sport Surgery Clinic, Dublin for their support, especially the Biomechanics team. We also acknowledge Ciaran Cosgrave, Director of the UPMC Sports Surgery Clinic, for his input in the study's conceptual development and for providing access to the dataset used in this research. This work was supported by Fundação para a Ciência e a Tecnologia (FCT): Grant numbers DOI: 10.54499/2020.07356.BD (PhD Grant), and DOI: 10.54499/UIDB/00447/2020, attributed to CIPER—Centro Interdisciplinar de Estudo da Performance Humana (unit 447).

Conflicts of Interest: Author Chris Richter was employed by the byte-Bavarian Digital Agency. The remaining authors declare that the research was conducted in the absence of any commercial or financial relationships that could be construed as a potential conflict of interest.

Abbreviations

The following abbreviations are used in this manuscript:

ACL	Anterior Cruciate Ligament
ACLR	ACL Reconstruction
CMC	Computed Muscle Control
CV	Coefficient of Variation
DOF	Degree of Freedom
DLDJ	Double Leg Drop Jump
EMG	Electromyography
GRF	Ground Reaction Forces
nRMSE	normalized Root Mean Square Errors
RRA	Residual Reduction Algorithm
RTP	Returning to play
SPM	Statistical Parametric Mapping

References

1. Kotsifaki, A.; Van Rossom, S.; Whiteley, R.; Korakakis, V.; Bahr, R.; D'Hooghe, P.; Papakostas, E.; Sideris, V.; Farooq, A.; Jonkers, I. Between-Limb Symmetry in ACL and Tibiofemoral Contact Forces in Athletes After ACL Reconstruction and Clearance for Return to Sport. *Orthop. J. Sport. Med.* **2022**, *10*, 232596712210847. [CrossRef]
2. Ardern, C.L.; Webster, K.E.; Taylor, N.F.; Feller, J.A. Return to Sport Following Anterior Cruciate Ligament Reconstruction Surgery: A Systematic Review and Meta-Analysis of the State of Play. *Br. J. Sports Med.* **2011**, *45*, 596–606. [CrossRef]
3. Ardern, C.L.; Webster, K.E.; Taylor, N.F.; Feller, J.A. Return to the Preinjury Level of Competitive Sport after Anterior Cruciate Ligament Reconstruction Surgery: Two-Thirds of Patients Have Not Returned by 12 Months after Surgery. *Am. J. Sports Med.* **2011**, *39*, 538–543. [CrossRef]
4. Spindler, K.P.; Huston, L.J.; Wright, R.W.; Kaeding, C.C.; Marx, R.G.; Amendola, A.; Parker, R.D.; Andrish, J.T.; Reinke, E.K.; Harrell, F.E.; et al. The Prognosis and Predictors of Sports Function and Activity at Minimum 6 Years after Anterior Cruciate Ligament Reconstruction: A Population Cohort Study. *Am. J. Sports Med.* **2011**, *39*, 348–359. [CrossRef]
5. Morris, R.C.; Hulstyn, M.J.; Fleming, B.C.; Owens, B.D.; Fadale, P.D. Return to Play Following Anterior Cruciate Ligament Reconstruction. *Clin. Sports Med.* **2016**, *35*, 655–668. [CrossRef]
6. Lynch, A.D.; Logerstedt, D.S.; Grindem, H.; Eitzen, I.; Hicks, G.E.; Axe, M.J.; Engebretsen, L.; Risberg, M.A.; Snyder-MacKler, L. Consensus Criteria for Defining "successful Outcome" after ACL Injury and Reconstruction: A Delaware-Oslo ACL Cohort Investigation. *Br. J. Sports Med.* **2015**, *49*, 335–342. [CrossRef]
7. Gokeler, A.; Hof, A.L.; Arnold, M.P.; Dijkstra, P.U.; Postema, K.; Otten, E. Abnormal Landing Strategies after ACL Reconstruction. *Scand. J. Med. Sci. Sport.* **2010**, *20*, 12–19. [CrossRef]
8. Oberländer, K.D.; Brüggemann, G.P.; Höher, J.; Karamanidis, K. Altered Landing Mechanics in ACL-Reconstructed Patients. *Med. Sci. Sports Exerc.* **2013**, *45*, 506–513. [CrossRef] [PubMed]
9. Paterno, M.V.; Ford, K.R.; Myer, G.D.; Heyl, R.; Hewett, T.E. Limb Asymmetries in Landing and Jumping 2 Years Following Anterior Cruciate Ligament Reconstruction. *Clin. J. Sport Med.* **2007**, *17*, 258–262. [CrossRef] [PubMed]
10. Paterno, M.V.; Rauh, M.; Schmitt, L.C.; Ford, K.R.; Hewett, T.E. Incidence of Second Anterior Cruciate Ligament (ACL) Injury 2 Years after Primary ACL Reconstruction and Return to Sport. *Orthop. J. Sport. Med.* **2013**, *1*, 116–121. [CrossRef]
11. Neitzel, J.A.; Kernozek, T.W.; Davies, G.J. Loading Response Following Anterior Cruciate Ligament Reconstruction during the Parallel Squat Exercise. *Clin. Biomech.* **2002**, *17*, 551–554. [CrossRef] [PubMed]
12. Wiggins, A.J.; Grandhi, R.K.; Schneider, D.K.; Stanfield, D.; Webster, K.E.; Myer, G.D. Risk of Secondary Injury in Younger Athletes after Anterior Cruciate Ligament Reconstruction. *Am. J. Sports Med.* **2016**, *44*, 1861–1876. [CrossRef] [PubMed]
13. King, E.; Richter, C.; Franklyn-Miller, A.; Daniels, K.; Wadey, R.; Moran, R.; Strike, S. Whole-Body Biomechanical Differences between Limbs Exist 9 Months after ACL Reconstruction across Jump/Landing Tasks. *Scand. J. Med. Sci. Sports* **2018**, *28*, 2567–2578. [CrossRef]
14. Thomas, A.C.; Villwock, M.; Wojtys, E.M.; Palmieri-Smith, R.M. Lower Extremity Muscle Strength after Anterior Cruciate Ligament Injury and Reconstruction. *J. Athl. Train.* **2013**, *48*, 610–620. [CrossRef]
15. King, E.; Richter, C.; Franklyn-Miller, A.; Wadey, R.; Moran, R.; Strike, S. Back to Normal Symmetry? Biomechanical Variables Remain More Asymmetrical Than Normal During Jump and Change-of-Direction Testing 9 Months After Anterior Cruciate Ligament Reconstruction. *Am. J. Sports Med.* **2019**, *47*, 1175–1185. [CrossRef] [PubMed]
16. He, X.; Leong, H.T.; Lau, O.Y.; Ong, M.T.Y.; Yung, P.S.H. Altered Neuromuscular Activity of the Lower-Extremities during Landing Tasks in Patients with Anterior Cruciate Ligament Reconstruction: A Systematic Review of Electromyographic Studies. *J. Sport Rehabil.* **2020**, *29*, 1194–1203. [CrossRef]
17. Laughlin, W.A.; Weinhandl, J.T.; Kernozek, T.W.; Cobb, S.C.; Keenan, K.G.; O'connor, K.M. The Effects of Single-Leg Landing Technique on ACL Loading. *J. Biomech.* **2011**, *44*, 1845–1851. [CrossRef]
18. Mokhtarzadeh, H.; Yeow, C.H.; Hong Goh, J.C.; Oetomo, D.; Malekipour, F.; Lee, P.V.S. Contributions of the Soleus and Gastrocnemius Muscles to the Anterior Cruciate Ligament Loading during Single-Leg Landing. *J. Biomech.* **2013**, *46*, 1913–1920. [CrossRef]
19. Peel, S.A.; Schroeder, L.E.; Weinhandl, J.T. Lower Extremity Muscle Contributions to ACL Loading during a Stop-Jump Task. *J. Biomech.* **2021**, *121*, 110426. [CrossRef]
20. Navacchia, A.; Ueno, R.; Ford, K.R.; Dicesare, C.A.; Gregory, D.; Hewett, T.E. EMG-Informed Musculoskeletal Modelling to Estimate Realistic Knee Anterior Shear Force during Drop Vertical Jump in Female Athletes. *Ann. Biomed. Eng.* **2019**, *47*, 2416–2430. [CrossRef]
21. Mazumder, O.; Chakravarty, K.; Chatterjee, D.; Sinha, A.; Poduval, M. Musculoskeletal Modelling to Predict and Reduce Antetrior Cruciate Ligament Injury during Single Leg Drop Jump Activity: Synergistic Muscle Co-Activation Approach. In Proceedings of the 2019 41st Annual International Conference of the IEEE Engineering in Medicine and Biology Society (EMBC), Berlin, Germany, 23–27 July 2019; pp. 4108–4112. [CrossRef]

22. Thomson, A.; Einarsson, E.; Hansen, C.; Bleakley, C.; Whiteley, R. Marked Asymmetry in Vertical Force (but Not Contact Times) during Running in ACL Reconstructed Athletes <9 Months Post-Surgery despite Meeting Functional Criteria for Return to Sport. *J. Sci. Med. Sport* **2018**, *21*, 890–893. [CrossRef]
23. Oberhofer, K.; Hosseini Nasab, S.H.; Schütz, P.; Postolka, B.; Snedeker, J.G.; Taylor, W.R.; List, R. The Influence of Muscle-Tendon Forces on ACL Loading during Jump Landing: A Systematic Review. *Muscles Ligaments Tendons J.* **2017**, *7*, 125–135. [CrossRef]
24. Ehara, Y.; Fujimoto, H.; Miyazaki, S.; Mochimaru, M.; Tanaka, S.; Yamamoto, S. Comparison of the Performance of 3D Camera Systems II. *Gait Posture* **1997**, *5*, 251–255. [CrossRef]
25. Collins, T.D.; Ghoussayni, S.N.; Ewins, D.J.; Kent, J.A. A Six Degrees-of-Freedom Marker Set for Gait Analysis: Repeatability and Comparison with a Modified Helen Hayes Set. *Gait Posture* **2009**, *30*, 173–180. [CrossRef] [PubMed]
26. Marshall, B.M.; Franklyn-Miller, A.; King, E.; Moran, K.; Strike, S.; Falvey, É.C. Biomechanical Factors Associated with Time to Complete a Change of Directions Cutting Maneuver. *J. Strength Cond. Res.* **2014**, *28*, 2845–2851. [CrossRef] [PubMed]
27. McGinley, J.L.; Baker, R.; Wolfe, R.; Morris, M.E. The Reliability of Three-Dimensional Kinematic Gait Measurements: A Systematic Review. *Gait Posture* **2009**, *29*, 360–369. [CrossRef] [PubMed]
28. Lu, T.W.; O'Connor, J.J. Bone Position Estimation from Skin Marker Co-Ordinates Using Global Optimisation with Joint Constraints. *J. Biomech.* **1999**, *32*, 129–134. [CrossRef]
29. Dempster, W.T. *Space Requirements of the Seated Operator: Geometrical, Kinematic, and Mechanical Aspects of the Body, with Special Reference to the Limbs*; 1955. Available online: https://deepblue.lib.umich.edu/items/a4869201-0cec-41ee-851c-2b0df68db9a3 (accessed on 6 February 2025).
30. Hanavan, E.P. A Mathematical Model of the Human Body. *AMRL-TR. Aerosp. Med. Res. Lab.* **1964**, *64*, 1–149.
31. Delp, S.L.; Anderson, F.C.; Arnold, A.S.; Loan, P.; Habib, A.; John, C.T.; Guendelman, E.; Thelen, D.G. OpenSim: Open-Source Software to Create and Analyze Dynamic Simulations of Movement. *IEEE Trans. Biomed. Eng.* **2007**, *54*, 1940–1950. [CrossRef]
32. Anderson, F.C.; Pandy, M.G. A Dynamic Optimization Solution for Vertical Jumping in Three Dimensions. *Comput. Methods Biomech. Biomed. Engin.* **1999**, *2*, 201–231. [CrossRef] [PubMed]
33. Thelen, D.G.; Anderson, F.C. Using Computed Muscle Control to Generate Forward Dynamic Simulations of Human Walking from Experimental Data. *J. Biomech.* **2006**, *39*, 1107–1115. [CrossRef]
34. Hicks, J.L.; Uchida, T.K.; Seth, A.; Rajagopal, A.; Delp, S.L. Is My Model Good Enough? Best Practices for Verification and Validation of Musculoskeletal Models and Simulations of Movement. *J. Biomech. Eng.* **2015**, *137*, 020905. [CrossRef] [PubMed]
35. Pataky, T.C. Generalized N-Dimensional Biomechanical Field Analysis Using Statistical Parametric Mapping. *J. Biomech.* **2010**, *43*, 1976–1982. [CrossRef]
36. Kipp, K.; Kim, H.; Cross, J.; Geiser, C. Muscle Force Contributions To Ground Reaction Force Profiles During Basketball Related Tasks. *ISBS Proc. Arch.* **2019**, *37*, 300–303.
37. Shi, H.; Huang, H.; Ren, S.; Yu, Y.; Liang, Z.; Wang, Q.; Hu, X.; Ao, Y. The Relationship between Quadriceps Strength Asymmetry and Knee Biomechanics Asymmetry during Walking in Individuals with Anterior Cruciate Ligament Reconstruction. *Gait Posture* **2019**, *73*, 74–79. [CrossRef]
38. Salem, G.J.; Salinas, R.; Harding, F.V. Bilateral Kinematic and Kinetic Analysis of the Squat Exercise after Anterior Cruciate Ligament Reconstruction. *Arch. Phys. Med. Rehabil.* **2003**, *84*, 1211–1216. [CrossRef] [PubMed]
39. Kim, H.J.; Lee, J.H.; Ahn, S.E.; Park, M.J.; Lee, D.H. Influence of Anterior Cruciate Ligament Tear on Thigh Muscle Strength and Hamstring-to-Quadriceps Ratio: A Meta-Analysis. *PLoS ONE* **2016**, *11*, e0146234. [CrossRef]
40. Cibulka, M.T.; Bennett, J. How Weakness of the Tensor Fascia Lata and Gluteus Maximus May Contribute to ACL Injury: A New Theory. *Physiother. Theory Pract.* **2020**, *36*, 359–364. [CrossRef]
41. Delp, S.L.; Loan, J.P.P.; Hoy, M.G.G.; Zajac, F.E.; Topp, E.L.; Rosen, J.M. An Interactive Graphics-Based Model of the Lower Extremity to Study Orthopedic Surgical Procedures. *IEEE Trans. Biomed. Eng.* **1990**, *37*, 757–767. [CrossRef]
42. Correa, T.A.; Baker, R.; Kerr Graham, H.; Pandy, M.G. Accuracy of Generic Musculoskeletal Models in Predicting the Functional Roles of Muscles in Human Gait. *J. Biomech.* **2011**, *44*, 2096–2105. [CrossRef]

applied sciences

MDPI

Article

Effects of Diaphragmatic Therapy on Pelvic Floor Muscle Activity, Stress Levels, and Sexual Life Satisfaction in Polish Women

Joanna Golec [1], Sara Gamrot [1,*], Monika Michalik [2], Iwona Sulowska-Daszyk [1], Monika Nowak [3] and Joanna Balicka-Bom [3]

1 Institute of Clinical Rehabilitation, University of Physical Culture in Kraków, 31-571 Kraków, Poland; joanna.golec@awf.krakow.pl (J.G.); iwona.sulowska@awf.krakow.pl (I.S.-D.)
2 School of Scientific-Behavioral Psychotherapy, Faculty in Katowice, SWPS University, 40-053 Katowice, Poland; monika7.michalik@gmail.com
3 Faculty of Health Science, Andrzej Frycz Modrzewski University in Kraków, 30-705 Kraków, Poland; mprzybytek@uafm.edu.pl (M.N.); jbalicka-bom@uafm.edu.pl (J.B.-B.)
* Correspondence: sara.gamrot@awf.krakow.pl

Abstract: Pelvic floor muscles (PFMs) in women play a key role, and their proper functioning depends on the coordinated interaction with other anatomical structures, particularly the diaphragm and deep abdominal muscles, which together constitute the so-called core stabilizing unit. The aim of this study was to evaluate the effects of diaphragmatic breathing therapy on pelvic floor muscle function and stress levels in healthy women. The randomized, controlled, parallel-group trial (allocation 1:1) included 42 women aged 21–30 years who met the inclusion and exclusion criteria. The experimental group received diaphragmatic breathing therapy. The following assessment tools were used: Surface Electromyography (sEMG), the Sexual Satisfaction Questionnaire in Close Relationships (KSS) by M. Plopa, and the Perception of Stress Questionnaire (KPS) by M. Plopa and R. Makarowski. In the experimental group, a significant reduction in resting PFM activity was observed in the final stage of the measurement protocol, along with a tendency toward decreased activity during relaxation phases. A trend toward increased amplitude during phasic and tonic contractions was also noted, more pronounced after therapy than in the control group, although not statistically significant. No significant associations between stress dimensions and sexual satisfaction were found in the control group, whereas in the experimental group, higher worry, reduced sense of meaning, low agency and pessimism correlated with lower sexual satisfaction and difficulties achieving orgasm. These findings suggest that diaphragmatic breathing therapy may reduce resting pelvic floor muscle activity and perceived emotional stress.

Keywords: pelvic floor muscles; diaphragmatic breathing therapy; stress; sexual satisfaction; Polish women

1. Introduction

Pelvic floor muscles (PFMs) in women are integral to maintaining urinary and fecal continence, supporting pelvic organs, contributing to lumbopelvic stability, and enhancing sexual function [1]. The proper function of these muscles relies on their synergistic relationship with other anatomical structures, particularly the diaphragm and deep abdominal muscles, collectively referred to as the "core stabilizing unit" [2]. Effective coordination between the diaphragm and PFM is critical for regulating intra-abdominal pressure (IAP) and maintaining postural stability [3].

During inhalation, the diaphragm contracts and moves downward, increasing IAP and inducing reflexive relaxation of the PFM. During exhalation, the diaphragm ascends while the PFMs contract, thus supporting expiration and trunk stabilization [3]. Disruptions in this coordination—often associated with respiratory dysfunction—may lead to compensatory tension within the pelvic floor, potentially resulting in pelvic pain, urinary incontinence, and sexual dysfunction [2].

Impaired diaphragmatic function is not confined to individuals with diagnosed medical conditions but can also be observed in seemingly healthy individuals. This is largely attributed to the widespread prevalence of upper chest (thoracic) breathing patterns that fail to properly engage the diaphragm. Such dysfunctional breathing is increasingly common among individuals experiencing chronic stress, fast-paced lifestyles, and prolonged sedentary behavior [4].

In recent years, growing attention has been paid to the role of breathing techniques and diaphragmatic work in reducing stress and improving the functioning of the musculoskeletal and reproductive systems in women. Research suggests that interventions combining physiotherapy with psychological support can contribute to the normalization of cortisol levels and enhance overall well-being in women with endometriosis [5]. These findings may serve as a basis for further investigation into the efficacy of breathing techniques, such as diaphragmatic therapy, in relation to pelvic floor muscle function, stress levels, and sexual quality of life.

There is evidence suggesting that even a single session of diaphragmatic therapy may have beneficial effects on psychological stress responses or reduce depressive symptoms [6,7]. Furthermore, it has been observed that a single relaxation session induced by diaphragmatic breathing can enhance the antioxidant defense capacity in athletes following exhaustive exercise. These effects are associated with a decrease in cortisol levels [8].

A positive correlation between perceived stress levels and pain intensity suggests that interventions aimed at reducing stress may substantially alleviate pain symptoms [9,10]. These effects may be linked to dysregulation of the hypothalamic–pituitary–adrenal (HPA) axis and the action of glucocorticoids [11]. Studies have shown that women experiencing high levels of stress exhibit reduced genital response to sexual stimuli and elevated cortisol levels, both of which may negatively impact sexual functioning [12].

Despite increasing clinical interest, comprehensive studies investigating the interactions between diaphragmatic breathing, pelvic floor muscle activity, psychological stress, and sexual satisfaction in women remain scarce. This knowledge gap underscores the need for an integrative therapeutic approach that incorporates both musculoskeletal and psychosocial dimensions. By exploring these multidimensional relationships, we may enhance our understanding of how non-invasive myofascial techniques can contribute to holistic care strategies in women's health.

In light of these physiological and psychological interconnections, the present study aimed to determine whether targeted diaphragm therapy could affect pelvic floor muscle activity, perceived stress levels, and sexual satisfaction in women.

The research question that guided this study was: Does diaphragm therapy improve PFM activity, reduce stress levels, and enhance sexual satisfaction in women?

2. Materials and Methods

2.1. Participants

A total of 76 women were initially recruited for the project. Based on the inclusion and exclusion criteria, 42 Polish women aged 21–30 years (mean age: 24.79 ± 2.75) were ultimately included in the study. Participants' body mass ranged from 48 to 89 kg (mean: 61.04 ± 9.31), and their height ranged from 155 to 178 cm (mean: 167.17 ± 5.1). The body mass index (BMI) of the participants ranged from 17.16 to 32.69 (mean: 22.18 ± 3.03). A detailed characterization of the study group is provided in Table 1.

Table 1. Characteristics of the participants.

Parameter		Total (n = 42)
Age [years]	Mean (SD)	24.79 (2.75)
	Median (quartiles)	24 (23–26)
	Range	21–30
	n	42
Education level	Vocational	1 (2.38%)
	Secondary	16 (38.10%)
	University	25 (59.52%)
Type of relationship	Marriage	6 (14.29%)
	Informal	36 (85.71%)
Relationship duration [months]	Mean (SD)	40.29 (28.48)
	Median (quartiles)	36 (13.5–64.5)
	Range	2–96
	n	42

SD—standard deviation.

The inclusion criteria were as follows: provision of written informed consent, age between 20 and 30 years, no history of pregnancy, childbirth, or miscarriage (confirmed nulliparity), sexual activity, and absence of self-reported urogenital dysfunction, neurological disorders, visual impairments, or musculoskeletal conditions. Additionally, participants were required to have had no physiotherapy targeting the diaphragm or pelvic floor muscles within the previous six months.

The exclusion criteria were as follows: pregnancy, hypertension, diagnosed urogenital dysfunctions, pathological changes within the pelvic region, history of surgical procedures or injuries involving the abdominal cavity or spine within six months prior to the study, neurological or musculoskeletal disorders, systemic diseases, use of hormonal contraception, and participation in physiotherapy within six months prior to the commencement of the study.

Participant recruitment was conducted through advertisements targeting the young adult population in Kraków and the surrounding areas (Poland), posted on social media platforms (Instagram, Facebook). Recruitment took place between 11 June and 15 July 2024, and all measurements were completed by 30 October 2024.

The study was conducted in the Functional Diagnostics Laboratory, which is part of the Central Scientific Research Laboratory at the Bronisław Czech University of Physical Education in Kraków. Ethical approval for the study was obtained from the Bioethics Committee (KBKA/43/O/2024). The project was funded by the Ministry of Science under the program Regional Initiative of Excellence for the years 2024–2027 (project number RID/SP/0027/2024/01; total funding: PLN 4,053,904.00).

Participants were randomly assigned to two groups in a 1:1 ratio using simple randomization. Group allocation was determined by drawing a sealed, opaque envelope containing the group designation (I or II), which had been prepared by a researcher who was not involved in either the assessment or the intervention. Group I (n = 20) received diaphragmatic therapy, which included selected techniques targeting the diaphragm (diaphragm normalization, work on the central tendon, and work on the crura of the diaphragm). Group II (n = 22) served as a control group with no intervention.

A total of six therapeutic sessions were applied to the diaphragm over a two-week period. Therapy was administered three times per week (Monday, Wednesday, and Friday). To evaluate both groups, assessments were performed before and after the six therapy sessions in Group I, and at equivalent time points in Group II without intervention.

The study was a randomized controlled trial with parallel groups. The investigator responsible for the measurements was blinded to the participants' group allocation. The participants were not informed of the exact research hypothesis but were told that both groups would be evaluated in terms of physiological responses associated with diaphragm function.

Prior to participation, all individuals were informed about the study aims, procedures, and the therapy protocol, as well as their right to withdraw from the study at any time without consequences. Written informed consent was obtained from all participants prior to inclusion.

2.2. Research Tools

To assess the effectiveness of the applied therapy, standardized research instruments were employed, including surface electromyography (sEMG) to measure the bioelectrical activity of the pelvic floor muscles, the Perceived Stress Questionnaire (PSQ) developed by Mieczysław Plopa and Mieczysław Markowski to evaluate the level of psychological stress, and the Sexual Satisfaction Questionnaire in Close Relationships (SSQ) by Mieczysław Plopa to analyze the level of satisfaction with sexual life.

2.2.1. Surface Electromyography (sEMG)

Surface electromyography (sEMG) was employed to assess the bioelectrical activity of the PFM, using the Noraxon Ultium device (Noraxon U.S.A. Inc., Scottsdale, AZ, USA) and theLifecare PR-02 vaginal probe (Everyway Medical Instruments Co., Ltd., New Taipei City, Taiwan). Findings from other studies suggest that differences in probe placement during PFM activation did not have a significant impact on the recorded sEMG signals [13]. The signal was recorded with 16-bit resolution at a sampling frequency of 1500 Hz. sEMG signal processing was performed using the MyoResearch XP software version 1.0 (Noraxon USA, Inc., Scottsdale, AZ, USA). The vaginal probe, measuring 76 mm in length and 28 mm in diameter, featured two longitudinal stainless steel and nickel plates.

Prior to each measurement, the sEMG system was calibrated in accordance with the manufacturer's recommendations to ensure signal accuracy and minimize baseline noise. Signal quality was visually inspected before recording to verify stable electrode impedance and the absence of artifacts.

The electromyographic data were band-pass filtered between 10 and 500 Hz, rectified, and smoothed using RMS filtering with a 100 ms window. The signals were expressed in microvolts (µV) and as a function of frequency (Hz). In addition, the sEMG data were normalized to the maximum recorded RMS EMG amplitude of the pelvic floor muscles. performed at the beginning of the protocol.

Measurements were not conducted during menstrual bleeding. Prior to the assessment, participants were asked to completely empty their bladders. After donning protective gloves and applying conductive gel, the participant independently inserted the vaginal probe, positioning the electrodes toward the iliac crests. The physiotherapist verified the correct placement of the probe. The examination was carried out in a supine position, with a cushion placed under the knees to allow for slight flexion of the hip and knee joints. The spine was maintained in a neutral position, and the upper limbs were positioned parallel to the torso.

The study followed the assessment protocol described by Oleksy et al. [14], which consisted of five stages:

1. Participants were instructed to relax the pelvic floor muscle (PFM) following a 60-s rest period prior to the assessment, which was divided into three segments: I—5 s, II—5 s, and III—50 s.
2. Participants were then asked to contract the PFM as quickly as possible and to immediately and fully relax it. This phase included five phasic (rapid) contractions, each lasting 2 s, with 10-s rest intervals between contractions.
3. In the next phase, participants were instructed to contract the PFM as strongly as possible, maintain the contraction for 10 s, and then fully relax. This consisted of five tonic contractions, each followed by a 10-s rest.
4. Subsequently, participants were asked to contract the PFM to a submaximal level that could be sustained for 60 s without changes in muscle tension. This phase involved one 60-s endurance contraction.
5. Finally, participants were instructed to relax the PFM during a 60-s resting period following the exercises.

2.2.2. The Sexual Satisfaction Questionnaire in Close Relationships (SSQ) by M. Plopa

To assess the level of sexual satisfaction before and after the series of therapy sessions, the Sexual Satisfaction Questionnaire in Close Relationships (Poland) was used. This Polish psychometric instrument was developed for research on the subjective perception of sexual quality of life. The questionnaire consists of 10 items rated on a 7-point Likert scale (from 1—"definitely not" to 7—"definitely yes"). The SSQ comprises three subscales: intimacy, caresses, and sex. The sex subscale refers to physical aspects such as the level of satisfaction with sexual intercourse and the experience of orgasm, for example, "I am satisfied with the quality of our sexual intercourse". The intimacy subscale concerns the depth of the romantic relationship, communicative openness, mutual trust, the feeling of joy from being close to one's partner, cuddling, the attractiveness of the partner's scent, and the readiness for erotic intimacy, as illustrated by the statement, "I can openly talk with my partner about my feelings and desires". The caress subscale includes the assessment of satisfaction with foreplay, behaviors leading to intimacy, mutual touching, and other forms of erotic contact outside of sexual intercourse, for instance, "I like when my partner touches me affectionately" [15].

The questionnaire allows for the assessment of overall sexual satisfaction as well as specific dimensions of sexual quality of life. It was developed and validated within the Polish cultural context using data collected in Poland. Participants were randomly divided into two subsamples: one for Exploratory Factor Analysis (EFA) and another for Confirmatory Factor Analysis (CFA). Both analyses supported a clear three-factor structure, with excellent model fit indices (CFI = 0.96, RMSEA = 0.048, SRMR = 0.043), confirming the theoretical model's consistency with the empirical data. The instrument demonstrated high internal consistency, with Cronbach's alpha coefficients ranging from 0.83 to 0.88 for subscales and 0.91 for the total score, indicating strong reliability. Convergent validity was confirmed through positive correlations with the Marital Adjustment Test (MAT), the Satisfaction with Life Scale (SWLS), and M. Rosenberg's Self-Esteem Scale (SES) [15].

Discriminant validity analyses further supported the measure, revealing equivalence across gender and consistent relationships with age and relationship status. Overall, the questionnaire exhibits robust psychometric properties and serves as a reliable and valid tool for assessing sexual satisfaction in both research and clinical contexts [15].

2.2.3. The Perceived Stress Questionnaire (PSQ) by M. Plopa and R. Makarowski

To measure the subjective level of stress, the Perceived Stress Questionnaire (Poland), developed by M. Plopa and R. Makarowski in 2010, was used. This instrument is designed to assess individual differences in the experience and perception of stress in everyday functioning. The questionnaire consists of 27 test items, divided into the following subscales: emotional tension (e.g., "I often feel irritated for no apparent reason"), intrapsychic stress (e.g., "I feel that I am unable to meet my own expectations"), external stress (e.g., "I have the impression that too much is expected of me"), and a lie scale (e.g., "I never make promises I cannot keep"). Respondents rate each statement on a five-point Likert scale, ranging from "definitely not" to "definitely yes" [16].

In addition to subscale scores, it is possible to obtain a general stress index, which serves as a synthetic measure of psychological burden and represents the sum of the three subscales. The questionnaire also includes a lie scale, which makes it possible to identify individuals who present themselves in an overly favorable light by attributing to themselves socially desirable behaviors that are in fact rare. Such a tendency may stem from a lack of self-criticism, limited insight, or deliberate misrepresentation [16].

The PSQ demonstrates good psychometric properties. The authors report internal consistency coefficients (Cronbach's alpha) ranging from 0.70 to 0.81, indicating that it is a reliable tool for measuring stress levels in both men and women. The reliability coefficients for women in individual scales are as follows: for Emotional Tension—Cronbach's alpha = 0.82, for External Stress—Cronbach's alpha = 0.70, for Intrapsychic Stress—Cronbach's alpha = 0.70, and for the Lie Scale—Cronbach's alpha = 0.57 [16].

The validity of the questionnaire is satisfactory and has been confirmed based on correlation coefficients with other psychological instruments: the NEO-FFI, C. D. Spielberger's Questionnaire, and the BDI—Beck Depression Inventory. The normalization of the PSQ was carried out on the basis of studies conducted among individuals representing both urban and rural populations in Poland. The sample was diverse in terms of gender, age, marital status, education, place of upbringing, occupation, and health condition. The questionnaire is widely used in psychological and clinical research, as well as in diagnostic practice [16].

2.3. *Intervention*

In the enrolled participants, therapy was performed using manual techniques aimed at normalizing diaphragmatic [17]. Diaphragmatic tension was assessed manually by a qualified physiotherapist. The therapeutic protocol included three manual techniques:

- Normalization of the costal arches (bilateral work on the costal margins): During this technique, participants were instructed to take a deep inhalation, while the therapist facilitated the upward and outward movement of the costal arches. During exhalation, the therapist maintained the achieved movement and prevented regression.
- Work on the central tendon of the diaphragm: The therapist placed the heel of the hand above the umbilicus and followed the participant's inhalation and exhalation. During exhalation, the heel of the hand was gradually pressed deeper into the tissue and rotated clockwise.
- Work on the diaphragmatic crura: This technique was performed with the participant in the prone position. The therapist placed one hand on the participant's thigh in the popliteal fossa region and the other hand over the diaphragmatic crura at the level of L1–L2, lateral to the spine, on the side where the therapist was positioned. The

participant was instructed to inhale deeply and then exhale. During exhalation, the participant was asked to maximally relax the entire body. With subsequent repetitions, the distance between the therapist's hands was gradually increased.

Each technique was performed in synchronization with the participant's breathing cycle, with inhalation and exhalation repeated several times. The duration of each therapeutic session was not predetermined but depended on the achievement of improved diaphragmatic mobility.

2.4. Statistical Analysis

The statistical analysis was performed using STATISTICA 12.0 Pl (Statsoft Polska, Krakow, Polska). Qualitative variables were presented as counts and percentages, while quantitative variables were expressed as mean and standard deviation. In the analysis of questionnaire results, comparisons of quantitative variables between two repeated measurements were performed using the Wilcoxon test. In the analyses based on questionnaire data, no correction for multiple testing was applied, as these analyses were exploratory. To assess differences in pelvic floor muscle activity assessed by electromyography, a two-way ANOVA was conducted, with one main factor between groups (experimental vs. control) and the other main factor being the repeated measure (time: pre- and post-intervention or pre- and post-break). For the analysis of PFM activity, post hoc comparisons were conducted using Tukey's test to control for multiple comparisons. The Shapiro–Wilk test was used to assess the normality of variable distribution. The homogeneity of variances across groups was evaluated using Levene's test. Effect sizes were determined with Cohen's d and classified as negligible (<0.1), small (0.1–0.3), medium (0.3–0.5), or large (>0.5) [18]. Spearman's correlation coefficient was used to assess correlation between two quantitative variables. Differences were considered statistically significant if the test probability level was lower than the assumed significance level ($p < 0.05$). Based on data from another study including the bioelectrical activity of the pelvic floor muscles, a power analysis for the two-way ANOVA determined that a minimum of 16 participants per group was required to achieve a statistical power of 0.80 at a two-tailed significance level of 0.05, assuming an effect size of d = 0.8. This analysis was based on data derived from previous literature on the primary outcome variables such as PFM bioelectrical activity [19].

3. Results

3.1. Bioelectrical Activity of the PFM

3.1.1. Resting Activity of the PFM During a 1-min Rest Period (First Phase of the Measurement Protocol—Pre-Exercise)

In the second measurement, a decrease in pre-exercise resting activity of the pelvic floor muscles was observed in both groups; however, these changes were not statistically significant ($p > 0.05$). The calculated effect size in the experimental group indicated a medium effect. The results are presented in Table 2.

Table 2. Resting activity of the PFM during a 1-min rest period (first phase of the measurement protocol—pre-exercise).

	Measurement	Experimental Group (n = 20)		p	ES a	Control Group (n = 22)		p	ES a	p *	ES b
		Mean ± SD	%MVC			Mean ± SD	%MVC				
Average mean amplitude (μV)	Before	9.57 ± 5.01	11.14 ± 5.77	0.395	−0.347	9.56 ± 6.49	11.60 ± 6.83	0.136	−0.248	0.995	−0.002
	After	8.03 ± 3.78	9.58 ± 5.04			7.94 ± 6.60	9.45 ± 8.56			1.000	−0.017

SD—standard deviation: p—between measurements; p *—between groups; ES a—effect size (Cohen's d) within each group; ES b—effect size (Cohen's d) between study groups; %MVC—percent of maximal voluntary contraction; μV—microvolts; %—percent.

3.1.2. Activity of the PFM During 2-s Phasic Contractions

During the assessment of 2-s phasic contractions, both groups showed an increasing trend in amplitude during contraction and a decreasing trend in resting pelvic floor muscle activity between contractions. However, observed changes were not statistically significant ($p > 0.05$). The data are presented in Table 3.

Table 3. Activity of the PFM during 2-s phasic contractions.

	Measurement	Experimental Group (n = 20)		p	ES a	Control Group (n = 22)		p	ES a	p *	ES b
		Mean ± SD	%MVC			Mean ± SD	%MVC				
Average peak amplitude (μV)	Before	82.51 ± 33.35	88.02 ± 16.79	0.923	0.109	74.96 ± 32.89	86.03 ± 17.60	0.984	0.054	0.859	−0.228
	After	85.71 ± 24.98	94.03 ± 12.48			76.72 ± 31.80	92.19 ± 25.23			0.784	−0.314
Time before peak (s)	Before	45.07 ± 15.46	-	0.922	0.192	52.61 ± 13.34	-	0.820	−0.257	0.355	0.522
	After	47.93 ± 14.24	-			48.84 ± 15.89	-				0.060
Average resting tension between phasic contractions (μV)	Before	12.74 ± 6.73	14.45 ± 6.78	0.623	−0.257	12.59 ± 7.11	15.93 ± 10.77	0.437	−0.279	1.000	−0.022
	After	11.17 ± 5.40	12.57 ± 5.16			10.72 ± 6.26	13.54 ± 8.20			0.996	−0.077

SD—standard deviation; p—between measurements; p *—between groups; ES a—effect size (Cohen's d) within each group; ES b—effect size (Cohen's d) between study groups; %MVC—percent of maximal voluntary contraction; μV—microvolts; %—percent; s—second.

3.1.3. Activity of the PFM During 10-s Tonic Contractions

Consistent with the findings for phasic contractions, the analysis of 10-s tonic contractions in the second assessment demonstrated an increase in contraction amplitude. Following the applied therapy, the value of amplitude normalized to the maximum voluntary contraction (MVC) increased by 11.55% in the experimental group, while in the control group, the increase was 7.41%. However, these changes were not statistically significant in either group (Table 4).

Table 4. Activity of the PFM during 10-s tonic contractions.

	Measurement	Experimental Group (n = 20)		p	ES a	Control Group (n = 22)		p	ES a	p *	ES b
		Mean ± SD	%MVC			Mean ± SD	%MVC				
Average mean amplitude (μV)	Before	47.52 ± 21.02	50.55 ± 12.37	0.734	0.211	47.44 ± 25.59	53.06 ± 11.31	1.000	0.010	1.000	−0.003
	After	51.47 ± 16.15	56.39 ± 10.48			47.67 ± 20.54	56.99 ± 18.43			0.938	−0.206
Time before peak (s)	Before	41.79 ± 18.92	-	0.883	−0.187	43.12 ± 21.66	-	0.958	−0.113	0.996	0.065
	After	38.47 ± 16.58	-			40.96 ± 16.17	-			0.972	0.152
Average resting tension between tonic contractions (μV)	Before	10.13 ± 4.74	11.93 ± 5.84	0.796	0.189	10.67 ± 6.51	13.08 ± 7.63	0.987	−0.105	0.996	0.095
	After	11.96 ± 12.83	12.40 ± 7.74			10.03 ± 5.65	12.90 ± 8.37			0.863	−0.195

SD—standard deviation; p—between measurements; p *—between groups; ES a—effect size (Cohen's d) within each group; ES b—effect size (Cohen's d) between study groups; %MVC—percent of maximal voluntary contraction; μV—microvolts; %—percent; s—second.

3.1.4. Activity of the PFM During 60-s Endurance Contractions

In the experimental group, during endurance contractions following the applied therapy, an upward trend in amplitude was observed, whereas the control group showed a slight decrease in amplitude in the second assessment. These changes were not statistically significant. Analysis of the fatigue index showed that the experimental group achieved a higher score following the applied therapy, indicating a medium effect size, whereas a decrease in this index was observed in the control group. The data are presented in Table 5.

Table 5. Activity of the PFM during 60-s endurance contraction.

	Measurement	Experimental Group (n = 20)		p	ES a	Control Group (n = 22)		p	ES a	p *	ES b
		Mean ± SD	%MVC			Mean ± SD	%MVC				
Average Mean Amplitude (µV)	Before	22.63 ± 8.54	25.50 ± 8.03	0.869	0.160	25.67 ± 17.28	28.98 ± 10.57	0.929	−0.098	0.896	0.223
	After	24.62 ± 15.38	26.05 ± 7.83			24.17 ± 13.01	28.68 ± 7.86			1.000	−0.032
Median Frequency (Hz)	Before	53.45 ± 6.84	-	0.679	−0.195	56.69 ± 10.73	-	0.992	−0.043	0.696	0.360
	After	51.68 ± 10.85	-			56.26 ± 9.38	-			0.421	0.452
Mean Frequency (Hz)	Before	65.74 ± 7.05	-	0.777	−0.190	70.61 ± 11.78	-	0.997	−0.034	0.448	0.502
	After	63.91 ± 11.62	-			70.23 ± 10.86	-			0.225	0.562
F ratio	Before	1.02 ± 0.13	-	0.709	0.320	1.04 ± 0.12	-	0.211	−0.660	0.967	0.160
	After	1.06 ± 0.12	-			0.97 ± −0.09	-			0.065	−0.849

SD—standard deviation; p—between measurements; p *—between groups; ES a—effect size (Cohen's d) within each group; ES b—effect size (Cohen's d) between study groups; %MVC—percent of maximal voluntary contraction; µV—microvolts; %—percent; Hz—hertz, f ratio—ratio of the median frequency of the third interval to the second interval.

3.1.5. Resting Activity of the PFM During a 1-min Rest Period (Last Phase of the Measurement Protocol—Post-Exercise)

In the experimental group, during the 1-min rest period in the last stage of the measurement protocol, a significant reduction in resting PFM activity was observed after therapy ($p < 0.05$). No significant changes were noted in the control group in the second assessment. The results are presented in Table 6.

Table 6. Resting activity of the PFM during a 1-min rest period (last phase of the measurement protocol—post-exercise).

	Measurement	Experimental Group (n = 20)		p	ES a	Control Group (n = 22)		p	ES a	p *	ES b
		Mean ± SD	%MVC			Mean ± SD	%MVC				
Average mean amplitude (µV)	Before	7.93 ± 4.00	9.40 ± 5.54	0.025	−0.431	8.79 ± 6.57	10.40 ± 6.33	0.419	−0.179	0.950	0.158
	After	6.15 ± 4.25	5.45 ± 3.86			7.71 ± 5.41	9.69 ± 7.23			0.767	0.321

SD—standard deviation; p—between measurements; p *—between groups; ES a—effect size (Cohen's d) within each group; ES b—effect size (Cohen's d) between study groups; %MVC—percent of maximal voluntary contraction; µV—microvolts; %—percent; the red color—statistically significant.

3.2. The Sexual Satisfaction Questionnaire in Close Relationships (SSQ) by M. Plopa and the Perceived Stress Questionnaire (PSQ) by M. Plopa and R. Makarowski

In the intervention group, statistically significant changes were observed in two stress-related variables: emotional tension and overall stress level, as measured by the Perceived Stress Questionnaire (PSQ). The emotional tension score was significantly lower after the intervention compared to baseline ($p = 0.006$), showing a 23% reduction. The overall stress level also decreased significantly by 17% ($p = 0.012$). The scores on the PSQ lie scale did not reach statistical significance in either group (Table 7).

Table 7. Comparison of the Perceived Stress Questionnaire (PSQ) and the Sexual Satisfaction Questionnaire in Close Relationships (SSQ) scores before and after the intervention in the therapy group.

Parameter	Measurement	N	Mean	SD	Median	Min	Max	Q1	Q3	p
SSQ: intimacy [sten]	Before intervention	20	5.20	1.82	5.5	1	8	4.75	6.25	$p = 1$
	After intervention	20	5.15	1.66	5.0	1	8	4.00	7.00	
SSQ: caresses [sten]	Before intervention	20	4.40	1.85	5.0	1	8	3.00	5.00	$p = 0.918$
	After intervention	20	4.45	1.73	4.5	1	8	3.00	5.00	
SSQ: sex [sten]	Before intervention	20	5.15	2.06	6.0	1	8	3.75	6.25	$p = 0.051$
	After intervention	20	6.00	1.97	6.5	1	8	5.00	7.25	
SSQ: total [sten]	Before intervention	20	5.60	1.93	6.0	1	9	4.00	7.00	$p = 0.253$
	After intervention	20	5.95	1.85	6.0	1	10	5.00	7.00	

Table 7. *Cont.*

Parameter	Measurement	N	Mean	SD	Median	Min	Max	Q1	Q3	p
PSQ: emotional tension [sten]	Before intervention	20	5.15	1.95	5.5	1	8	3.75	6.25	$p = 0.006$ *
	After intervention	20	3.95	1.88	3.5	1	8	3.00	5.00	
PSQ: external stress [sten]	Before intervention	20	2.85	1.46	2.5	1	6	2.00	4.00	$p = 0.112$
	After intervention	20	2.45	1.28	2.0	1	5	1.00	3.25	
PSQ: intrapsychic stress [sten]	Before intervention	20	3.35	1.69	3.0	1	8	2.00	4.00	$p = 0.821$
	After intervention	20	3.30	1.92	3.0	1	9	2.00	4.00	
PSQ: lie scale [sten]	Before intervention	20	5.80	2.95	5.0	1	10	3.75	9.00	$p = 0.566$
	After intervention	20	5.50	2.61	5.0	1	10	3.75	7.00	
PSQ: total [sten]	Before intervention	20	3.75	1.59	4.0	1	7	3.00	5.00	$p = 0.012$ *
	After intervention	20	3.10	1.59	3.0	1	7	2.00	4.00	

p—Wilcoxon paired test * and the red color—statistically significant ($p < 0.05$).

In the control group, no statistically significant changes ($p > 0.05$) were observed in any of the assessed dimensions, either in the SSQ or the PSQ questionnaires. Similarly, the scores on the PSQ lie scale were not statistically significant ($p > 0.05$). The results are presented in Table 8.

Table 8. Comparison of the Perceived Stress Questionnaire (PSQ) and the Sexual Satisfaction Questionnaire in Close Relationships (SSQ) scores before and after the intervention in the control group.

Parameter	Measurement	N	Mean	SD	Median	Min	Max	Q1	Q3	p
SSQ: intimacy [sten]	Before intervention	22	5.68	1.32	5.5	3	8	5.00	6.00	$p = 0.851$
	After intervention	22	5.73	1.35	6.0	3	9	5.00	6.00	
SSQ: caresses [sten]	Before intervention	22	4.73	1.80	5.0	1	8	3.25	6.00	$p = 0.376$
	After intervention	22	4.95	1.70	5.0	1	8	4.00	5.75	
SSQ: sex [sten]	Before intervention	22	5.82	1.68	6.0	2	9	5.00	7.00	$p = 0.08$
	After intervention	22	6.14	1.67	6.0	2	9	5.25	7.00	
SSQ: total [sten]	Before intervention	22	6.36	1.50	6.0	3	9	6.00	7.00	$p = 0.212$
	After intervention	22	6.59	1.40	6.0	4	10	6.00	7.00	
PSQ: emotional tension [sten]	Before intervention	22	3.91	1.93	3.5	1	7	3.00	5.75	$p = 0.503$
	After intervention	22	3.77	1.63	3.5	1	7	3.00	5.00	
PSQ: external stress [sten]	Before intervention	22	2.45	1.60	2.0	1	7	1.00	3.00	$p = 0.39$
	After intervention	22	2.64	1.65	2.0	1	7	1.25	3.00	
PSQ: intrapsychic stress [sten]	Before intervention	22	2.95	1.76	3.0	1	6	1.25	4.00	$p = 1$
	After intervention	22	2.95	1.62	2.0	1	6	2.00	4.00	
PSQ: lie scale [sten]	Before intervention	22	6.32	3.36	7.0	1	10	3.00	9.75	$p = 0.495$
	After intervention	22	6.09	3.10	7.0	1	10	3.00	9.00	
PSQ: total [sten]	Before intervention	22	3.00	1.72	3.0	1	7	1.25	4.00	$p = 0.803$
	After intervention	22	3.05	1.73	3.0	1	7	2.00	3.75	

p—Wilcoxon paired test.

Correlation analysis in the intervention group revealed significant and consistent associations. The SSQ intimacy subscale score after the intervention showed a significant positive correlation with the PSQ lie scale (r = 0.628; $p = 0.003$). A similar pattern was observed for the SSQ caress subscale, which was significantly associated with the PSQ lie scale (r = 0.461; $p = 0.041$). Furthermore, the SSQ lie scale positively correlated with the overall SSS score (r = 0.570; $p = 0.009$). Statistically significant correlations are presented in Table 9. Detailed results of the analysis are provided in Supplementary Table S1.

Table 9. Correlation between the Sexual Satisfaction Questionnaire in Close Relationships (SSQ) and the Perceived Stress Questionnaire (PSQ)—therapy group.

	R	*p*
SSS: intimacy [sten]—after intervention & SSQ: lie scale [sten]—after intervention	r = 0.628	*p* = 0.003 *
SSS: caresses [sten]—after intervention & SSQ: lie scale [sten]—after intervention	r = 0.461	*p* = 0.041 *
SSS: total [sten]—after intervention & SSQ: lie scale [sten]—after intervention	r = 0.570	*p* = 0.009 *
SSQ: intrapsychic stress [sten]—after intervention & SSS: sex [sten]—after intervention	r = −0.457	*p* = 0.043 *
SSQ:lie scale [sten]—after intervention & SSS: intimacy [sten]—after intervention	r = 0.628	*p* = 0.003 *
SSQ:lie scale [sten]—after intervention & SSS: caresses [sten]—after intervention	r = 0.461	*p* = 0.041 *
SSQ:lie scale [sten]—after intervention & SSS: total [sten]—after intervention	r = 0.570	*p* = 0.009 *
	R	*p*
SSQ: intimacy [sten]—after intervention & PSQ: lie scale [sten]—after intervention	r = 0.628	*p* = 0.003 *
SSQ: caresses [sten]—after intervention & PSQ: lie scale [sten]—after intervention	r = 0.461	*p* = 0.041 *
SSQ: total [sten]—after intervention & PSQ: lie scale [sten]—after intervention	r = 0.570	*p* = 0.009 *
PSQ: intrapsychic stress [sten]—after intervention & SSQ: sex [sten]—after intervention	r = −0.457	*p* = 0.043 *
PSQ:lie scale [sten]—after intervention & SSQ: intimacy [sten]—after intervention	r = 0.628	*p* = 0.003 *
PSQ:lie scale [sten]—after intervention & SSQ: caresses [sten]—after intervention	r = 0.461	*p* = 0.041 *
PSQ:lie scale [sten]—after intervention & SSQ: total [sten]—after intervention	r = 0.570	*p* = 0.009 *

Spearman's correlation coefficient * and the red color—statistically significant ($p < 0.05$).

In the control group, unlike in the intervention group, no significant correlations ($p > 0.05$) were found between PSQ stress dimensions and SSQ intimacy or sexual satisfaction indicators. Similarly, no significant correlations ($p > 0.05$) were observed for the PSQ lie scale. Detailed results of the conducted analysis are provided in Supplementary Table S2.

3.3. *Correlations Between Standardized Questionnaire Outcomes and Pelvic Floor Muscles*

In the assessment of correlations between the results obtained from standardized questionnaires and the analysis of pelvic floor muscle function prior to the intervention, several associations were observed in the experimental group. A significant ($p < 0.05$) positive correlation was noted between the following variables: Average peak amplitude during phasic and tonic contractions [%MVC] and the SSQ intimacy scale score, Average resting tension between phasic contractions [μV] and the PSQ external stress scale score, Post-exercise resting activity [%MVC] and the PSQ intimacy scale score, as well as between Maximal voluntary contraction and the PSQ external stress scale score. Following the intervention, a significant ($p < 0.05$) negative correlation was observed in the experimental group between Pre-exercise resting activity [%MVC] and the SSQ caress scale score. In contrast, the variable Average peak amplitude during phasic contractions [μV] showed a significant ($p < 0.05$) positive correlation with the SSQ sex scale score. Statistically significant correlations are presented in Table 10. Detailed results of the analysis are provided in Supplementary Tables S3 and S4.

In the control group, during the first measurement, significant ($p < 0.05$) negative correlations were observed between the following variables: Average peak amplitude during phasic contractions [%MVC] and the overall SSQ score, Median frequency during endurance contraction [Hz] and both the PSQ emotional tension scale and the PSQ overall score, as well as Mean frequency during endurance contraction [Hz] and the PSQ overall score. In the second assessment within the control group, significant ($p < 0.05$) positive correlations were found between the following variables: Average peak amplitude during phasic contractions [μV] and both the PSQ external stress scale and the PSQ overall score; Average resting tension between phasic contractions [μV] and the PSQ emotional tension scale, PSQ external stress scale, and PSQ overall score; Average peak amplitude during tonic contractions [μV] and the PSQ external stress scale; as well as Average resting tension between tonic contractions [μV] and both the PSQ emotional tension and overall scales. Additionally, a significant ($p < 0.05$) negative correlation was observed between Mean frequency during endurance contraction [Hz] and the PSQ emotional tension scale. Statistically significant correlations are presented in Table 11. Detailed results of the analysis are provided in Supplementary Tables S5 and S6.

Table 10. Correlations between standardized questionnaire outcomes and pelvic floor muscles—Experimental group.

		R	p
Before intervention	Average peak amplitude during phasic contractions [%MVC]—before intervention & SSQ: intimacy [sten]—before intervention	r = 0.6	p = 0.005 *
	Average resting tension between phasic contractions [μV]—before intervention & PSQ: external stress [sten]—before intervention	r = 0.565	p = 0.009 *
	Average peak amplitude during tonic contractions [%MVC]—before intervention & SSQ: intimacy [sten]—before intervention	r = 0.545	p = 0.013 *
	Post-exercise resting activity [%MVC]—before intervention & SSQ: intimacy [sten]—before intervention	r = 0.474	p = 0.035 *
After intervention	Pre-exercise resting activity [%MVC]—after intervention & SSQ: caresses [sten]—after intervention	r = −0.555	p = 0.017 *
	Average peak amplitude during phasic contractions [μV]—after intervention & SSQ: sex [sten]—after intervention	r = 0.458	p = 0.042 *

Spearman's correlation coefficient * and the red color—statistically significant ($p < 0.05$).

Table 11. Correlations between standardized questionnaire outcomes and pelvic floor muscles—Control group.

		R	p
Before intervention	Average peak amplitude during phasic contractions [%MVC]—before intervention & SSQ: total [sten]—before intervention	r = −0.428	p = 0.047 *
	Median frequency during endurance contraction [Hz]—before intervention & PSQ: emotional tension [sten]—before intervention	r = −0.434	p = 0.043 *
	Median frequency during endurance contraction [Hz]—before intervention & PSQ: total [sten]—before intervention	r = −0.471	p = 0.027 *
	Mean frequency during endurance contraction [Hz]—before intervention & PSQ: intrapsychic stress [sten]—before intervention	r = −0.462	p = 0.031 *
	Mean frequency during endurance contraction [Hz]—before intervention & PSQ: total [sten]—before intervention	r = −0.522	p = 0.013 *
	F ratio during endurance contraction—before intervention & PSQ: emotional tension [sten]—before intervention	r = −0.631	p = 0.002 *
	F ratio during endurance contraction—before intervention & PSQ: intrapsychic stress [sten]—before intervention	r = −0.428	p = 0.047 *
	F ratio during endurance contraction—before intervention & PSQ: total [sten]—before intervention	r = −0.539	p = 0.01 *
After intervention	Average peak amplitude during phasic contractions [μV]—after intervention & PSQ: external stress [sten]—after intervention	r = 0.509	p = 0.015 *
	Average peak amplitude during phasic contractions [μV]—after intervention & PSQ: total [sten]—after intervention	r = 0.429	p = 0.046 *
	Average resting tension between phasic contractions [μV]—after intervention & PSQ: external stress [sten]—after intervention	r = 0.459	p = 0.032 *
	Average resting tension between phasic contractions [μV]—after intervention & PSQ: total [sten]—after intervention	r = 0.574	p = 0.005 *
	Average peak amplitude during tonic contractions [μV]—after intervention & PSQ: external stress [sten]—after intervention	r = 0.438	p = 0.041 *
	Average resting tension between tonic contractions [μV]—after intervention & PSQ: total [sten]—after intervention	r = 0.482	p = 0.023 *
	Mean frequency during endurance contraction [Hz]—after intervention & SSQ:emotional tension [sten]—after intervention	**r = −0.462**	p = 0.031 *

Spearman's correlation coefficient * and the red color—statistically significant ($p < 0.05$).

4. Discussion

The respiratory diaphragm plays a pivotal role not only in the mechanics of breathing but also in trunk stabilization, regulation of intra-abdominal pressure, and coordination of postural muscle activity [20]. In the present study, we aimed to evaluate the extent to which targeted diaphragm therapy may influence pelvic floor muscle activity, perceived stress levels, and sexual life satisfaction in women.

This relationship can be explained by the anatomical and fascial continuity between the diaphragm, the transversus abdominis muscle, and the pelvic floor muscles. The transversalis fascia originates from the cervical and mediastinal fascia and extends downward to the pubic symphysis, where it connects with the pelvic fascia. The transversus abdominis muscle, as a continuation of the endothoracic fascia, functionally links the diaphragm with the pelvic floor, forming an integrated fascial system responsible for trunk stabilization and the regulation of intra-abdominal pressure [21]. Any tension or restriction along this pathway may affect diaphragm mobility and impair the coordination between the diaphragm and the pelvic floor muscles [22]. Some studies have shown that applying manual therapy to one structure along a fascial line can have a significant effect on other structures within the same line. Likewise, Marizeiro et al. [23] reported that two sessions of manual diaphragm release significantly improved chest wall expansion, respiratory muscle strength, posterior chain muscle flexibility, and lumbar spine range of motion.

The most novel finding of this study was that the applied diaphragm therapy significantly reduced the resting activity of the pelvic floor muscles. Analysis of the results obtained from standardized questionnaires revealed that the intervention significantly reduced emotional tension and overall stress levels. In both groups, at both measurement points, a higher level of stress was associated with greater resting tension of the pelvic floor muscles, and this correlation was statistically significant.

In recent years, knowledge regarding the functioning of the PFM has expanded considerably, allowing for a more holistic perspective on the complex synergies between the PFM and respiratory mechanisms. Studies by other authors emphasize that the PFMs do not function in isolation but rather cooperate with the muscles of the anterolateral abdominal wall, and that these two muscle groups form a system closely linked to respiratory processes [3,24–27]. Moreover, the superficial and deep layers of the PFM do not always exhibit simultaneous contractile activity and can be activated separately, both during voluntary activation and during various tasks related to respiratory function [28–30]. Similar observations were reported by Park et al. [2], who confirmed in their studies that pelvic floor muscle contraction is associated with diaphragmatic mobility and influences pulmonary function.

Scientific literature emphasizes that breathing techniques may influence both brain function and the autonomic nervous system, which is likely one of the primary mechanisms underlying their stress-reducing effects. Appropriately selected breathing techniques increase the activity of the parasympathetic nervous system, thereby helping to balance the excessive sympathetic arousal characteristic of stress and anxiety states [31]. Moreover, the synchronization of brain rhythms with respiration promotes modulation of neuronal activity and affects mood regulation and cognitive processes [32]. Some sources highlight that, on the one hand, psychological load shapes breathing patterns [33–35], while on the other, an altered breathing pattern may influence the intensity of emotional experiences [36–38].

Hopper et al. [39], in a systematic review, concluded that diaphragmatic breathing may be a widely applicable intervention for reducing both psychological and physiological stress. Evidence presented in the systematic review indicates that diaphragmatic breathing can decrease physiological stress measured by blood pressure, respiration, and cortisol levels, as well as psychological stress assessed by the stress subscale of the DASS-21. They also noted a lack of evidence that unequivocally measures the effects of diaphragmatic breathing on stress reduction using hard biomarkers [39].

In the literature, surface electromyography (sEMG) is described as a simple and non-invasive method for assessing the PFM [14,40]. In our study, following diaphragmatic breathing therapy, a significant decrease in the resting activity of the pelvic floor muscles was observed during the final phase of the measurement protocol. In addition to assessing the ability of the PFM to relax, their activation level during phasic and tonic contractions was also evaluated. The second measurement revealed a tendency toward increased signal amplitude, indicating greater PFM activation during contractions. This change was more pronounced in the experimental group following the applied therapy than in the control group; however, it did not reach statistical significance. Studies with larger sample sizes may help to identify potential differences; therefore, further research in this area is warranted.

Several studies in the literature have investigated the hypopressive technique in relation to pelvic floor muscle function. López-Torres et al. [41] demonstrated that a structured two-month abdominal hypopressive training program significantly improved PFM strength. Other studies suggest that diaphragmatic breathing exercises may serve as an alternative to pelvic floor muscle training in the management of urinary incontinence [42,43].

In our study, the intervention protocol consisted of six sessions conducted over a two-week period (three sessions per week). This schedule is consistent with international studies on manual diaphragm therapy, in which short-term interventions have demonstrated measurable physiological and clinical effects. Rocha et al. [44] applied a manual diaphragm release technique to patients with chronic obstructive pulmonary disease and observed significant improvements in diaphragmatic mobility, inspiratory capacity, and exercise tolerance after six to eight sessions conducted over a two- to three-week period. Similarly, Cheng et al. [21] implemented a two-week intervention in individuals with chronic neck pain, achieving improvements in pain intensity, disability, and diaphragmatic function.

Analyzing the correlations between the results of standardized questionnaires and PFM activity revealed the presence of certain relationships. In the experimental group before the intervention, a positive correlation was observed between PFM activation levels and the score on the intimacy scale. A similar relationship was noted between PFM resting tension and stress level. Following the applied therapy, there was a tendency for greater PFM activation during phasic contractions to accompany higher scores on the sex scale, and for lower PFM resting tension to be associated with caressing. In the control group, both positive and negative correlations were observed, primarily concerning variables related to emotional tension and stress, although the direction and strength of these relationships were not consistent. In summary, the correlation analysis showed that in both groups, a higher level of stress was associated with greater resting tension of the PFM. This observation is consistent with previous reports indicating that PFM overactivity may represent a chronic response to stress [45].

The results of our study show convergence in certain areas with the findings of Petrelluzzi et al. [5], who implemented a comprehensive therapeutic program in women with endometriosis, combining physiotherapy with psychological intervention. This program consisted of breathing techniques, relaxation techniques aimed at reducing PFM tension, and elements of cognitive–behavioral therapy. After the eight-week program, the authors observed a significant decrease in cortisol levels, improvement in subjective vitality, as well as a reduction in pain and stress symptoms. In our study, the use of diaphragm therapy may be regarded as part of a similar approach in which regulation of muscle tension plays an important role. This effect translates not only into improved control of PFM but may also influence the psychosexual aspects of women's lives, including perceived sexual satisfaction. The findings of Petrelluzzi et al. therefore point to potential mechanisms through which breathing techniques may support the overall improvement of women's quality of life [5].

In our study, following diaphragmatic breathing therapy, women exhibited a reduction in overall perceived stress, particularly in the dimension related to emotional tension manifested by anxiety, nervousness, and difficulty relaxing. The decrease in emotional tension was also associated with reduced irritability, especially in interpersonal relationships, as well as with a lower sense of fatigue.

Donatti L. et al. [46] demonstrated a positive relationship between stress levels and the severity of pelvic pain in women suffering from deep endometriosis, reflecting the connection between emotional state and physiological processes.

Psychological and physiological stress are associated with a range of biochemical and physiological responses that exceed the body's resting level. Exposure to stressors leads, among other effects, to the release of cortisol and adrenaline, increased heart rate, elevated electrodermal activity, blood pressure, and muscle tension [47]. One of the interventions used to reduce stress is breathing exercises, including diaphragmatic breathing. Experimental studies show that even a few minutes of such exercises can influence emotional state. Breathing at a rate of approximately six breaths per minute is associated with decreased arousal and an increased sense of positive energy. Prolonging the exhalation phase relative to inhalation further promotes a sense of control, relaxation, mindfulness, and stress reduction [48]. Other scientific studies likewise indicate that individuals practicing specific breathing techniques exhibit lower levels of tension and achieve mental clarity more easily [49,50].

Stress is a complex construct whose sources and consequences may take various forms. As noted by Plopa [15], its internal and external dimensions include, among others, intrapsychic, emotional, and environmental stress. A distinction between acute and chronic stress is also widely accepted. Acute stress is a short-term reaction to a sudden threat (e.g., abrupt vehicle braking), whereas chronic stress has long-term effects and negatively impacts quality of life, including functioning and sexual satisfaction [51,52]. Kanner et al. [53] emphasize that everyday minor stressors may produce more harmful health consequences than rare but more severe events. Wrześniewski, in turn, points out that it is the interpretation of a situation rather than the stressful event itself that plays a key role [54,55].

In our study, the Perceived Stress Questionnaire (PSQ) based on the cognitive–transactional theory of Lazarus and Folkman [56], in which stress is defined as a relationship between the individual and the environment, was used [56,57]. This instrument enables the assessment of subjective stress perception, understood as internal or external pressure. An analogous approach is represented by the Perceived Stress Scale (PSS) in the Polish adaptation by Juczyński and Ogińska-Bulik [58,59]. In turn, the CISS developed by S. Cohen, T. Kamarck, and R. Mermelstein and Mini-COPE developed by Charles S. Carver allow for the evaluation of task-oriented, emotion-oriented, and avoidance coping styles. In the present study, significant relationships were found between these strategies and tendencies to distort responses [57,58].

The relationship between stress and sexual functioning is bidirectional: stress hinders sexual life, and sexual difficulties increase stress levels [60]. The view that psychological stress inhibits sexual responses is supported, among others, by the studies of Bancroft [61], who points to the natural occurrence of inhibition under conditions of fatigue or emotional tension. Particularly relevant in this context is the Basson model [62], according to which the female sexual response is nonlinear and strongly conditioned by emotional and relational factors. This model emphasizes the importance of emotional closeness, relationship satisfaction, and openness to sexual contact even in the absence of spontaneous desire. Consistent with Basson's framework, our findings indicate that women's sexual functioning is more susceptible to emotional burden and perceived stress than to purely biological factors [62].

In the control group, no significant relationships were found between stress dimensions and sexual satisfaction. In contrast, among women undergoing therapy, higher levels of worry, decreased sense of life meaning, low agency, and pessimism correlated with lower sexual satisfaction and difficulties in achieving orgasm. These findings partly reflect the observations of Kanner et al. [53], which emphasize the importance of minor, everyday stressors. Furthermore, in this group, a tendency to lie or present oneself in a favorable light was observed, which can be interpreted as a manifestation of avoidance or emotion-focused strategies. This possibility should be taken into account when interpreting the results, particularly with regard to the dimensions of emotional closeness and caressing.

The obtained results are consistent with the findings of Bodenmann et al. [63], who demonstrated that higher levels of daily stress in women correlate with a greater number of sexual problems and lower sexual satisfaction. Similar conclusions were presented by Ter Kuile et al. [64], who showed that women with high stress levels experienced lower genital arousal, although they did not differ from others in terms of psychological arousal. This mechanism can be explained by the cognitive consequences of stress—distraction and a limited ability to focus on erotic stimuli [65]. In our study, a negative correlation was also found between daily stress, sexual satisfaction, and genital arousal.

Several limitations of the present study should be noted. First, although the required minimum number of participants was met, the sample size was relatively small, which may have limited the ability to detect the effects of the therapy. The study is probably underpowered to identify moderate differences. Furthermore, no assessment of the long-term effects of the therapy was conducted, which would certainly be valuable for evaluating the effectiveness of the applied intervention. Another limitation of the study is the short duration of the therapy and the lack of assessment of the long-term sustainability of the observed effects. A total of six therapeutic sessions were conducted over a two-week period. It is possible that extending the duration of the therapy could have resulted in additional beneficial effects in the experimental group. In future studies, follow-up assessments are planned at 1 and 3 months to determine whether the beneficial effects are maintained over time. Another limitation of the study is the relatively narrow inclusion criteria. On the one hand, this selection allowed for the formation of a homogeneous group, thereby increasing the reliability of the obtained results. However, restricting the study group to such specific criteria limits the generalizability of the findings to other populations. Further research is needed in this area, including studies involving other groups, such as women with stress urinary incontinence or those in older age categories. Another limitation of the present study is the absence of a sham or placebo control group. The lack of a sham group may have influenced subjective outcomes, particularly those related to perceived stress levels and satisfaction, which should be considered when interpreting the results. Another limitation of the present study is the lack of a direct assessment of diaphragmatic mobility, for example, using ultrasonography. Although surface electromyography (sEMG) provides valuable information on muscle activity, it does not allow for the direct evaluation of diaphragm displacement or its mechanical efficiency. Future studies should consider employing objective tools for assessing respiratory function, such as ultrasonography or spirometry, to obtain a more comprehensive understanding of diaphragm function and its relationship with pelvic floor muscle activity.

Another limitation of the present study is the lack of complete blinding of the participants. Although the assessor performing the measurements was blinded to group allocation, the participants were aware of whether they were receiving the intervention, which may have influenced their responses. An additional limitation is that only healthy, asymptomatic women were included, which limits the generalizability of the findings to clinical populations. The inclusion of women without urogenital or musculoskeletal disorders allowed for the formation of a homogeneous sample and minimized potential confounding factors. Another limiting factor is the lack of control over the menstrual cycle phase of the participants, which may have affected both physiological and psychological parameters, including stress and satisfaction levels. Furthermore, future studies should consider the potential influence of cultural and social factors on the interpretation of results obtained using psychometric tools such as the SSQ and PSQ. Moreover, it should be noted that no correction for multiple testing was applied to the analyses based on questionnaire data, which increases the risk of type I error. Therefore, these findings should be interpreted with caution. Another limitation of the study is the lack of multivariate analysis to adjust for potential confounding factors. However, the experimental and control groups were comparable at baseline in terms of age, body mass, height, BMI, education level, type of relationship, and relationship duration, which likely minimized the influence of these variables on the results.

Despite the limitations of the study related to sample size and the lack of long-term follow-up, the results indicate the clinical potential of diaphragm therapy as a complementary physiotherapeutic method supporting both psychophysical health and women's sexual quality of life. The obtained findings suggest that diaphragmatic breathing therapy may help improve PFM activity. Further studies on larger populations are necessary to clearly determine the clinical potential of this method in physiotherapy practice.

5. Conclusions

The findings of the present study suggest that therapy targeting the respiratory diaphragm reduces the resting activity of the PFM and perceived stress levels, particularly in the emotional dimension. The observed association between higher stress levels and greater resting tension of the PFM suggests a bidirectional relationship between emotional state and physiological functioning. Further research in this area is certainly needed to confirm these effects.

From a clinical perspective, diaphragm-focused therapy may serve as a valuable adjunct to physiotherapeutic interventions aimed at reducing pelvic floor muscle activity as well as stress-related tension in women. Integrating diaphragmatic techniques into pelvic floor rehabilitation programs could further enhance therapeutic effectiveness.

Supplementary Materials: The following supporting information can be downloaded at: https://www.mdpi.com/article/10.3390/app152212055/s1, Table S1: Correlation of subscales (intimacy, caresses, sex, overall score) of the Sexual Satisfaction Questionnaire in Close Relationships (SSQ) with the subscales (emotional tension, internal stress, intrapsychic stress, lie scale, overall stress) of the Perceived of Stress Questionnaire (PSQ)—therapy group; Table S2: Correlation of subscales (intimacy, caresses, sex, overall score) of the Sexual Satisfaction Questionnaire in Close Relationships (SSQ) with the subscales (emotional tension, internal stress, intrapsychic stress, lie scale, overall stress) of the Perceived of Stress Questionnaire (PSQ)—control group; Table S3: Correlations between standardized questionnaire outcomes and pelvic floor muscles—Experimental group before intervention; Table S4: Correlations between standardized questionnaire outcomes and pelvic floor muscles—Experimental group after intervention; Table S5: Correlations between standardized questionnaire outcomes and pelvic floor muscles—Control group before intervention; Table S6: Correlations between standardized questionnaire outcomes and pelvic floor muscles—Control group after intervention.

Author Contributions: Conceptualization, J.G. and S.G.; methodology, J.G., S.G., M.M. and I.S.-D.; formal analysis, I.S.-D., S.G. and M.M.; investigation, J.G., S.G., I.S.-D., M.M., M.N. and J.B.-B.;

resources, J.G., S.G., I.S.-D., M.M., M.N. and J.B.-B.; data curation, I.S.-D., M.M. and S.G.; writing—original draft preparation, J.G., S.G., I.S.-D. and M.M.; writing—review and editing, J.G., S.G., I.S.-D., M.M., M.N. and J.B.-B. All authors have read and agreed to the published version of the manuscript.

Funding: This research was funded by the Ministry of Science under the program Regional Initiative of Excellence for the years 2024–2027 (project number RID/SP/0027/2024/01; total funding: PLN 4,053,904.00).

Institutional Review Board Statement: The study was conducted in accordance with the Declaration of Helsinki and approved by the Ethics Committee of the Andrzej Frycz Modrzewski Krakow University (approval number KBKA/43/O/2024).

Data Availability Statement: All data generated or analyzed during this study are included in this published article.

Conflicts of Interest: The authors declare no conflicts of interest. The funders had no role in the design of the study; in the collection, analyses, or interpretation of data; in the writing of the manuscript; or in the decision to publish the results.

References

1. Baessler, K.; Burgio, K.L.; Norton, P.A.; Schüssler, B.; Moore, K.H.; Stanton, S.L. (Eds.) *Pelvic Floor Re-Education. Principles and Practice*; Springer: London, UK, 2008; ISBN 978-1-85233-968-5.
2. Park, H.; Han, D. The Effect of the Correlation between the Contraction of the Pelvic Floor Muscles and Diaphragmatic Motion during Breathing. *J. Phys. Ther. Sci.* **2015**, *27*, 2113–2115. [CrossRef]
3. Neumann, P.; Gill, V. Pelvic Floor and Abdominal Muscle Interaction: EMG Activity and Intra-Abdominal Pressure. *Int. Urogynecol. J.* **2002**, *13*, 125–132. [CrossRef] [PubMed]
4. Bieniek, M.; Zaleski, A. The Impact of the Action of the Diaphragm on the Selected Organs of Human Body. *Man. Med.* **2018**, *1*, 13–18. [CrossRef]
5. Friggi Sebe Petrelluzzi, K.; Garcia, M.C.; Petta, C.A.; Ribeiro, D.A.; de Oliveira Monteiro, N.R.; Céspedes, I.C.; Spadari, R.C. Physical Therapy and Psychological Intervention Normalize Cortisol Levels and Improve Vitality in Women with Endometriosis. *J. Psychosom. Obstet. Gynecol.* **2012**, *33*, 191–198. [CrossRef]
6. Okado, Y.; De Pace, D.; Ewing, E.; Rowley, C. Brief Relaxation Training for the Prevention of Stress-Related Difficulties: A Pilot Study. *Int. Q. Community Health Educ.* **2020**, *40*, 193–200. [CrossRef] [PubMed]
7. Conlon, A.; Arnold, R.; Preatoni, E.; Moore, L.J. Pulling the Trigger: The Effect of a 5-Minute Slow Diaphragmatic Breathing Intervention on Psychophysiological Stress Responses and Pressurized Pistol Shooting Performance. *J. Sport. Exerc. Psychol.* **2022**, *44*, 206–219. [CrossRef]
8. Martarelli, D.; Cocchioni, M.; Scuri, S.; Pompei, P. Diaphragmatic Breathing Reduces Exercise-Induced Oxidative Stress. *Evid.-Based Complement. Altern. Med.* **2011**, *2011*, 932430. [CrossRef]
9. Schrepf, A.; O'Donnell, M.A.; Luo, Y.; Bradley, C.S.; Kreder, K.J.; Lutgendorf, S.K. Inflammation and Symptom Change in Interstitial Cystitis or Bladder Pain Syndrome: A Multidisciplinary Approach to the Study of Chronic Pelvic Pain Research Network Study. *Urology* **2016**, *90*, 56–61. [CrossRef]
10. Anderson, R.U.; Orenberg, E.K.; Morey, A.; Chavez, N.; Chan, C.A. Stress Induced Hypothalamus-Pituitary-Adrenal Axis Responses and Disturbances in Psychological Profiles in Men with Chronic Prostatitis/Chronic Pelvic Pain Syndrome. *J. Urol.* **2009**, *182*, 2319–2324. [CrossRef]
11. Stephens, M.A.C.; Wand, G. Stress and the HPA Axis: Role of Glucocorticoids in Alcohol Dependence. *Alcohol. Res.* **2012**, *34*, 468–483. [CrossRef]
12. Hamilton, L.D.; Meston, C.M. Chronic Stress and Sexual Function in Women. *J. Sex. Med.* **2013**, *10*, 2443–2454. [CrossRef]
13. Halski, T.; Ptaszkowski, K.; Słupska, L.; Dymarek, R. The Evaluation of Bioelectrical Activity of Pelvic Floor Muscles Depending on Probe Location: A Pilot Study. *Biomed. Res. Int.* **2013**, *2013*, 238312. [CrossRef] [PubMed]
14. Oleksy, Ł.; Mika, A.; Sulowska-Daszyk, I.; Rosłoniec, E.; Kielnar, R.; Stolarczyk, A. The Reliability of Pelvic Floor Muscle Bioelectrical Activity (SEMG) Assessment Using a Multi-Activity Measurement Protocol in Young Women. *Int. J. Environ. Res. Public Health* **2021**, *18*, 765. [CrossRef]
15. Plopa, M. *Sexual Satisfaction in Close Relationships: Research Method (KSS)*; Publishing House of the University of Economics and Human Sciences: Warsaw, Poland, 2021.
16. Plopa, M.; Makarowski, R. *Perceived Stress Questionnaire*; Vizja Press & IT: Warsaw, Poland, 2010.
17. Tyszko-Bury, E. The Use of Osteopathic Techniques in Patients with Pelvic Floor Pain. *Pract. Physiother. Rehabil.* **2015**, *66*, 6–12.

18. Cohen, J. A Power Primer. *Psychol. Bull.* **1992**, *112*, 155–159. [CrossRef]
19. Błudnicka, M.; Piernicka, M.; Kortas, J.; Biernacka, B.D.; Szumilewicz, A. Effects of a One-Time Biofeedback EMG Session on Neuromuscular Activity of the Pelvic Floor Muscles in Pregnant Women. *Neurophysiology* **2020**, *52*, 438–445. [CrossRef]
20. Yu, X.; Jiang, H.; Zhang, C.; Jin, Z.; Gao, L.; Wang, R.; Fang, J.; Su, Y.; Xi, J.; Fang, B. The Role of the Diaphragm in Postural Stability and Visceral Function in Parkinson's Disease. *Front. Aging Neurosci.* **2021**, *13*, 785020. [CrossRef] [PubMed]
21. Cheng, K.-C.; Hii, E.Y.X.; Lin, Y.-N.; Kuo, Y.-L.; Tsai, Y.-J. Effects of Manual Diaphragm Release on Pain, Disability and Diaphragm Function in Patients with Chronic Neck Pain: A Pilot Randomized Controlled Trial. *BMC Complement. Med. Ther.* **2025**, *25*, 349. [CrossRef]
22. Ajimsha, M.S.; Shenoy, P.D.; Gampawar, N. Role of Fascial Connectivity in Musculoskeletal Dysfunctions: A Narrative Review. *J. Bodyw. Mov. Ther.* **2020**, *24*, 423–431. [CrossRef]
23. Marizeiro, D.F.; Florêncio, A.C.L.; Nunes, A.C.L.; Campos, N.G.; Lima, P.O.d.P. Immediate Effects of Diaphragmatic Myofascial Release on the Physical and Functional Outcomes in Sedentary Women: A Randomized Placebo-Controlled Trial. *J. Bodyw. Mov. Ther.* **2018**, *22*, 924–929. [CrossRef]
24. Talasz, H.; Kremser, C.; Kofler, M.; Kalchschmid, E.; Lechleitner, M.; Rudisch, A. Phase-Locked Parallel Movement of Diaphragm and Pelvic Floor during Breathing and Coughing—A Dynamic MRI Investigation in Healthy Females. *Int. Urogynecol. J.* **2011**, *22*, 61–68. [CrossRef] [PubMed]
25. Hodges, P.W.; Sapsford, R.; Pengel, L.H.M. Postural and Respiratory Functions of the Pelvic Floor Muscles. *Neurourol. Urodyn.* **2007**, *26*, 362–371. [CrossRef]
26. Sapsford, R. Rehabilitation of Pelvic Floor Muscles Utilizing Trunk Stabilization. *Man. Ther.* **2004**, *9*, 3–12. [CrossRef]
27. Bordoni, B.; Zanier, E. Anatomic Connections of the Diaphragm: Influence of Respiration on the Body System. *J. Multidiscip. Healthc.* **2013**, *6*, 281–291. [CrossRef]
28. Aljuraifani, R.; Stafford, R.E.; Hall, L.M.; van den Hoorn, W.; Hodges, P.W. Task-Specific Differences in Respiration-Related Activation of Deep and Superficial Pelvic Floor Muscles. *J. Appl. Physiol.* **2019**, *126*, 1343–1351. [CrossRef]
29. Betschart, C.; Kim, J.; Miller, J.M.; Ashton-Miller, J.A.; DeLancey, J.O.L. Comparison of Muscle Fiber Directions between Different Levator Ani Muscle Subdivisions: In Vivo MRI Measurements in Women. *Int. Urogynecol. J.* **2014**, *25*, 1263–1268. [CrossRef] [PubMed]
30. Vodušek, D.B. Neural Control of Pelvic Floor Muscles. In *Pelvic Floor Re-Education*; Springer: London, UK, 2008; pp. 22–35.
31. Bentley, T.G.K.; D'Andrea-Penna, G.; Rakic, M.; Arce, N.; LaFaille, M.; Berman, R.; Cooley, K.; Sprimont, P. Breathing Practices for Stress and Anxiety Reduction: Conceptual Framework of Implementation Guidelines Based on a Systematic Review of the Published Literature. *Brain Sci.* **2023**, *13*, 1612. [CrossRef]
32. Ashhad, S.; Kam, K.; Del Negro, C.A.; Feldman, J.L. Breathing Rhythm and Pattern and Their Influence on Emotion. *Annu. Rev. Neurosci.* **2022**, *45*, 223–247. [CrossRef]
33. Nardi, A.E.; Freire, R.C.; Zin, W.A. Panic Disorder and Control of Breathing. *Respir. Physiol. Neurobiol.* **2009**, *167*, 133–143. [CrossRef]
34. Wilhelm, F.H.; Gevirtz, R.; Roth, W.T. Respiratory Dysregulation in Anxiety, Functional Cardiac, and Pain Disorders. *Behav. Modif.* **2001**, *25*, 513–545. [CrossRef] [PubMed]
35. Martinez, J.M.; Kent, J.M.; Coplan, J.D.; Browne, S.T.; Papp, L.A.; Sullivan, G.M.; Kleber, M.; Perepletchikova, F.; Fyer, A.J.; Klein, D.F.; et al. Respiratory Variability in Panic Disorder. *Depress. Anxiety* **2001**, *14*, 232–237. [CrossRef]
36. Bhimani, N.T.; Kulkarni, N.B.; Kowale, A.; Salvi, S. Effect of Pranayama on Stress and Cardiovascular Autonomic Function. *Indian J. Physiol. Pharmacol.* **2011**, *55*, 370–377.
37. Sellakumar, G.K. Effect of Slow-Deep Breathing Exercise to Reduce Anxiety among Adolescent School Students in a Selected Higher Secondary School in Coimbatore, India. *J. Psychol. Educ. Res.* **2015**, *23*, 54–72.
38. Hunt, M.G.; Rushton, J.; Shenberger, E.; Murayama, S. Positive Effects of Diaphragmatic Breathing on Physiological Stress Reactivity in Varsity Athletes. *J. Clin. Sport Psychol.* **2018**, *12*, 27–38. [CrossRef]
39. Hopper, S.I.; Murray, S.L.; Ferrara, L.R.; Singleton, J.K. Effectiveness of Diaphragmatic Breathing for Reducing Physiological and Psychological Stress in Adults: A Quantitative Systematic Review. *JBI Database Syst. Rev. Implement. Rep.* **2019**, *17*, 1855–1876. [CrossRef]
40. Aukee, P.; Penttinen, J.; Immonen, P.; Airaksinen, O. Intravaginal Surface EMG Probe Design Test for Urinary Incontinence Patients. *Acupunct. Electro-Ther. Res.* **2002**, *27*, 37–44. [CrossRef]
41. López-Torres, O.; Álvarez-Sáez, M.; Lorenzo Calvo, J.; Carmona, L.; Soriano, L. Effects of an 8-Week Abdominal Hypopressive Technique Program on Pelvic Floor Muscle Contractility: An Assessor-Blinded Randomized Controlled Trial. *Appl. Sci.* **2025**, *15*, 5844. [CrossRef]
42. Zachovajeviene, B.; Siupsinskas, L.; Zachovajevas, P.; Venclovas, Z.; Milonas, D. Effect of Diaphragm and Abdominal Muscle Training on Pelvic Floor Strength and Endurance: Results of a Prospective Randomized Trial. *Sci. Rep.* **2019**, *9*, 19192. [CrossRef]

43. Toprak, N.; Sen, S.; Varhan, B. The Role of Diaphragmatic Breathing Exercise on Urinary Incontinence Treatment: A Pilot Study. *J. Bodyw. Mov. Ther.* **2022**, *29*, 146–153. [CrossRef] [PubMed]
44. Rocha, T.; Souza, H.; Brandão, D.C.; Rattes, C.; Ribeiro, L.; Campos, S.L.; Aliverti, A.; de Andrade, A.D. The Manual Diaphragm Release Technique Improves Diaphragmatic Mobility, Inspiratory Capacity and Exercise Capacity in People with Chronic Obstructive Pulmonary Disease: A Randomised Trial. *J. Physiother.* **2015**, *61*, 182–189. [CrossRef] [PubMed]
45. van Reijn-Baggen, D.A.; Han-Geurts, I.J.M.; Voorham-van der Zalm, P.J.; Pelger, R.C.M.; Hagenaars-van Miert, C.H.A.C.; Laan, E.T.M. Pelvic Floor Physical Therapy for Pelvic Floor Hypertonicity: A Systematic Review of Treatment Efficacy. *Sex. Med. Rev.* **2022**, *10*, 209–230. [CrossRef] [PubMed]
46. Donatti, L.; Podgaec, S.; Baracat, E.C. Efficacy of Cognitive Behavioral Therapy in Treating Women with Endometriosis and Chronic Pelvic Pain: A Randomized Trial. *J. Health Psychol.* **2025**, *30*, 1004–1016. [CrossRef]
47. Strelau, J. Temperament and Stress: Temperament as a Moderating Factor of Stressors, the State and Effects of Stress, and Coping with Stress. In *Man in a Stressful Situation: Theoretical and Methodological Issues*; University Publishing House: Katowice, Poland, 1996; pp. 88–132.
48. Ma, X.; Yue, Z.-Q.; Gong, Z.-Q.; Zhang, H.; Duan, N.-Y.; Shi, Y.-T.; Wei, G.-X.; Li, Y.-F. The Effect of Diaphragmatic Breathing on Attention, Negative Affect and Stress in Healthy Adults. *Front. Psychol.* **2017**, *8*, 874. [CrossRef]
49. Reiner, R. Integrating a Portable Biofeedback Device into Clinical Practice for Patients with Anxiety Disorders: Results of a Pilot Study. *Appl. Psychophysiol. Biofeedback* **2008**, *33*, 55–61. [CrossRef]
50. Conrad, A.; Müller, A.; Doberenz, S.; Kim, S.; Meuret, A.E.; Wollburg, E.; Roth, W.T. Psychophysiological Effects of Breathing Instructions for Stress Management. *Appl. Psychophysiol. Biofeedback* **2007**, *32*, 89–98. [CrossRef]
51. Bancroft, J.; Loftus, J.; Long, J.S. Distress About Sex: A National Survey of Women in Heterosexual Relationships. *Arch. Sex. Behav.* **2003**, *32*, 193–208. [CrossRef] [PubMed]
52. Kingsberg, S.A.; Janata, J.W. Female Sexual Disorders: Assessment, Diagnosis, and Treatment. *Urol. Clin. N. Am.* **2007**, *34*, 497–506. [CrossRef]
53. Kanner, A.D.; Coyne, J.C.; Schaefer, C.; Lazarus, R.S. Comparison of Two Modes of Stress Measurement: Daily Hassles and Uplifts versus Major Life Events. *J. Behav. Med.* **1981**, *4*, 1–39. [CrossRef]
54. Włodarczyk, D.; Wrześniewski, K. The Stress Appraisal Questionnaire. *Psychol. Rev.* **2010**, *53*, 479–496.
55. Wrześniewski, K. Stress Coping Styles and Strategies. Measurement Issues. In *Man in a Stressful Situation: Theoretical and Methodological Issues*; University Publishing House: Katowice, Poland, 1996; pp. 44–64.
56. Lazarus, R.S.; Folkman, S. *Stress: Appraisal and Coping*; Springer: New York, NY, USA, 1984.
57. Szczepaniak, P.; Strelau, J.; Wrześniewski, K. Diagnosis of Stress Coping Styles Using the Polish Version of the CISS Questionnaire by Endler and Parker. *Rev. Psychol.* **1996**, *1*, 187–210.
58. Cohen, S.; Kamarck, T.; Mermelstein, R. A Global Measure of Perceived Stress. *J. Health Soc. Behav.* **1983**, *24*, 385. [CrossRef] [PubMed]
59. Juczyński, Z.; Ogińska-Bulik, N. *Stress and Stress Management Tools*; Psychological Testing Laboratory of the Polish Psychological Society: Warsaw, Poland, 2009.
60. Zdrojewicz, Z.; Lelakowska, K. The Role of Stress in Sexual Problems, Disorders, and Preferences. *Pol. Sexol.* **2006**, *4*, 69–79.
61. Bancroft, J.; Janssen, E.; Strong, D.; Carnes, L.; Vukadinovic, Z.; Long, J.S. The Relation Between Mood and Sexuality in Heterosexual Men. *Arch. Sex. Behav.* **2003**, *32*, 217–230. [CrossRef] [PubMed]
62. Basson, R. The Female Sexual Response: A Different Model. *J. Sex. Marital. Ther.* **2000**, *26*, 51–65. [CrossRef]
63. Bodenmann, G.; Atkins, D.C.; Schär, M.; Poffet, V. The Association between Daily Stress and Sexual Activity. *J. Fam. Psychol.* **2010**, *24*, 271–279. [CrossRef]
64. Ter Kuile, M.M.; Vigeveno, D.; Laan, E. Preliminary Evidence That Acute and Chronic Daily Psychological Stress Affect Sexual Arousal in Sexually Functional Women. *Behav. Res. Ther.* **2007**, *45*, 2078–2089. [CrossRef]
65. Adams, A.E.; Haynes, S.N.; Brayer, M.A. Cognitive Distraction in Female Sexual Arousal. *Psychophysiology* **1985**, *22*, 689–696. [CrossRef]

applied sciences

MDPI

Article

A Robotic Gamified Framework for Upper-Limb Rehabilitation

Anahis Casanova †, Natalia Sempere †, Cristina Romero, Koralie Porcel, Andres Ubeda * and Carlos A. Jara

Human Robotics Group, University of Alicante, San Vicente del Raspeig s/n, 03690 Alicante, Spain; anahis.casanova@ua.es (A.C.); natalia.sempere@ua.es (N.S.); c.romero@ua.es (C.R.); koralie.porcel@gmail.com (K.P.); carlos.jara@ua.es (C.A.J.)

* Correspondence: andres.ubeda@ua.es

† These authors contributed equally to this work.

Abstract: Robotic devices have become increasingly important in upper-limb rehabilitation, as they assist therapists, improve treatment efficiency, and enable personalised therapy. However, the lack of standardised protocols and integrative tools limits their widespread adoption and effectiveness. To address these challenges, a robotic framework was developed for upper-limb rehabilitation in patients with acquired brain injury (ABI). The framework is designed to be adaptable to various ROS-compatible collaborative robots with admittance control and potentially adaptable to other types of control, and also integrates kinematic and electrophysiological (EMG) metrics to monitor patient performance and progress. It combines data acquisition through EMG and robot motion sensors, gamification elements to enhance engagement, and configurable robot control modes within a unified software platform. A pilot evaluation with eight healthy subjects performing upper limb movements on an ROS-compatible robot from the UR family demonstrated the feasibility of the framework's components, including robot control, EMG acquisition and synchronization, gamified interaction, and synchronised data collection. User performance through all levels remained below the controller limits of force and velocity thresholds even in the most resistive damping. These results support the potential of the proposed framework as a flexible, extensible, and integrative tool for upper-limb rehabilitation, providing a foundation for future clinical studies and multi-platform implementations.

Keywords: neurorehabilitation; collaborative robots; electromyography; physical assessment; gamification

1. Introduction

Every year, 15 million people suffer a stroke, with 5 million of the survivors requiring rehabilitation. Many of these individuals experience permanent motor disabilities such as hemiparesis, which is characterised by the weakening of one side of the body. Rehabilitation for these patients involves significant challenges due to the limited availability of human resources and clinical personnel [1].

To address these challenges, robotic systems have been increasingly employed in healthcare settings in the last two decades. From surgical to rehabilitation assistance, robots play a crucial role in easing the patient's recovery. In the rehabilitation domain, robotic systems assist therapists during treatment routines, thereby augmenting the number of patients that can be effectively managed by a single therapist. This capability is particularly useful during the early stages of the rehabilitation, where providing immediate and personalised assistance is decisive for the success of the treatment [2]. Furthermore, the need for rehabilitation due to acquired brain injury (ABI) has increased by over 63% in the last decades [3], promoting the works addressing this issue.

End-effector robots intended for upper-limb rehabilitation can accomplish active and passive motor skills training for the wrist, forearm, and shoulder; in addition, they can increase the intensity and repeatability of rehabilitation protocols. There have been a number of outstanding projects based on end-effector robots, including MIT-Manus [4], MIME [5], GENTLE/s [6], ARM Guide [7], and reachMAN2 [8], leading to a breakthrough in the field of rehabilitation robotics. In the last few years, collaborative robots with end-effector tools have been extensively used for rehabilitation, particularly for patients with upper limb disabilities [9,10]. These robots are specially prepared for human–robot interaction and can assist motion in different modalities depending on the limb's mobility (passive, active, active-assistive, etc). The robot end-effector attachment point is connected to the patient's limb, and can guide it over a fixed path or apply assistance-as-needed control for rehabilitation therapy [11].

In addition to robotic systems, the use of biosensors to evaluate rehabilitation performance has been shown to be very beneficial. Biosensors can monitor a wide range of information during therapy, including electrophysiological activity and motion performance. In this context, so-called neuromechanical biomarkers can be used as metrics for tracking motor function [12]. Effects on motor coordination [13], muscle strength [14], and others can be monitored by using electromyography (EMG), while kinematics can also be measured during robot-assisted rehabilitation [15]. From this information, the rehabilitation therapy can be reoriented to fulfill its goals more effectively. An example of this was presented in [16], where a motion sensor was used to capture hand kinematics and translate them into game inputs within a rehabilitation environment while adjusting the difficulty according to the patient's progress. More recently, a personalized EMG-based feedback training was applied together with a wearable device to adapt hand rehabilitation [17].

These therapeutic exercises are often integrated with serious games, which have gained popularity due to their ability to engage patients more effectively in their therapy. Gamification in rehabilitation typically includes features such as scoring systems, progress summaries, and thematic elements that can enhance the gaming experience [18]. In addition, the incorporation of robots and sensors along with serious games in rehabilitation contexts not only enhances patient engagement through stimulating experiences but also improves the accuracy and efficiency of data collection. Consequently, gamification is increasingly being implemented in clinical practice to improve rehabilitation outcomes [19].

Current Limitations of Available Technology

In recent years, exoskeletons and end-effector platforms have been introduced commercially. However, as summarized in Table 1, in upper-limb rehabilitation the market remains limited, with only a few available options. Commercial robots offer different training modes (passive, active, assistive); while some robots incorporate multiple modes, such as ReoGo [20], and Yidong-Arm1 [21], others such as ALEx S [22], ALEx RS [23] and ArmeoPower [24,25] focus primarily on a single therapy mode and are restricted to passive therapy. Even within similar modes, implementation strategies differ; Harmony SHR [21] requires therapists to manually select assistive force, while REAPlan [25] and InMotion Arm [26] adjust force adaptively during tasks. A significant drawback is that available robots often omit resistive modes, which are essential for strengthening muscles in intermediate and advanced therapy stages [27]. Moreover, robots such as Hocoma's Armeo series address specific rehabilitation stages, requiring multiple investments as the patient progresses [25].

Additionally, physical therapy assessment is often limited to kinematic parameters (range of motion, trajectory, speed, execution time), overlooking critical EMG parameters such as muscle fatigue and activation patterns. Currently, robots incorporating EMG are mainly research-oriented and seldom clinically available [28].

Table 1. Commercial upper-limb rehabilitation robots.

Robot	Aim [1]	Training [2] Modes	Gamification	Data [3]
ALEx S	S-E	P	✓	K
ALEx RS	S-E-FA-W	-	+VR	K
ArmeoPower	S-E-H *	P	+VR	-
ArmeoSpring	S-E-H *	-	✓	K
ArmeoSpring Pro	S-E-H *	-	✓	K
Harmony SHR	S-E	P + Ass	-	-
Nx-A2	S-E-H	-	+VR	K
ReoGo	S *-E-W	P + Ass	✓	K
Yidong-Arm1	S-E-H	P, A, M	-	-
REAPlan	S-E	Adap	✓	-
InMotion Arm	S-E-H	Adap	✓	IE-K

* If requested. [1] (S) Shoulder, (E) elbow, (H) hand, (FA) forearm, (W) wrist. [2] (P) Passive, (A) active, (M) mixed, (Ass) assistive, (Adp) adaptative. [3] (K) Kinematic, (IE) intelligent evaluation.

Personalisation of therapy is a critical area for innovation that seeks to enhance patient outcomes by adapting robotic systems to individual needs, abilities, and preferences. While some systems support customisation, limitations such as restricted exercise durations or limited degrees of freedom (DoF) continue to persist. For example, ReoGo supports three-joint movement but has limited joint DoF, while Armeo and InMotion Arm require additional wrist joints for full-range therapy [24,26].

Despite these advancements, a clear standardised protocol for rehabilitation robots remains lacking [29]. This work seeks to specifically address these gaps by developing a flexible, scalable, and interoperable robotic framework for upper-limb rehabilitation in ABI patients. The proposed framework integrates real-time monitoring and performance analysis using EMG systems and robot sensors, addressing limitations such as insufficient integration of EMG parameters in clinical practice and the lack of adaptive difficulty during therapy. It also incorporates gamification elements to enhance patient motivation and provides diverse end-effector tools accommodating different gripping requirements, thereby overcoming the need for multiple robot investments in different rehabilitation stages. As a first approach, admittance control was chosen specifically because it provides a natural interaction, is robust to varying levels of patient engagement, and allows for dynamic adjustment of robot resistance, contrasting positively against impedance, force, or hybrid control methods.

2. Robotic Framework Design

The software framework integrates an end-effector robotic platform based on the Universal Robot UR10e collaborative robot, which is equipped with an internal force/torque sensor necessary for implementing admittance control. The end-effector is an ergonomic handle specifically designed for ABI patients, and is synchronised with gamified activities displayed visually. For EMG acquisition, eight Noraxon Ultium electrodes are used, providing real-time muscle activity monitoring. Figure 1 illustrates the complete hardware setup, including the sensor placements and interactive interface.

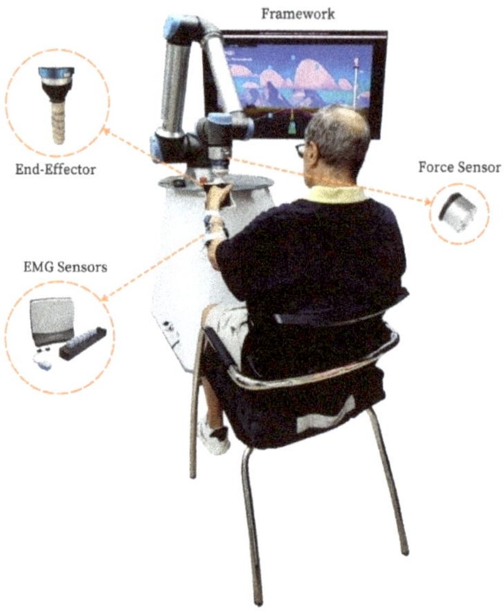

Figure 1. Setup with the robot, EMG sensors, screen, end-effector, and a participant.

2.1. Control System

Rehabilitation robots offer several advantages: they ensure consistent execution of exercises, eliminate the need for therapist guidance in directing movements, and enable collection of precise position and force data. To that end, a proper control needs to be implemented in order to fulfill the task requirements. In this work, the rehabilitation exercise involves manually gripping and moving the robot's end-effector in the specified directions. During the initial stages, the robot must not oppose any resistance to the trajectory, facilitating a free arm movement for the patient. As rehabilitation progresses, the robot should provide increasing resistance to movements for the purposes of adapting the patient's training and the game difficulty. To accomplish this, the control strategy is implemented based on admittance control, which is selected due to its suitability for providing safe and natural interactions during rehabilitation exercises [30].

Admittance control requires precise force/torque sensing, conditions that are fully satisfied by the UR10e's integrated sensors [30]. The force data from the robot's end-effector is processed using ROS, subtracting gravitational effects and computing the tool centre point (TCP) velocity via a mass-damper-spring transfer function, as shown in the following equation:

$$\dot{x}_i^{desired} = \frac{1}{m + c_{user}} \cdot F_{net,i}. \tag{1}$$

This equation computes the desired TCP velocity ($\dot{x}_i^{desired}$), where $m = 10.9$ Kg is the maximum mass accepted by the robot, c_{user} is the damping value provided manually through the user interface (UI) or by specific levels (see Section 2.4), and $F_{net,i}$ is the total net force after the force threshold and validation logic:

$$F_{net,i} = \begin{cases} |F_{ext,i}| - F_{th} \cdot \gamma, & \text{if } F_{ext,i} \geq 0 \text{ and } |F_{ext,i}| > F_{th} \cdot \gamma \\ -(|F_{ext,i}| - F_{th} \cdot \gamma), & \text{if } F_{ext,i} < 0 \text{ and } |F_{ext,i}| > F_{th} \cdot \gamma \\ 0, & \text{if } |F_{ext,i}| \leq F_{th} \cdot \gamma \end{cases} \tag{2}$$

where the scaling factor $\gamma = \frac{F_{th}}{|F_{ext,x}|+|F_{ext,y}|+|F_{ext,z}|}$ distributes the force threshold (F_{th}) between the active axes of the external force ($F_{ext,i}$). To guarantee safe interaction, forces are validated within the range $2 \leq |F_{ext,i}| \leq 200$ N to remove low-level noise and saturating inputs that exceed safe limits. Finally, a dual-rate low-pass filter is implemented to prevent jolting motion:

$$\dot{x}_i^{filt} = \dot{x}_i^{prev} + \alpha(\dot{x}_i^{desired} - \dot{x}_i^{prev}) \quad (3)$$

where $\dot{x}_i^{filt}$ is the final filtered velocity at the TCP and $\dot{x}_i^{prev}$ is the previous TCP velocity. The filter employs two gains α depending on motion phase: during acceleration ($\frac{1}{32}$) and during deceleration ($\frac{1}{10}$). The complete system relationship can be represented as follows:

$$\dot{X}_i^{filt}(s) = G_{total}(s) \cdot F_{net,i}(s) \quad (4)$$

where $G_{total}(s) = \frac{\alpha}{s+\alpha} \cdot \frac{1}{m+c_{user}}$ represents the combined transfer function of the admittance controller and velocity smoothing filter. The first term $\frac{\alpha}{s+\alpha}$ is the first-order low-pass filter from Equation (3), while the second term represents the admittance gain. This cascade configuration ensures smooth velocity profiles by filtering high-frequency components while maintaining the force–velocity relationship dictated by the maximum mass accepted by the robot.

While demonstrated specifically with the UR10e robot, this approach can be adapted to other collaborative robots provided that they possess or integrate comparable force/torque sensing capabilities and ROS compatibility. Adjustments necessary for integrating this control strategy into other robot platforms primarily involve calibrating the sensor configurations and adapting the transfer function parameters according to robot-specific dynamics.

2.2. Data Acquisition and Processing

The developed software framework is designed to gather data from various sources, visualize these data, and create datasets to process them efficiently. To this end, the core manager of the framework runs in Unity and the robot data and control run in ROS. To visualise data in the UI and automatically create datasets, Unity and ROS require an intermediate communication interface. On the ROS side, two threads are in charge of the robot: one thread continuously publishes joint values, TCP position and speed, and detected force and torque to the corresponding topics; meanwhile, another is responsible for the robot control loop, which requires communication with Unity to update the robot's resistance to movement in real-time.

On the Unity side, the *Unity Robotics Hub* package has been imported, allowing for the creation of custom scripts that act as ROS nodes. Consequently, two components have been created: one for publishing data and another for subscribing data. This setup enables ROS and Unity to communicate through topics. Robot data are published at a frequency of 100 Hz concurrently with the execution of the control loop at 500 Hz, thereby regulating the velocity reference in the robot's TCP. Unity retrieves and processes these data in real time, displaying the values in the UI and storing them to construct a dataset with the robot information. This dataset includes details such as the patient's code, age, gender, arm to rehabilitate, arm length, and robot resistance value. For each batch of robot data obtained in Unity, information is stored in the following format: timestamp, joints, TCP position, TCP speed, force, and the active target, which varies according to the selected game and its corresponding set of targets.

Regarding the EMG sensors, they are managed through the official Noraxon software (Noraxon MR 3.20.68 version). To obtain EMG data in Unity, the sensors placed on the patient are activated to start publishing information about muscular activation in mV. These data are captured via HTTP streaming with the help of Noraxon software carrying out

HTTP requests. When a request is made, a batch of all the data stored between the previous and current requests is received. For this reason, although the sensors publish data at 2000 Hz, Unity does not receive data at this frequency due to the slower HTTP request process, necessitating further processing.

Synchronisation of both the robot and EMG datasets is achieved thanks to ROS timestamps. Each collected batch is labelled with the latest ROS timestamp immediately after the HTTP request, then resampled to 1500 Hz and interpolated to ensure sufficient temporal equidistance for performance analysis. After collecting and processing EMG data, a new dataset is automatically created. This dataset includes the original timestamp, the interpolated-resampled timestamp, and information from the eight channels, where the EMG signals are delivered in raw format to allow for maximum flexibility during postprocessing. This approach enables the offline application of customised filtering, rectification, and smoothing techniques tailored to the specific analysis objectives and patient characteristics.

As a result of the previous processes, two datasets in CSV format are delivered to the therapist after completing a rehabilitation routine, as depicted in Figure 2. The robot dataset provides real-time information on the patient's hand position, speed, and applied force, while the EMG dataset contains information on the patient's electrical activity. Furthermore, the framework allows the user to select the export path and the information to be included in the dataset. By default, all available information is exported.

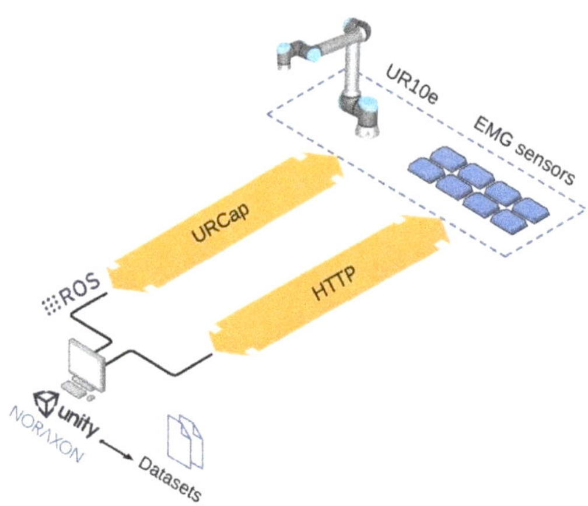

Figure 2. Diagram of data acquisition.

2.3. User Interaction

Developing accurate robot control and acquiring data from the hand's position, applied force, robot parameters, and EMG data would be pointless without proper interaction between the therapist, patient, and system. It is essential to equip the therapist with the necessary tools to understand how these data correlate with the conducted routine as well as to enable the system to receive inputs from the therapist to customise the exercises. Additionally, visual feedback is provided to the patient in order to facilitate the association between movements in the real space executed through the robot and the movements of objects in the games. Consequently, a comprehensive UI of the software framework has been developed in Unity.

2.3.1. User Interface

The main UI, integrated in C# within Unity, serves as the principal link between all the framework components: the robot, EMG sensors, therapist, and patient. It handles core data processing, synchronisation, and routine management in the back-end, while the front-end provides the therapist with all the necessary configurations to set up the routine.

Regarding the front-end, it has been designed with a minimalist and user-friendly style, as providing an intuitive interface can ease the therapist's experience while using the framework. It is divided into two main scenes. The first scene contains the login screen (Figure 3), where it is possible to register a new patient or initialise a session. The benefit of having a login interface is that it allows the storage of repetitive data across different sessions with various patients, such as name, age, exercise, or robot configuration. These data are only accessible by the therapist, who logs in with a unique ID and password for each patient. Additionally, because the proposed framework supports the use of various robots, the UR3 (as described in [31]) and UR10e models (available in the lab) can be seen in the image. Therefore, the login screen also offers the possibility of choosing a robot among the displayed options.

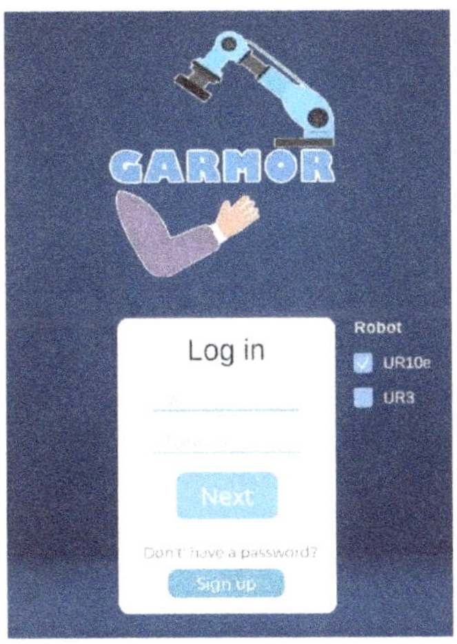

Figure 3. Login screen in the proposed framework.

After logging in, the second main scene displays all available configurations and games included in the framework. It is subdivided into four tabs according to the content section. The first tab (Figure 4a) presents a mosaic of all serious games for the therapist to choose from and start the rehabilitation process. Each game has been developed in a separate scene, allowing for future expansion of the serious games repository. Although the games are presented in separate scenes, they all share the components responsible for managing the robots and the sensors, which were developed intentionally to enable switching between games within the same session.

The second tab (Figure 4b) offers a real-time view of all data stored in the datasets, including data related to the robot's position in Cartesian and joint space, velocity, force at the TCP, and values from the EMG sensors. Additionally, the recording of data can be started and stopped through this tab without selecting a serious game; doing so initialises the admittance control and data exchange without gamification, leaving the therapist in charge of indicating the movements.

The third tab (Figure 4c) contains configurations related to the robot and admittance control. Here, the therapist can select the level of resistance the robot will apply, which communicates with the controller in ROS and adjusts the system's damping value. There is also a button for calibrating the robot. This button measures the force and torque while the robot is stationary in order to subtract it from the real force in the admittance control. Moreover, the default admittance control operates in free space; however, certain serious games may require restricting motion to one axis. Consequently, a drop-down menu has been added to this tab in order to select the plane in which the robot should move (XY, YZ, or XZ), using the robot's base as the reference system. For reference, a visual representation of the selected plane is shown just below the menu.

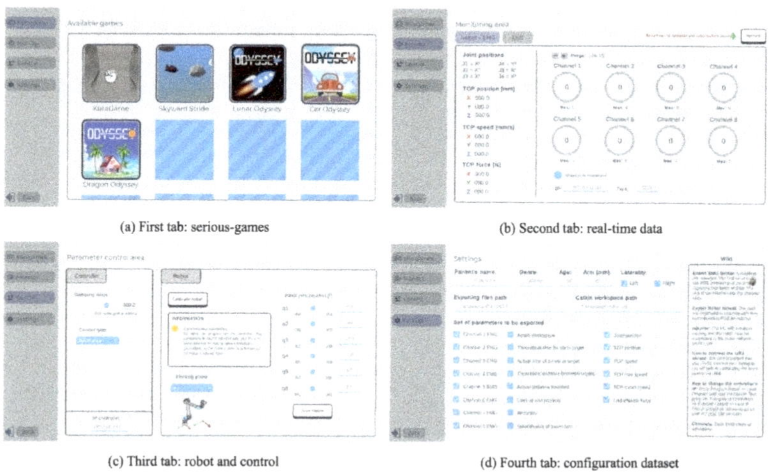

(a) First tab: serious-games

(b) Second tab: real-time data

(c) Third tab: robot and control

(d) Fourth tab: configuration dataset

Figure 4. Different tabs of the user interface.

The last tab (Figure 4d) allows for the configuration of parameters related to the patient's data and the datasets' export path. It is possible to add or remove information from these datasets prior to their creation; for example, the joint position or a certain muscle can be excluded from the dataset generation. These configurations are automatically stored for each patient, and do not need to be re-entered in subsequent sessions with that patient.

Using the previous tabs, the therapist can navigate through the framework, customise the exercises according to the patient's needs, and record data freely without following any game. There is no need to worry about starting the robot or storing the dataset, as these tasks are managed automatically. Regarding user interaction, all buttons and pressable items provide feedback, such as changing colour or enlarging. Additionally, messages inform the therapist about the current state of the systems. If the robot is not calibrated and a data recording is initiated, startup is not allowed and the therapist is informed of this through a pop-up message.

2.3.2. Gamification

Even though the main UI provides basic interaction with the framework, its primary objectives are to verify the correct functionality of the system, configure data related to the patient and routines, and select the serious game that will be used. The actual rehabilitation exercises are based on one of the provided serious games, which can be selected in the first tab and redirect the user to a new Unity scene (see Figure 4a). Currently, the application offers three games:

- **Odyssey** is based on a centre-out approach [32] in which the patient moves the end-effector to reach targets shown on the screen. Eight targets are placed along a circular path, with one target in the centre, making for a total of nine targets. The objective is to move from the centre target to one of the targets on the circumference and then return. Depending on the user, the order of these targets can be randomised or arranged in a clockwise sequence, an option that can be selected in the game's UI. If the user is a healthy individual, targets can be either randomised or set in a clockwise order. Randomising the targets prevents the user from learning patterns and keeps them constantly focused on the exercise. The target order is set clockwise if the user is an upper-limb rehabilitation patient, since a known sequence is easier to follow and ensures the same number of repetitions in all eight directions. There are three skins for this game: Lunar Odyssey, Car Odyssey, and Dragon Odyssey (Figure 5).

- **Skyward Stride** consists of an aeroplane that flies through an infinite world in a 2D environment with obstacles to be avoided. The user can change the height of the aeroplane to deal with the obstacles, obtain bonus items, and finish different levels with variable difficulty. For upper-limb rehabilitation, the movements can be controlled by linear displacements on a plane (arm reaching) or commanded by single joint movements of the wrist or elbow.
- **Kora Game** allows for rehabilitation of the upper limb as the patient moves the end-effector of the robot with admittance control. In this game, the movements of the robot are mapped to a 2D hand that collects apples and pears appearing randomly in a forest, which is as large as the robot's working space. Different difficulty levels can be set by changing the number of fruits that appear or the maximum time for the user to grasp them.

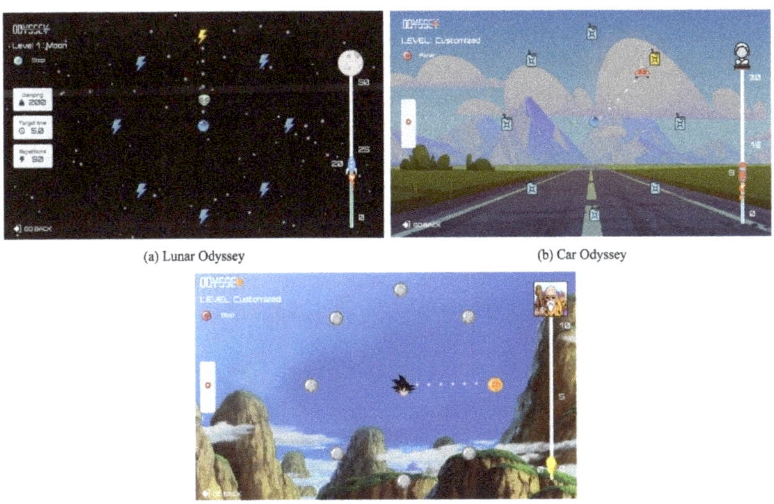

(a) Lunar Odyssey (b) Car Odyssey (c) Dragon Odyssey

Figure 5. Different tabs of the game interface.

2.4. Configuration

There are a total of six different parameters that can be used to properly tune the rehabilitation activities in Odyssey:

1. The first parameter is the user's **arm length**, which is essential for adjusting the robot's targets to match the user's maximum arm reach. This measurement is entered into the game, which automatically adjusts the robot's targets to ensure that the arm is fully extended and not flexed when reaching a target. The targets are mapped in an elliptic shape that allows for full arm extension when reaching them in the movement, where left–right movements have a greater range than forward–backward movements. This

design ensures that the patient performs the exercise and stretches the arm to the maximum range in order to reach each target.

2. The second personalisation parameter is the robot's **plane of work.** Odyssey can be used in two different planes, namely, the transverse and the frontal planes, with the human body as the frame of reference. This allows the rehabilitation exercise to be executed with either horizontal or vertical movements. In both planes, targets are adjusted according to the arm length.
3. Third, the **damping value** for the admittance control is adjusted for each patient. This parameter can be adjusted from 0 to 600 Ns/m in individual exercises or in consecutive exercises with a specific damping for each level. For healthy individuals, it can be set to a high value, causing the robot to provide significant resistance to movement. For rehabilitation patients, the resistance can initially be set to zero and gradually increased as rehabilitation progresses. This refinement and selection of the damping value is modified by the person in charge of the rehabilitation routine, and can be updated at any time during the exercises. When exporting data, the current damping value during the exercise is also stored.
4. Another relevant parameter is the **number of repetitions** required to complete the current game level. One repetition consists of two movements: one from the centre to the active target on the circumference, and a second from the target back to the centre. Visual feedback is implemented in a progress repetition bar (see Figure 5, right side of the visual interface). The number of repetitions depends on the exercise's purpose and the patient's resilience, and must be configured by the therapist.
5. The damping value and number of repetitions alone are not enough to define the specific speed at which the exercise should be performed. Consequently, the **reaching time** for each target is constantly measured, and can also be modified. If the user does not reach the target in time, it turns red and a message appears indicating the need to increase speed for the next repetition. When a target is not reached, it is not counted as a successful target. The total number of achieved or failed targets is stored in the generated robot dataset.
6. Keeping the patient motivated is an essential aspect of rehabilitation. For this reason, the sixth personalisation parameter is the **game skin**. Three different themes have been designed: a space adventure (Figure 5a), a driving journey (Figure 5b), and a skin inspired by the Dragon Ball anime show (Figure 5c). This allows the user to choose the theme they prefer before starting rehabilitation.

Furthermore, the game provides two modes of play: level-based and fully customised. The level-based mode consists of six levels with all parameters pre-set except for one, which changes with each level. For example, when evaluating the speed, the time allowed to reach the targets decreases with each level. The number of repetitions and the damping value can also be adjusted through the levels. It is the responsibility of the person in charge of the therapy to adapt the levels to the patient. The benefit of the level-based mode is its gamification, providing rewards for succeeding in each level. Therefore, the second gaming mode is the personalised mode, in which all parameters can be modified in real time.

2.5. Safety and Hardware Requirements

As the developed framework communicates via ROS and the robot controller, safety stops and parameter limits are configured directly on the robot controller, in this case the UR10e teach pendant, which is installed on the same computer as the framework. Certain parameter limits are configured directly in the robot controller program. These limits include workspace plane thresholds related to the user's arm length, the target location, and the maximum force and velocity thresholds (107 N and 250 mm/s). When any of these limits is reached, a safety stop occurs within 200 ms. The robot can be restarted through the controller, and the framework resumes when communication is restored. When a communication loss of more than 250 ms is detected or power loss occurs, the framework console displays warning and error messages and the robot stops at the last detected coordinate position. The robot can be resumed through the controller. The only safety stop that requires restarting both the robot and the framework is the one triggered manually by the Z-stop.

System requirements for installing the proposed framework include an Ethernet connection on the same network used for both the robot and the computer station. The computer station should run Ubuntu 20.04.6 LTS and ROS for robot control and visualization. For the EMG acquisition system, the software should be compatible with the operating system on the computer station. If this is not the case, another computer can be used to handle the communication. It also needs to be connected to the same network as the main computer station through Ethernet or WiFi.

3. Validation Methodology

To demonstrate the feasibility of the proposed framework, a pilot evaluation was conducted involving eight healthy participants performing upper limb movements using the UR10e robot. The primary goal of this pilot was to assess the correct functioning of the framework components, robot control, EMG acquisition, gamification interface, and data synchronisation rather than to test a particular clinical hypothesis.

3.1. Experiment Protocol

In this experiment, the participants (Table 2) performed reaching movements while seated upright with feet flat on the floor and the elbow positioned at 90°. The task involved moving an object from a central position to one of eight radial targets arranged in clockwise order, then returning to the centre in the x–y plane. Following a one-minute resting baseline and a familiarization trial at 100 Ns/m, which were excluded from the analysis, each participant completed one repetition set of the eight movements at three damping levels of 200, 400, and 600 Ns/m (D200, D400, D600 respectively), with a one-minute rest period between damping levels. Participants were asked to try to maintain consistent velocity throughout all movements and damping levels. The experiment started with measurement of the participant's dominant arm length. The chair was adjusted so that when the participant grasped the end effector, the arm was aligned with the central target of the centre-out tasks depicted in Figure 5.

Table 2. Description of the healthy subjects.

Subject ID	Sex	Age	Arm Length (mm)	Sleep Time
S01	F	22	540	7 h
S02	F	27	460	7 h
S03	M	24	560	7 h
S04	F	23	490	7 h
S05	M	39	560	7 h
S06	M	47	580	7 h
S07	M	21	530	7 h
S08	F	24	460	<5 h

All participants were right-handed.

Eight EMG sensors were placed on the skin surface over the following muscles: Middle Trapezius (MT), Anterior, Middle, and Posterior Deltoid (AD, AM, AP), Biceps Brachii

(BB), Brachioradialis (BR), Extensor Digitorum (ED), and Extensor Carpi Ulnaris (CUE). An example of recorded EMG signals for one set of movements is shown in Figure 6.

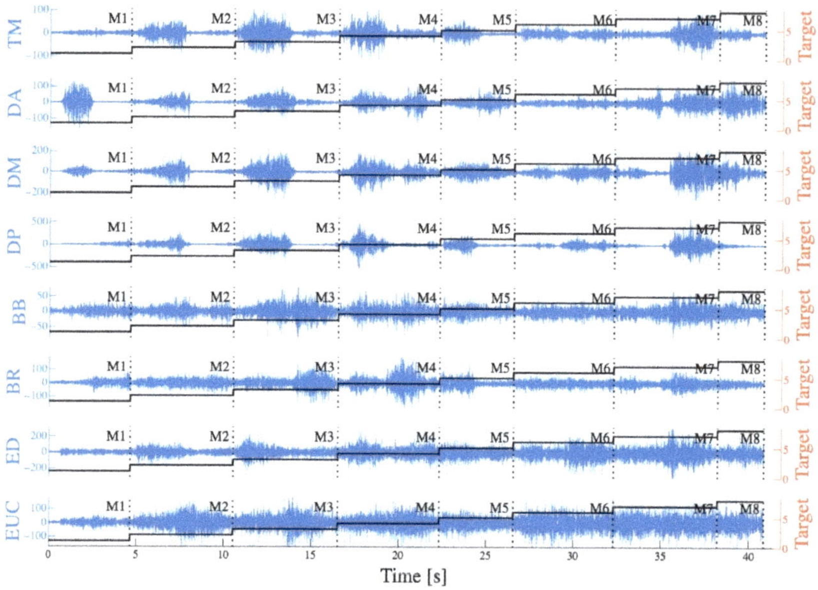

Figure 6. EMG data synchronization of eight different muscles activity (blue) during the movement execution of the eight different targets (black) on the first level.

The game was then configured according to each participant's data, particularly arm length. Each participant selected a preferred game skin to perform the task. The reaching time was set to 5 s; otherwise, it was considered a failure. Visual feedback of target failure was presented with a failure message appearing at the target to reach; the colour and shape of both the target and the avatar icon changed, the progress bar did not count the failed target, and at the end of the session the progress bar showed only the total number of targets reached. To minimize variability and initial fatigue, participants were asked to abstain from caffeine on test day, avoid upper limb exercise the day before, and ensure at least seven hours of sleep before testing (see Table 2). The protocol was approved by the Research Ethics Committee of the University of Alicante, with file number UA-2023-01-23 2.

3.2. Data Analysis

Because the framework provides two datasets, namely, kinematics and EMG, we performed two offline analyses. For the kinematic data, the system recorded the TCP positions, velocities, and forces in the x–y plane as well as the active target at each time point. The first milliseconds of each recording were discarded in order to remove initial adaptation to the robot's movement. The data were then segmented into eight movements (M) for each damping level. The absolute magnitudes of force and velocity were calculated for all segments. The comparison of mean force across participants was normalised to the 95th percentile of force across all damping conditions. Time was measured as the total duration required to complete a movement, from centre to target to centre. For TCP positions, trajectory segments were compared with an ideal path, defined as the shortest path between the centre point and the objective target. Based on this comparison, errors were calculated and normalised to the corresponding movement distance: mean absolute error (MAE), maximum error (MaxE), and percentage root mean square error (PRMSE).

For the EMG data, the signals underwent multistage processing. First, outliers exceeding three times the standard deviation were removed and replaced with the median value of the window using a Hampel filter with a temporal window of $0.05 \cdot fs$. Baseline noise was subtracted, then a fourth-order Butterworth notch filter was applied to remove power line interference at 50 Hz and its harmonics. Cardiac noise in the back and shoulder channels (MT, AD, MD, PD) was subtracted from the signal using an adaptive filter. Finally, a 20 to 350 Hz bandpass filter was applied to remove low and high frequency noise. To segment muscle activity for each movement, kinematic data were interpolated and then aligned with the corresponding timestamps. Muscle activation periods were detected from the signal envelope whenever the amplitude exceeded two times the standard deviation for at least 500 ms. Within the active periods of each repetition, the root mean square (RMS) and the mean frequency (MNF) were computed using a 250-ms sliding window with 50% overlap.

4. Results

Validation of the proposed rehabilitation platform was performed with the Odyssey game. To support personalized therapy for each individual, the system incorporates different personalisation parameters that can be updated in both the main UI and within the game. The developed framework was preliminary tested with eight healthy subjects in order to validate the data acquisition process and overall system functionality as well as to assess possible kinematic metrics for evaluating user performance.

4.1. System Usability

To validate system's usability in real-life scenarios, maximum forces and velocities were obtained. Figure 7 shows the maximum force (A) and velocity (B) values across all movement tasks. As mentioned in Section 2.5, force and velocity have limit thresholds to ensure correct robot functionality. Maximum force values for almost all subjects remained within safe limits. Notably, S06 (the subject with the longest arm length) and S07 (the youngest male subject) both showed peak values close to the force threshold. In both cases, the maximum force was observed at D400 and the lowest velocity at D600, while S06 also reported the highest velocity at D400 and S07's velocity decreased with increasing damping, as expected. In the case of S07, this may be because the highest velocity was observed at D200, which was considerably faster than any other subject; from this initial damping level, the subject tried to maintain the same velocity across all damping levels, inducing more force than necessary through the robot. In addition, all male subjects showed higher variability in velocity, which may be due to different velocities across movements. This phenomenon may improve with a reminder to the subjects to maintain constant velocities, thereby reducing force peaks.

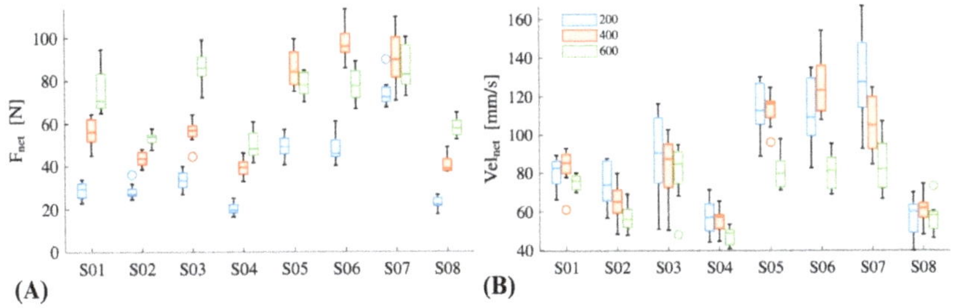

Figure 7. Boxplots showing the maximum net force (**A**) and net velocity (**B**) detected in all eight subjects for each damping level. The x-axis shows the subject ID, while the y-axis shows the corresponding metric and units. All plots show D200 in blue, D400 in orange, and D600 in green.

Figure 6 shows muscle synchronization with kinematic targets for all eight muscles. Certain targets activate muscles at the beginning of the task (centre-target), while others produce more activation at the end of the task (target-centre) and still others maintain muscle activation through the entire reach. For example, as expected, DM and DP activated at the beginning of M5, as the subject performed a downward rotation of the shoulder, while AD became more active in the middle of the task as the user performed a shoulder abduction.

4.2. User Performance

To show the trajectory performance, Figure 8 displays the expected and real TCP trajectories for all eight movements at each damping condition in the x–y plane for S02. The blue plot corresponds to the first damping level (D200), the yellow plot corresponds to the middle level (D400), and the green plot corresponds to the last level (D600). In all three plots, the dark line is the ideal trajectory. To differentiate the eight movements, each direction is represented with a gradient of its corresponding colour. In this subject, at D200 the trajectories appear inaccurate in almost all directions, while at D400 they show more precise movements with straighter paths.

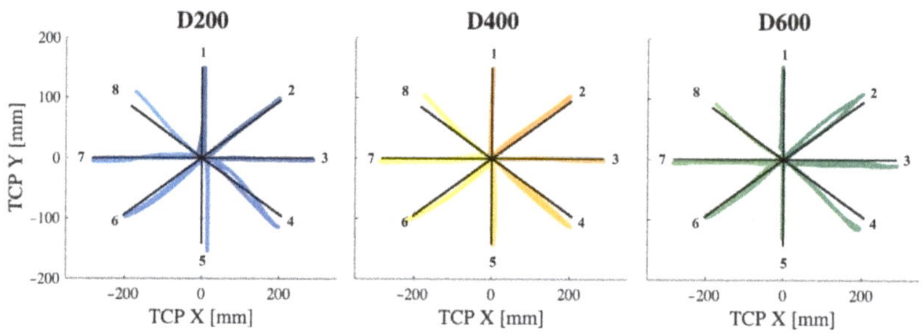

Figure 8. End-effector trajectories at all three damping levels for user A.

To measure the movement performance in the eight trajectories, kinematic metric plots were generated to compare movements, as shown in Figure 9. To maintain consistency between figures, results for D200 are shown in blue, for D400 in orange, and for D600 in green. Figure 9A shows a bar plot of the mean total time for each movement across all subjects. In general, time increases with higher movement resistance generated by the robot, as expected. It is important to note that while M5 and M8 have the lowest times, the difference between these two is nearly double. This is because the movement performed in the last direction was only centre-to-target, resulting in approximately half the time relative to the other movements, which were centre–target–centre. In addition, the movements with the highest times were M3 and M7, which correspond to the lateral targets, as described in Figure 8.

Regarding force values, Figure 9B shows a box plot of the normalised force across all movements. Force medians are clearly separated by damping condition, as expected: D600 > D400 > D200, which confirms that the framework applies graded mechanical resistance. The results for force also shows a directional pattern, with lateral and diagonal targets (M2, M3, M4, M6, M7, and M8) having higher values compared to vertical targets.

Performance was also analysed using movement errors. Figure 9C shows the plots for the PRMSE, mean error, and maximum error. The PRMSE values are small overall, typically between approximately 5 to 15% of the movement distance. In most directions, the PRMSE tends to be slightly lower at D600 and the movement accuracy varies between directions, with vertical targets showing the largest errors. The mean error is clustered around 0.05 to 0.12, with minimal separation between resistance levels. Interestingly, all subjects reported that M3 and M7 were the most difficult movements, yet these have the lowest errors across metrics, while M1 and M5 were reported as the easiest but have the highest errors and highest variability between subjects.

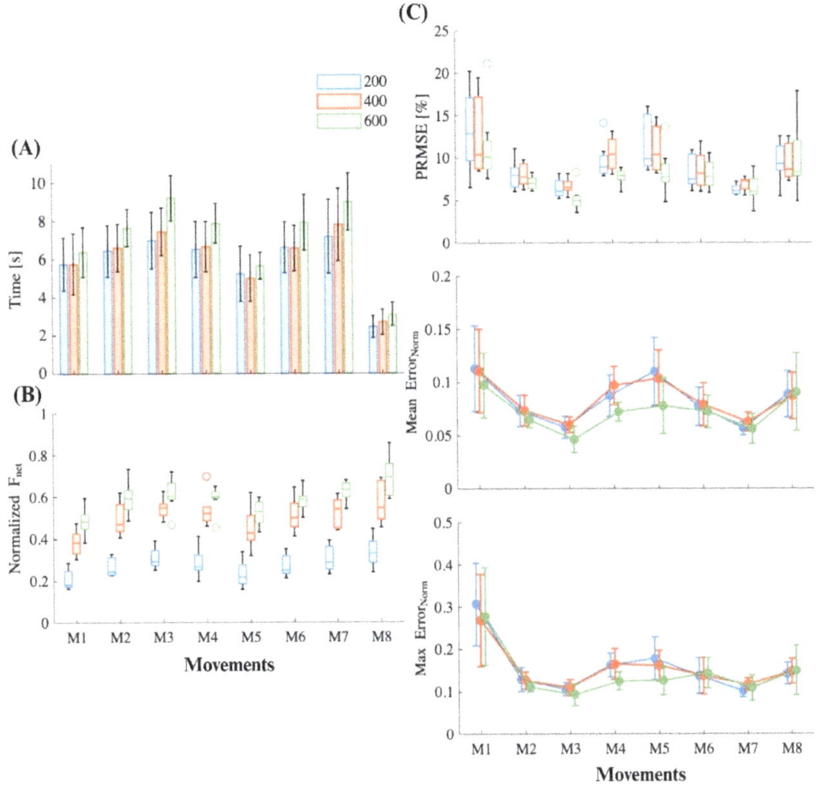

Figure 9. Kinematics plots of the subjects' performance across the different movements and damping conditions, showing the average time to complete a target movement (**A**), the normalised net force used in each movement (**B**), and the mean normalised errors (PRMSE, mean error, and maximum error) obtained in each movement (**C**). The x-axis represents the movement number, and all plots show D200 in blue, D400 in orange, and D600 in green.

For the eight-muscle channel EMG data, normalised RMS values between 0 and 1 were computed, with values near 1 representing the session maximum. Figure 10 shows heatmaps for the three damping conditions of the normalised RMS values. The x-axis corresponds to the movement number and the y-axis to the muscles. As expected, the normalised RMS increases with damping across all directions. The proximal shoulder muscles show anatomically consistent directional tuning. The AD peaks in anterior and left anterolateral reaches (M1, M7, and M8), while the MD and PD peak in lateral movements (M3, M4, M6, and M7). The MT peaks in right lateral reaches (M2, M3, and M4) and in the left lateral movement (M7). This pattern is preserved as the load increases. Elbow flexors (BB, BR) show moderate diagonal biased recruitment that scales with damping, with the exception of M8. Distal extensors (ED, ECU) increase broadly, which indicates wrist and hand stabilization through co-contraction. Overall, M3, M6, and M7 activate all muscles more in the higher damping conditions. Figure 11 shows the mean MNF values for S02 across the three damping conditions, scaled to the maximum MNF value recorded during the protocol for each muscle. As the protocol was not sufficiently demanding to induce fatigue, no significant changes in MNF values were consistently observed in all subjects.

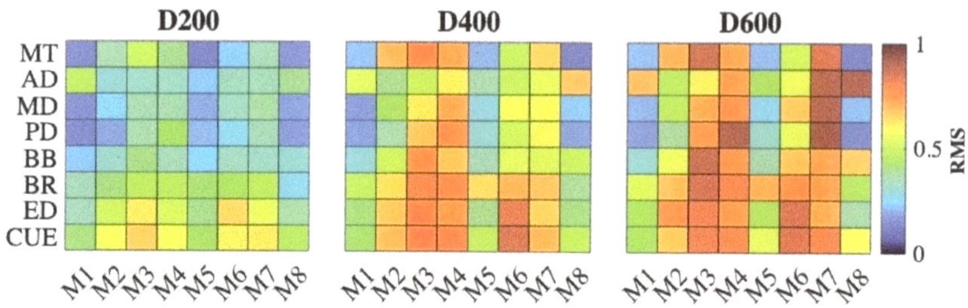

Figure 10. Mean normalised RMS values for all subjects across the eight muscles, eight movements, and three damping levels.

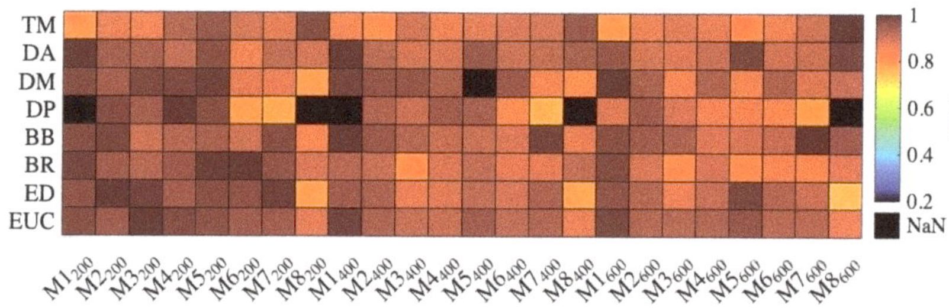

Figure 11. Mean MNF values for subject S02, scaled to the maximum value for each muscle across all eight movements and all three damping levels. The x-axis corresponds to the movement number and associated damping level. Black zones indicate periods with no detected muscle activation.

5. Discussion

The primary objective of the pilot study was to assess the feasibility of the proposed framework in a real environment with different subjects and subject characteristics. For this reason, the analysis focused primarily on the visualisation and interpretation of the data generated by the system rather than on a detailed evaluation of the participants' performance. Within the context of the present study, the most relevant aspects were the correct synchronisation of data, effective communication between the various system modules (Unity, ROS, and EMG), and functional validation of the complete framework. Preliminary results, particularly the RMS values obtained from the EMG signals, demonstrated that the system responds appropriately: as damping increases, there is a proportional rise in muscle activation, accompanied by an increase in the applied forces and an expected reduction in execution speed. With respect to trajectory errors, a decreasing trend was observed as the damping level increased. However, because only one set of eight movements was performed per damping level, this trend could be attributed either to motor adaptation or to the possibility that higher damping facilitates the execution of more stable and linear trajectories in the absence of fatigue.

The developed framework enables linear movements in the x–y and y–z planes, allowing tasks with up to four degrees of freedom. This primarily targets the shoulder and elbow, although the wrist may also benefit to some extent. Moreover, the system can be extended to support three-dimensional movements provided that compatible skins are developed to accommodate this functionality. In comparison with other robotic rehabilitation platforms for the upper limb, the present system offers capabilities similar to recent technologies that support 3D interaction, such as the Armeo family [25], Gentle/s, and Proficio devices. In contrast, earlier systems such as ARM-Guided and MIT-Manus are limited to movements with one or two degrees of freedom, respectively [33]. A key aspect to consider is post-session feedback. Most current rehabilitation robots provide only basic kinematic metrics; some offer performance indicators at the end of the therapy session, as is the case with the Proficio [34]. However, EMG signals are typically excluded from this type of performance analysis, being used mainly for device control rather than as quantifiable indicators of the patient's condition or progression [28].

Future Work and Current Limitations

While this work has demonstrated the technical feasibility of the proposed framework in controlled settings with healthy participants and ROS-compatible collaborative robots from the UR family, further studies are required to fully validate its applicability in broader rehabilitation contexts. Future work should focus on conducting clinical trials with at least ten ABI patients with a minimum of three hours per week and a total of five sessions [33] in order to assess the effectiveness of the framework in real rehabilitation scenarios. This study would include assessment of impact on motor recovery outcomes, user engagement, and therapist usability. Analysis of both kinematic data and EMG recordings should be addressed in order to provide patients with a complete overview of the therapy sessions, which should include movement performance with error metrics such as PRMSE, MaxE, and MAE along with muscle strengthening with force and velocity metrics for the kinematic analysis and fatigue and muscle power for the EMG analysis. At the end of the therapy, a complete report including patient score and improvement should be provided.

Additionally, further experiments are planned to implement and validate the framework on other ROS-compatible robots beyond the Universal Robots family. This would help to confirm the framework's adaptability to diverse hardware platforms and configurations. To support wider adoption, collaborative efforts with external research groups are envisioned, allowing independent teams to deploy and evaluate the system in different environments. Furthermore, we intend to expand the gamification module with more personalized and adaptive tasks as well as to integrate biosignal processing pipelines for user feedback, since one of the main limitations of the current framework is the lack of real-time feedback. At present, only time feedback is provided through changes in target colour, a failure message, and a progress bar that does not count missed targets. Markers such as fatigue detection, large increases in error, and messages about force and velocity when values approach their limits are key objectives for future online analysis. A kinematic report should also be provided at the end of the session instead of raw kinematic and EMG data. Finally, regarding data privacy and long-term management, measures should include implementation of end-to-end encryption and multifactor authentication across data capture, transmission, and storage as well as integration of automated pseudonymization and legally compliant retention policies supported by immutable audit trails for full traceability.

6. Conclusions

This paper has presented a robotic framework designed for upper-limb rehabilitation in ABI patients. The proposed framework offers an adaptable solution for rehabilitation exercises integrating data acquisition, serious games to enhance patient engagement, and flexible robot control options. A technical validation of the framework was carried out in a controlled setting with eight healthy participants and three different damping conditions with an ROS-compatible collaborative robot from the UR family, demonstrating the feasibility and functionality of the proposed framework as a proof of concept.

Author Contributions: A.C. and N.S. contributed to data curation, formal analysis, investigation, and writing. Both also participated in reviewing and editing along with C.R. Study methodology and software development was the principal contribution of N.S., with the help of K.P. in software development. A.U. and C.A.J. provided supervision and contributed to writing and reviewing of the article. All authors have read and agreed to the published version of the manuscript.

Funding: This work is part of the GARMOR project, reference PID2022139105OB-I00, funded by the Ministry of Science, Innovation and Universities (MCIN/AEI/10.13039/501100011033).

Institutional Review Board Statement: This study has ethical approval from the Research Ethics Committee of the University of Alicante, file number UA-2023-01-23 2.

Informed Consent Statement: Informed consent was obtained from all subjects involved in the study.

Data Availability Statement: The datasets and materials are available upon request to the corresponding author.

Conflicts of Interest: The authors declare no conflicts of interest.

Abbreviations

The following abbreviations are used in this manuscript:

ABI	Acquired Brain Injury
ROS	Robot Operating System
EMG	Electromyography
UR	Universal Robots
DoF	Degrees of Freedom
UI	User Interface
HTTP	Hypertext Transfer Protocol
CSV	Comma-Separated Values

References

1. Zorowitz, R.D.; Chen, E.; Tong, K.B.; Laouri, M. Costs and Rehabilitation Use of Stroke Survivors: A Retrospective Study of Medicare Beneficiaries. *Top. Stroke Rehabil.* **2009**, *16*, 309–320. [CrossRef]
2. Aprile, I.; Germanotta, M.; Cruciani, A.; Loreti, S.; Pecchioli, C.; Cecchi, F.; Montesano, A.; Galeri, S.; Diverio, M.; Falsini, C.; et al. Upper Limb Robotic Rehabilitation after Stroke: A Multicenter, Randomized Clinical Trial. *J. Neurol. Phys. Ther.* **2020**, *44*, 3–14. [CrossRef] [PubMed]
3. Kennard, M.; Hassan, M.; Shimizu, Y.; Suzuki, K. Max Well-Being: A Modular Platform for the Gamification of Rehabilitation. *Front. Robot. AI* **2024**, *11*, 1382157. [CrossRef]
4. Krebs, H.I.; Hogan, N.; Aisen, M.L.; Volpe, B.T. Robot-aided Neurorehabilitation. *IEEE Trans. Rehabil. Eng.* **1998**, *6*, 75–87. [CrossRef]
5. Lum, P.S.; Burgar, C.G.; Shor, P.C.; Majmundar, M.; Van der Loos, M. Robot-assisted Movement Training Compared with Conventional Therapy Techniques for the Rehabilitation of Upper-limb Motor Function after Stroke. *Arch. Phys. Med. Rehabil.* **2002**, *83*, 952–959. [CrossRef]
6. Richardson, R.; Brown, M.; Bhakta, M.; Levesley, M.C. Design and Control of a Three Degree of Freedom Pneumatic Physiotherapy Robot. *Robotica* **2003**, *21*, 589–604. [CrossRef]
7. Reinkensmeyer, D.J.; Kahn, L.E.; Averbuch, M.; McKenna-Cole, A.N.; Schmit, B.D.; Rymer, W.Z. Understanding and Treating Arm Movement Impairment after Chronic Brain Injury: Progress with the ARM Guide. *J. Rehabil. Res. Dev.* **2000**, *37*, 653–662.
8. Zhu, T.L.; Klein, J.; Dual, S.A.; Leong, T.C.; Burdet, E. ReachMAN2: A Compact Rehabilitation Robot to Train Reaching and Manipulation. In Proceedings of the 2014 IEEE/RSJ International Conference on Intelligent Robots and Systems, Chicago, IL, USA, 14–18 September 2014; pp. 2107–2113. [CrossRef]

9. Chiriatti, G.; Palmieri, G.; Palpacelli, M.C. A Framework for the Study of Human-robot Collaboration in Rehabilitation Practices. In *Advances in Service and Industrial Robotics, Proceedings of the International Conference on Robotics in Alpe-Adria Danube Region, Kaiserslautern, Germany, 19 June 2020*; Springer: Cham, Switzerland, 2020; pp. 190–198. [CrossRef]
10. Rodrigues, J.C.; Menezes, P.; Restivo, M.T. An Augmented Reality Interface to Control a Collaborative Robot in Rehab: A Preliminary Usability Evaluation. *Front. Digit. Health* **2023**, *5*, 1–16. [CrossRef]
11. Molteni, F.; Gasperini, G.; Cannaviello, G.; Guanziroli, E. Exoskeleton and End-Effector Robots for Upper and Lower Limbs Rehabilitation: Narrative Review. *Innov. Influenc. Phys. Med. Rehabil.* **2018**, *10*, 174–188. [CrossRef] [PubMed]
12. Garro, F.; Chiappalone, M.; Buccelli, S.; De Michieli, L.; Semprini, M. Neuromechanical Biomarkers for Robotic Neurorehabilitation. *Front. Neurorobot.* **2021**, *15*, 742163. [CrossRef] [PubMed]
13. Lencioni, T.; Fornia, L.; Bowman, T.; Marzegan, A.; Caronni, A.; Turolla, A.; Jonsdottir, J.; Carpinella, I.; Ferrarin, M. A Randomized Controlled Trial on the Effects Induced by Robot-assisted and Usual-care Rehabilitation on Upper Limb Muscle Synergies in Post-stroke Subjects. *Sci. Rep.* **2021**, *11*, 5323. [CrossRef] [PubMed]
14. Huang, C.; Chen, M.; Zhang, Y.; Li, S.; Zhou, P. Model-based Analysis of Muscle Strength and EMG-force Relation with Respect to Different Patterns of Motor Unit Loss. *Neural Plast.* **2021**, *2021*, 5513224. [CrossRef] [PubMed]
15. Goffredo, M.; Proietti, S.; Pournajaf, S.; Galafate, D.; Ciota, M.; Le Pera, D.; Posterato, F.; Francesichini, M. Baseline Robot-measured Kinematic Metrics Predict Discharge Rehabilitation Outcomes in Individuals with Subacute Stroke. *Front. Bioeng. Biotechnol.* **2022**, *10*, 1012544. [CrossRef] [PubMed]
16. Afyouni, I.; Rehman, F.U.; Qamar, A.M.; Ghani, S.; Hussain, S.O.; Sadiq, B.; Rahman, M.A.; Murad, A.; Basalamah, S. A Therapy-driven Gamification Framework for Hand Rehabilitation. *User Model. User-Adapt. Interact.* **2017**, *27*, 215–265. [CrossRef]
17. Simić, M.; Stojanović, G.M. Wearable Device for Personalized EMG Feedback-based Treatments. *Results Eng.* **2024**, *23*, 102472. [CrossRef]
18. Alfieri, F.M.; da Silva Dias, C.; de Oliveira, N.C.; Battistella, L.R. Gamification in Musculoskeletal Rehabilitation. *Curr. Rev. Musculoskelet. Med.* **2022**, *15*, 629–636. [CrossRef]
19. Tuah, N.M.; Ahmedy, F.; Gani, A.; Yong, L.N. A Survey on Gamification for Health Rehabilitation Care: Applications, Opportunities, and Open Challenges. *Information* **2021**, *12*, 91. [CrossRef]
20. Faran, S.; Einav, O.; Yoeli, D.; Kerzhner, M.; Geva, D.; Magnazi, G.; van Kaick, S.; Mauritz, K.-H. Reo Assessment to Guide the ReoGo Therapy: Reliability and Validity of Novel Robotic Scores. In Proceedings of the 2009 Virtual Rehabilitation International Conference, Haifa, Israel, 29 June–2 July 2009; [CrossRef]
21. Exoskeleton Report, "Upper Body Fixed Rehabilitation: Exoskeleton Catalog Category for Upper Body Fixed (or Stationary) Wearable Exoskeletons". Available online: https://exoskeletonreport.com/product-category/exoskeletoncatalog/medical/upper-body-fixed-rehabilitation/ (accessed on 19 November 2023).
22. Nexum by Wearable Robotics. Available online: https://nexumrobotics.it/ (accessed on 19 November 2023).
23. D'Antonio, E.; Galofaro, E.; Patane, F.; Casadio, M.; Masia, L. A Dual Arm Haptic Exoskeleton for Dynamically Coupled Manipulation. In Proceedings of the 2021 IEEE/ASME International Conference on Advanced Intelligent Mechatronics (AIM), Delft, The Netherlands, 12–16 July 2021; IEEE: Delft, The Netherlands, 2021; pp. 1237–1242.
24. Calabrò, R.S.; Russo, M.; Naro, A.; Milardi, D.; Balletta, T.; Leo, A.; Filoni, S.; Bramanti, P. Who May Benefit From Armeo Power Treatment? A Neurophysiological Approach to Predict Neurorehabilitation Outcomes. *PM&R* **2016**, *8*, 971–978... [CrossRef]
25. Calabrò, R.S. (Ed.) *Translational Neurorehabilitation: Brain, Behavior and Technology*; Springer International Publishing: Cham, Switzerland, 2024. [CrossRef]
26. Hsieh, Y.; Lin, K.; Wu, C.; Shih, T.; Li, M.; Chen, C. Comparison of Proximal versus Distal Upper-Limb Robotic Rehabilitation on Motor Performance after Stroke: A Cluster Controlled Trial. *Sci. Rep.* **2018**, *8*, 2091. [CrossRef]
27. Zhang, L.; Guo, S.; Sun, Q. An Assist-as-Needed Controller for Passive, Assistant, Active, and Resistive Robot-Aided Rehabilitation Training of the Upper Extremity. *Appl. Sci.* **2021**, *11*, 340. [CrossRef]
28. Guatibonza, A.; Solaque, L.; Velasco, A.; Peñuela, L. Assistive Robotics for Upper Limb Physical Rehabilitation: A Systematic Review and Future Prospects. *Chin. J. Mech. Eng.* **2024**, *37*, 69. [CrossRef]
29. Bessler, J.; Prange-Lasonder, G.B.; Schaake, L.; Saenz, J.F.; Bidard, C.; Fassi, I.; Valori, M.; Lassen, A.B.; Buurke, J.H. Safety Assessment of Rehabilitation Robots: A Review Identifying Safety Skills and Current Knowledge Gaps. *Front. Robot. AI* **2021**, *8*, 602878. [CrossRef] [PubMed]
30. Shoaib, M.; Asadi, E.; Cheong, J.; Bab-Hadiashar, A. Cable Driven Rehabilitation Robots: Comparison of Applications and Control Strategies. *IEEE Access* **2021**, *9*, 110396–110420. [CrossRef]
31. Mamani, W.; Sempere, N.; Casanova, A.; Morell, V.; Jara, C.A.; Ubeda, A. Upper Limb EMG-based Fatigue Estimation During End Effector Robot-assisted Activities. In *Converging Clinical and Engineering Research on Neurorehabilitation V, Proceedings of the International Conference on NeuroRehabilitation, La Granja, Spain, 4–7 November 2024*; Springer: Cham, Switzerland, 2024; pp. 441–445. [CrossRef]

32. Rohrer, B.; Fasoli, S.; Krebs, H.I.; Hughes, R.; Volpe, B.; Frontera, W.R.; Stein, J.; Hogan, N. Movement Smoothness Changes During Stroke Recovery. *J. Neurosci.* **2002**, *22*, 8297–8304. [CrossRef]
33. Mahfouz, D.M.; Shehata, O.M.; Morgan, E.I.; Arrichiello, F. A Comprehensive Review of Control Challenges and Methods in End-Effector Upper-Limb Rehabilitation Robots. *Robotics* **2024**, *13*, 181. [CrossRef]
34. Abdel Majeed, Y.; Awadalla, S.; Patton, J.L. Effects of Robot Viscous Forces on Arm Movements in Chronic Stroke Survivors: A Randomized Crossover Study. *J. NeuroEng. Rehabil.* **2020**, *17*, 156. [CrossRef] [PubMed]

applied sciences

MDPI

Article

Pilot Study of an Online Exercise Therapy Programme for Home Office Workers in Terms of Musculoskeletal and Mental Health

Tünde Szilágyi [1,*], Ilona Veres-Balajti [2], Balázs Lukács [2], Enikő Király [1], Anna Mária Laczkó [1] and Zoltán Jenei [1]

[1] Department of Medical Rehabilitation and Physical Medicine, Faculty of Medicine, University of Debrecen, 4032 Debrecen, Hungary

[2] Department of Physiotherapy, Faculty of Health Sciences, University of Debrecen, 4032 Debrecen, Hungary

* Correspondence: szilagyine.tunde@med.unideb.hu

Abstract: (1) Background: Working from home is becoming increasingly common and has many advantages, but also negative consequences such as reduced physical activity, poor ergonomics and stress. Many people find it difficult to attend in-person rehabilitation sessions due to transportation barriers and limited access to rehabilitation facilities. Therefore, the aim of our study was to investigate whether our online preventive exercise programme effectively improves the musculoskeletal and mental health of sedentary workers. (2) Methods: The study participants worked from home on a long-term basis. Our research group consisted of 30 people, 16 in the intervention group and 14 in the control group. The intervention group members participated in a complex preventive exercise programme three times a week for 10 weeks. The training took place online. Before and after the exercise programme, various tests were used to assess the functional status of the participants' spine, the strength and flexibility of their core muscles (Core, Kempf and Kraus–Weber tests) and their mental state (Beck Depression Inventory). (3) Results: Significant improvement was observed in the intervention group in the Core, Kempf, Kraus–Weber and Beck tests. (4) Conclusions: Our online complex training programme proved to be effective during the evaluation. It is crucial to tailor remote rehabilitation services to the individual needs and preferences of patients and healthcare professionals.

Keywords: core muscles; patient education; prevention; home office; movement therapy

1. Introduction

1.1. Background

Online gymnastics is a distance-based exercise programme using communication technologies. In recent years, online technology has developed rapidly, but the restrictions imposed by COVID-19 have helped to accelerate the development of this system [1]. Biddle and Mutrie studied the psychological impact of physical activity. Their research has proven that regular exercise reduces anxiety and depression [2]. Peter Salmon, a psychologist at the University of Liverpool, also considers physical activity as a potential way of maintaining mental well-being [3].

The aim of a study by Özel and colleagues was to examine the effect of telerehabilitation-based, remote supervised or unsupervised structured exercise therapy on pain, disability and quality of life in patients with chronic, non-specific neck pain [4]. Moretti and colleagues in Italy described the exacerbation of low back and neck pain in the remote workers they surveyed, but it is not known how teleworking affected the prevalence of musculoskeletal pain [5]. The Department of Physiotherapy and Rehabilitation at Muğla Sıtkı

Koçman University involved 50 patients with chronic low back pain syndrome. The 8-week gymnastics programme was delivered to the patients through a web-based telerehabilitation platform called Fizyoweb. The telerehabilitation group reported improvements in pain, function, quality-of-life, kinesiophobia, satisfaction and motivation [6]. Cooper and colleagues tested a 6-week online multimodal exercise and health education program involving 23 women focusing on menopause education and lifestyle advice. The effectiveness of the program was evaluated in terms of the physical performance, menopausal symptoms, well-being and self-efficacy of participants in Laois, Ireland. Exercise sessions were held twice a week for 40 min, while the health education sessions were held once a week, also for 40 min. Participants were divided into two groups: perimenopausal and postmenopausal. It was found that the 6-week online exercise program was time-efficient and that multimodal exercise and health education tailored to menopause significantly improved the physical performance, menopausal symptoms, well-being and physical self-efficacy of perimenopausal and postmenopausal women [7].

Henriëtte A. Meijer and colleagues systematically reviewed qualitative studies providing evidence on patients' responses to videoconferencing (VC) services in an orthopaedic setting. Videoconferencing (VC) can play an important role in the management of orthopaedic pathologies. Despite positive reports regarding telemedicine studies, the uptake in clinical practice remains low. Acceptability by patients is an important element in the adoption of telemedicine into the system, and a focus on qualitative methodology may reveal the reasons behind the acceptability of using VC. Further qualitative research is needed to investigate acceptability for both patients and clinicians using a theoretical framework that allows for replicability and generalizability [8].

Malinska and her colleagues conducted a study among computer workers. They used a questionnaire to collect data on demographics, musculoskeletal disorders, lifestyle characteristics, computer work ergonomics and psychosocial and physical work characteristics. Almost half of the respondents complained about neck pain and low back pain. Prolonged computer time, increased job demand and older age were identified as risk factors [9].

Studies to date suggest that exercise protects against the adverse effects of stress and has an anti-stress and antidepressant effect [10]. The positive mental health effects of regular exercise are particularly important for people who work long hours at home [11].

Work from home (WFH), also referred to as home office, emerged in the early 2000s with the advancement of teleworking technologies. These developments enabled employees to perform their professional duties remotely, reducing the need for commuting, allowing for greater flexibility in work schedules and promoting a better work–life balance [12]. Moreover, working from home offers advantages not only for employers but also for employees. Most importantly, it allows employees to save the time otherwise spent commuting and to achieve a more balanced work–life relationship [13]. Working from home enables employees to schedule their work during periods of peak productivity and to avoid many of the workplace distractions typically associated with American-style open offices [13]. In addition, with WFH, employees have more control over environmental factors, internal environmental quality (IEQ) factors such as lighting, temperature, humidity, noise and air quality [14]. These IEQ factors have a significant impact on employee comfort, which is strongly associated with job satisfaction [15].

Despite these numerous benefits, full-time home-office work has also been associated with certain disadvantages. Remote workers tend to experience reduced social interaction with colleagues and lower levels of physical activity (e.g., the absence of walking between meeting locations) [12]. In addition, prolonged screen exposure due to work can lead to fatigue, headaches and eye-related symptoms [16]. For individuals living alone, the absence of face-to-face interaction and daily social support may contribute to mental health problems, including depression and anxiety, as well as a blurring of the boundaries between work and personal life. Such difficulties in psychological detachment from work can elevate stress levels, while ongoing work–family conflict may lead to emotional exhaustion [17].

The home office has also drawn attention to the relational and ergonomic needs of the remote working environment. Many employees perform their tasks at temporary workstations such as kitchen tables, counters, sofas, beds or coffee tables. With reduced business travel and commuting, they spend considerably more time in front of a computer each day [18].

Shariat A. et al. reported a high prevalence of musculoskeletal pain among office workers [19]. Musculoskeletal pain can have a significant negative impact on quality of life, work efficiency and job performance according to Wu S. et al. [20].

Based on the studies discussed above, we aimed to develop an online preventive exercise programme adapted to home office workers and to examine its effects on participants' musculoskeletal function and mental health. The effects were assessed using various validated tests by comparing the intervention and control groups at the beginning and end of the study.

In Hungary, a significant number of people who work from home rely on access to the internet for their work. By February 2020, the number of people working from home was only a few hundred thousand. The peak was in May 2020, when 17% of the domestic workforce, or around 760,000 people, worked from home [20]. According to the Central Statistical Office, the vast majority of home workers (77%) have a tertiary education, 21% have a school leaving certificate, 1.6% have a vocational secondary school qualification and 0.4% have a primary education [21].

The following is the recommended correct sitting position in front of a computer: The recommended joint angles are 0–10° at the shoulder, 90–100° at the elbow, 90–100° at the hip, 90–110° at the knee, and 100–120° at the ankle. While sitting, care should be taken to keep the pelvis stable and centred, avoid posterior pelvic tilt and maintain the physiological curves of the spine—lumbar lordosis, thoracic kyphosis and cervical lordosis [22]. The ideal layout of the workspace is important. It is determined by the frequency of use and placement of office equipment and documents. On this basis, we can divide our working environment into three parts:

1 Usual work
2 Occasional work
3 Non-working area

1.2. Low Back Pain (LBP)

Low back pain refers to pain located between the lower edge of the ribs and the buttocks. The pain can be of short duration (acute), slightly longer duration (subacute) or long duration (chronic). It can occur in any job but is most common in sedentary workers. LBP makes it difficult to move around and can affect quality of life and mental wellbeing. These factors in turn can affect work and relationships with family and friends. Research suggests that 619 million people live with LBP, and it is said to be the leading cause of disability worldwide [23,24]. According to the Global Burden of Diseases, Injuries, and Risk Factors Study (GBD), published in 2023, the global number of LBP cases is expected to reach 619 million in 2020, rising to 843 million by 2050 [23]. There are significant geographical variations in prevalence (cases per 100,000 population), with the highest prevalence in Central and Eastern Europe and Hungary having the highest prevalence per 100,000 population. The already high prevalence of low back pain may be exacerbated by the negative effects of teleworking (including reduced physical activity, poor ergonomics, stress etc.). In this context, preventive interventions targeting the musculoskeletal system play a particularly important role.

Core muscle stability therapy has become increasingly popular in recent years, with research by Wang and colleagues [25] finding that core exercises produced better results in the treatment of low back pain compared with general spinal exercise in the first three months of treatment for LBP. Further studies have shown that core-specific interventions reduce pain and improve movement limitations [26].

1.3. The Aim of Our Study

Our goal was to develop a new experimental online preventive exercise programme for home office workers, aiming to assess and improve the functional and mental condition of participants. In our study, the primary endpoint was the musculoskeletal status of the participants (strength and flexibility of core muscles), while the secondary endpoint was the mental health of the subjects.

2. Materials and Methods

- Thirty patients participated in the study on a voluntary basis. We asked a small local IT company to participate in the online training programme we had developed, which we then confirmed using a validated questionnaire and physical examination.
- Participants signed a consent form in which we informed them about the purpose and ethical background of our programme.
- We divided the participants into two groups. Sixteen people participated in the intervention group and 14 in the control group. The inclusion of the control group was essential for validating the results.
- All participants were examined in their own homes. The assessments were carried out before and after the exercise programme.
- Members of the intervention group participated in a 45 min online exercise programme three times a week for 10 weeks at prearranged times, which took the form of live lectures supervised by a physical therapist.
- Members of the control group did not participate in online therapy sessions. We asked the participants to continue their usual habits and activities and not to start any new activities or other things. This approach was considered important to ensure that the comparison between the control and intervention groups was not influenced by external factors.
- Those who did not meet the above criteria were excluded from the study.
- The data were analysed using statistical methods.

2.1. Target Group

The survey involved young people who work from home for the IT company we contacted and who responded quickly to our online call for applications.

2.2. Study Process

Our study was conducted between 10 March 2024 and 22 November 2024. Registration for our survey was completed online. The inclusion criteria for participants were as follows:

1. Working from home.
2. Consent required measurements and participation in the exercise programme.
3. Attendance at the pre- and post-measurements before and after the 10-week programme.
4. Internet access and use of Microsoft Teams application.
5. Attendance at all sessions.

6. Availability of the necessary equipment (half-litre PET bottles, rubber bands, chair and sofa).
7. Sufficient space to carry out the exercise.
8. Sufficient physical and health condition to participate in the training programme.

Presentation of the research structure:

1. Registration for our research was completed online.
2. Inclusion criteria were completed by 30 participants. The intervention group (n = 16) and control group (n = 14).
3. A validated questionnaire and physical examination were carried out in the participants' homes with the personal assistance of the physiotherapist (n = 30).
4. The 10-week online exercise programme took place in the presence of a physiotherapist (n = 16).
5. To measure back the results obtained during the study, the previously used validated questionnaires and physical examinations were re-recorded in the personal presence of the physiotherapist (n = 30).

2.3. Questionnaires

We wanted to know the mental status of the home office workers so we asked the participants to fill in a form consisting of a Beck depression test at the beginning and end of the study. The test taken at the beginning of the study allows us to assess the presence and severity of depressive symptoms, with a back test at the end to show changes in symptoms. The Hungarian abridged version consists of nine statements and is equivalent to the original, with acceptable reliability [27]. The scores obtained after completion of the test allow us to differentiate between four levels of depression: 0–9 normal, 10–18 mild, 19–25 moderate and ≥26 severe.

2.4. Functional Ability Test

Kraus–Weber (K–W) test: this test is designed to test the strength of the core muscles. The test consists of 6 sets of exercises, which are scored separately and evaluated on the basis of the total score.

The maximum score is 60 points. Scores between 0 and 40 are considered unsatisfactory, scores above 40 are considered good and scores above 50 are considered very good [28].

Core test: an endurance test that has been proposed for the trunk flexors; the subject is in the push-up position in the forearm position. The whole torso is straight; it is important that the hips are neither elevated nor bent. The head is in the line of the spine with shoulders above the elbows. The correct posture should be maintained as long as they can. Meanwhile, measure the number of seconds held with a stopwatch. The exercise continues until the subject gives up or cannot maintain control of the lumbar spine [29].

The Kempf test: named after Hans-Dieter Kempf, the test system consists of 8 exercises and is an excellent way of testing the strength and flexibility of the core muscles [30].

2.5. Mobility Programme

Under the supervision of a qualified physiotherapist, participants in the programme performed targeted posture-strengthening and core muscle-strengthening spinal exercises three times a week for 10 weeks, using a variety of tools. These tools were carefully selected and incorporated into the exercise programme based on a preliminary assessment to ensure that they were available to all participants in their area of residence. We used half-litre PET bottles to replace dumbbells, rubber bands for strengthening, and home equipment such as a chair and a sofa to carry out the exercises. The exercises were performed on carpet or polyfoam mattresses. Each session consisted of a warm-up, main part and a cool-down. Occasionally and on demand, the exercises were supplemented with a short relaxation session. The exercises were always held at pre-arranged times using the Microsoft Teams communication platform. The exercise programme was put together by physical therapists. The physical therapist discussed the tasks with the participant during personal assessments and corrected them during the live programme if necessary. The online programme also allowed the physical therapist to monitor participation and activity during the exercises.

2.5.1. The Warm-Up

We started the exercises with a general warm-up, moving through the joints from cranial to cauda, with more emphasis on the shoulder, hip, knee and ankle joints. The warm-up consisted of 10 min in total.

2.5.2. The Main Part

Based on the individual data obtained during the initial assessment, a physiotherapist compiled a set of exercises for the intervention group that were suitable for all participants and enabled them to progress. The tasks changed weekly. The physiotherapist also determined the number of sets and repetitions based on the group dynamics, ensuring that the training sessions were not one-sided [31]. Each session lasted 45 min.

In the first phase of this stage of training, we teach the correct position of the lumbo-pelvic-hip complex and the formation of normal lordosis, with the aim of establishing proper pelvic control and maintaining it in different body positions.

In the second phase, to achieve effective co-contraction of the transversus abdominis and multifidi muscles, we teach the isolated active tensioning of these muscle groups in different body positions, checking the correct execution by palpation.

Subsequently, to develop the endurance of these muscles, we practice holding the tension for 10 s while breathing continuously and calmly, repeating each exercise ten times, with a one-minute break between training different muscle groups. This is followed by integrated training of the local and global muscle systems, where, following the principle of gradual progression, we increase the intensity of the exercises in several body positions, first in a closed and then in an open kinetic chain, requiring increasing concentration.

The most important exercises used in the exercise programme:

- Squats:
    - Squats → Squats with side steps → Chair-assisted trunk support squats;
- Burpees:
    - One-sided burpees
- Pelvic lifts:
    - Pelvic lifts with stable and unstable support
- Other functional exercises:
- Kneeling push-ups → Push-ups
- Abdominal crunches on the floor and on a chair
- Trunk extension on the floor
- Plank with knee support → Traditional plank → Side plank with knee support

Based on the individual data obtained during the initial assessment, a physiotherapist compiled exercises for the intervention group that were suitable for all participants and allowed them to progress. The tasks varied from week to week. The number of sets and repetitions was also determined by the physiotherapist according to the group dynamics.

The main parts of the exercise programme are shown in Table 1.

Table 1. Exercise programme main part.

Time (week)	Body Position	Device Used	Exercises
1	Supine	Polyfom	Easier spinal mobilisation and isometric stretching
2–3	Variable (lying, sitting)	Half litre bottled water, chair, polyfom	Upper and lower abdominal exercises, deep back muscle strengthening
4–5	Variable (lying, sitting, standing)	Half litre bottled water, chair, polyfom, rubber band	Shoulder girdle and trunk dynamic exercises
6	Variable (lying, sitting, standing, on all fours)	Half litre bottled water, chair, polyfom, rubber band	Shoulder girdle and trunk stabilisation exercises
7–8	Variable (lying, sitting, standing, on all fours)	Half litre bottled water, chair, polyfom, rubber band	Heavy back and abdominal muscle strengthening exercises
9–10	Variable (lying, sitting, standing, on all fours)	Half litre bottled water, chair, polyfom, rubber band	Difficult strain stabilisation exercises

The exercise is illustrated in Figure 1.

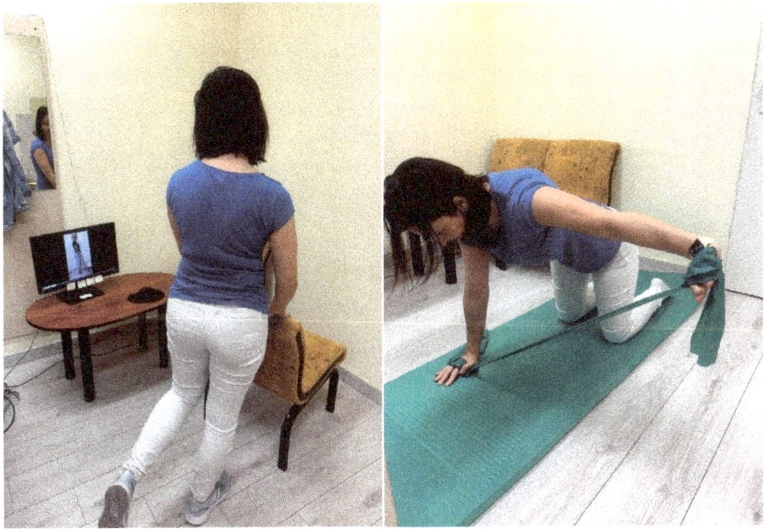

Figure 1. The picture shows one of the subjects doing gymnastics.

2.5.3. The Cool Down

After each exercise we did outstretching, with the main focus on the muscle groups worked. The recovery took 10 min per session. We also supplemented the stretching with short breathing exercises.

2.5.4. The Relaxation Part

Participants lay on a mat or mat in a comfortable position. Relaxation was performed with eyes closed. Once they had managed to focus on the sensations of their body, they started with slow, deep inhalations and exhalations. Following the conscious breathing, they said to themselves, "I am completely relaxed and at ease". These cycles were repeated three times and ended with stretching [32]. The relaxation sessions were 5–10 min each.

2.6. Statistical Methods Used

The results obtained during the physical survey were processed according to a database created in Microsoft Excel spreadsheet software (Microsoft Office LTSC Professional Plus 2024; Excel 2408 version). Our results are measured using median and interquartile ranges (IQR1-IQR3). To implement the statistical tests, in addition to the Microsoft Excel spreadsheet software mentioned above. The two paired (before/after intervention) data were analysed using Wilcoxon signed rank tests and between-group comparisons (intervention group vs. control group) Mann-Whitney U test was used. To evaluate categorical data Fisher's exact test was used. For statistical analysis Stata 13 software was used. The result was considered significant if the *p*-value was less than 0.05. The tests were used to examine whether there was a significant change in the parameters assessed in our case group during the 10-week exercise programme. Strength and extensibility of the spine and core muscles (Core, Kempf and Kraus-Weber test) and mental state (Beck depression test) were identified as dependent variables and time as an independent variable.

2.7. Ethics Licence

The study was approved by the Research Ethics Committee of the University of Debrecen (6774-2024), and participants gave informed consent to the collection and processing of data.

3. Results

A total of thirty people participated in the study, with 14 people in the control group (8 women with a mean age of 29.25 years and 6 men with a mean age of 27.50 years) and 16 people in the intervention group (5 women with a mean age of 27.80 years and 11 men with a mean age of 29.63 years). Table 2 compares the characteristics of the intervention group and the control group at baseline (e.g., age, gender, baseline scores on all tests). There were no differences in age, gender ratio, or test scores between the intervention and control groups. Although the Kraus-Weber Total score was significantly higher in the intervention group before the intervention, the exercise programme still managed to improve the score, while the control group showed no significant change in Kraus-Weber Total score, which also confirms the effectiveness of the exercise programme.

Table 2. Baseline results.

	Control Group		**Intervention Group**		
	male (%)	*female (%)*	*male (%)*	*female (%)*	*p-value (control* vs. *intervention)*
gender	69.00	31.00	43.00	57.00	0.153
	Median	*IQR*	*Median*	*IQR*	*p-value (control* vs. *intervention)*
age	28.00	27.00–30.25	27.00	26.25–32.00	0.461
Kraus-Weber Total	57.00	54.75–57.00	58.00	56.75–60.00	0.022
Kempf 1	15.50	13.00–17.25	16.50	13.75–20.75	0.449
Kempf 2	16.50	13.25–17.50	17.00	13.25–19.25	0.583
Kempf 3	11.50	10.00–15.00	13.50	13.50–17.50	0.482
Kempf 4	13.50	10.00–16.00	13.50	10.00–16.00	0.985
Core	49.50	38.00–58.00	59.00	46.75–72.75	0.104
Beck total	22.00	22.00–23.00	22.00	20.25–24.00	1.000

3.1. Results of the Kraus-Weber Test

Based on the total scores of the Kraus-Weber test, a significant change ($p = 0.003$) was observed after the exercise programme in the intervention group. At the start of the study, the median was 58 (56.75; 60.00), increasing to 60 (59.00; 60.00). No significant ($p = 0.082$) change was observed in the control group.

Looking at the first part of the Kraus-Weber test (Kraus-Weber 1), it can be seen that there was a significant change ($p = 0.015$) from the start of the study (median = 10.00; iqr = 8.75; 10.00). For the second test part (Kraus-Weber 2), there was no significant change ($p = 0.084$) in the effect of the exercise programme. Both at baseline and after the exercise programme, the median was 10.00 (iqr = 10.00; 10.00). The results of the third part (Kraus-Weber 3) (before and after median = 10.00; iqr = 10.00; 10.00) also showed no significant change ($p = 0.157$). Also, no significant change was found for the fourth (Kraus-Weber 4) ($p = 0.083$) (before and after median = 10.00; iqr = 10.00; 10.00) test part. When examining the fifth part (Kraus-Weber 5), it can be seen that there was a significant change ($p = 0.046$) from the beginning of the test (median = 10.00; iqr = 9.75; 10.00) to the end (median = 10.00; iqr = 10.00; 10.00). Finally, a significant change ($p = 0.025$) was also observed for the sixth part (Kraus-Weber 6) (median = 10.00; iqr = 9.00; 10.00) compared to the pre-exercise condition (median = 10.00; iqr = 10.00; 10.00). There were no significant changes in the control group in any of the tests (Kraus-Weber 1–6) (Table 3).

Table 3. Kraus-Weber test results.

		Control Group			Intervention Group			
Test	*Time (Before/After Intervention)*	*Median*	*IQR*	*p-Value*	*Median*	*IQR*	*p-Value*	*p-Value* (*Control* vs. *Intervention*)
Kraus-Weber Total	*Before*	57.00	54.75–57.00	0.082	58.00	56.75–60.00	0.003	0.020
	After	57.00	54.75–58.00		60.00	59.00–60.00		<0.001
Kraus-Weber 1	*Before*	9.00	8.75–10.00	-	10.00	8.75–10.00	0.015	0.639
	After	9.00	8.75–10.00		10.00	10.00–10.00		0.078
Kraus-Weber 2	*Before*	10.00	10.00–10.00	-	10.00	10.00–10.00	0.084	0.978
	After	10.00	10.00–10.00		10.00	10.00–10.00		0.279
Kraus-Weber 3	*Before*	9.50	9.00–10.00	-	10.00	10.00–10.00	0.157	0.035
	After	9.50	9.00–10.00		10.00	10.00–10.00		0.007
Kraus-Weber 4	*Before*	9.50	8.00–10.00	-	10.00	10.00–10.00	0.083	0.052
	After	9.50	8.00–10.00		10.00	10.00–10.00		0.006
Kraus-Weber 5	*Before*	10.00	10.00–9.00	0.082	10.00	9.75–10.00	0.046	0.737
	After	10.00	10.00–10.00		10.00	10.00–10.00		0.551
Kraus-Weber 6	*Before*	9.00	8.75–10.00	-	10.00	9.00–10.00	0.025	0.079
	After	9.00	8.75–10.00		10.00	10.00–10.00		0.001

3.2. Results of the Core Test

For the intervention group. a significant change ($p = 0.015$) was observed in the Core test comparing the results before (median = 59.00; iqr = 46.75; 72.75) and after (median = 59.5; iqr = 50.50; 73.50) the exercise programme. There was no significant change in the control group (median$_{before}$ = 49.50; iqr = 38.00; 58.00. median$_{after}$ = 49.50; iqr = 39.75; 55.75. $p = 0.327$).

3.3. Kempf Test Results

A significant improvement ($p < 0.001$) was observed for the first part of the Kempf test (Kempf 1). At the start of the test the median was 16.50 (iqr = 13.75; 20.25) and after the exercise programme the median was 23.50 (iqr = 18.25; 25.50). The results of the second part of the test (Kempf 2). also showed a significant improvement ($p < 0.001$) (median = 17.00; iqr = 13.25; 19.25) compared to the initial value (median = 20.00; iqr = 16.50; 26.00). The third part of the test (Kempf 3) showed a similar positive result ($p < 0.001$). The median was 13.50 (iqr = 13.50; 17.50) for the first measurement and 17.00 (iqr = 15.00; 20.25) after the exercise programme. Finally. the results of the fourth part of the Kempf test (Kempf 4) (median = 13.50; iqr = 10.00; 16.00) also showed a significant improvement ($p < 0.001$) after the gymnastics programme (median = 15.00; iqr = 12.50; 22.00). In contrast. there were no significant changes in the control group in any of the tests (Kempf 1–4). (Table 4).

Table 4. Results of the first four parts of the Kempf test (Kempf 1–4).

		Control Group			Intervention Group			
Test	*Time (Before/After Intervention)*	*Median*	*IQR*	*p-Value*	*Median*	*IQR*	*p-Value*	*p-Value (Control vs. Intervention)*
Kempf 1	*Before*	15.50	13.00–17.25	-	16.50	13.75–20.75	<0.001	0.449
	After	15.50	13.00–17.25		23.50	18.25–25.50		0.001
Kempf 2	*Before*	16.50	13.25–17.50	-	17.00	13.25–19.25	<0.001	0.583
	After	16.50	13.25–17.50		20.00	16.50–26.00		0.047
Kempf 3	*Before*	11.50	10.00–15.00	-	13.50	13.50–17.50	<0.001	0.482
	After	11.50	10.00–15.00		17.00	15.00–20.25		0.003
Kempf 4	*Before*	13.50	10.00–16.00	-	13.50	10.00–16.00	<0.001	0.984
	After	13.50	10.00–16.00		15.00	12.50–22.00		0.205

Improvements were observed in the other parts of the Kempf test (K5–K8) but none of our results were significant. In the fifth part of the test, two respondents showed improvement in both their right and left legs ($p = 0.157$). When examining the seventh test part. a positive change was found in 1 respondent ($p = 0.317$). Finally, for the last part, there was an improvement for both right and left leg in 3 participants ($p = 0.083$). In the control group, no significant difference was found between the results before and after the programme when the Kempf test was administered again.

3.4. Beck Questionnaire

Based on the total scores of the Beck questionnaire. there was a significant change ($p < 0.001$) in the effect of the exercise programme. At the beginning of the study, the median was 22.00 (iqr = 20.25; 24.00) and at the end of the study the median decreased to 15.00 (iqr = 13.75; 17.00). In our study significant reduction was also evaluated in the control group ($p < 0.001$). but the improvement in the intervention group ($median_{intervention}$ = 15.00 iqr = 13.75; 17.00) was significantly greater ($p = 0.012$) than in the control group ($median_{control}$ = 18.00 iqr = 16.50; 19.00) after the intervention. The results of the questionnaire are shown in Table 5 broken down by question (Table 5).

Table 5. Beck test results.

		Control Group			Intervention Group			
Test	*Time (Before/After Intervention)*	*Median*	*IQR*	*p-Value*	*Median*	*IQR*	*p-Value*	*p-Value (Control vs. Intervention)*
Beck Total	*Before*	22.00	22.00–23.00	<0.001	22.00	20.25–24.00	<0.001	1.000
	After	18.00	16.50–19.00		15.00	13.75–17.00		0.012
Beck 1	*Before*	2.00	1.25–3.00	0.001	2.00	1.75–3.25	0.002	0.863
	After	1.50	1.00–2.00		1.50	1.00–2.00		1.000
Beck 2	*Before*	2.00	1.00–3.00	0.003	2.00	1.00–3.25	0.003	0.619
	After	1.00	1.00–2.00		2.00	1.00–2.25		0.616
Beck 3	*Before*	3.00	2.25–3.75	0.018	3.00	2.00–4.00	0.001	0.878
	After	3.00	2.00–3.00		2.00	1.00–2.25		0.024
Beck 4	*Before*	3.00	3.00–4.00	<0.001	3.50	2.75–4.00	<0.001	0.946
	After	2.50	2.00–3.00		2.00	1.00–3.00		0.140
Beck 5	*Before*	3.00	2.00–3.75	0.018	3.00	2.00–3.00	0.003	0.527
	After	3.00	2.00–3.00		2.00	1.00–2.00		0.042
Beck 6	*Before*	1.50	1.00–2.00	0.018	1.50	1.00–2.00	0.014	0.982
	After	1.00	1.00–1.75		1.00	1.00–1.25		0.849
Beck 7	*Before*	3.00	2.00–3.00	0.003	3.00	2.00–3.00	<0.001	0.927
	After	2.00	2.00–3.00		2.00	1.00–2.00		0.023
Beck 8	*Before*	2.00	2.00–3.00	0.027	2.00	2.00–3.00	0.009	0.661
	After	2.00	1.00–2.00		2.00	1.00–2.00		0.980
Beck 9	*Before*	1.50	1.00–2.00	0.04	2.00	1.00–2.00	0.015	0.528
	After	1.00	1.00–1.00		1.00	1.00–1.25		0.842

4. Discussion

The aim of our study was to create a new experimental online spinal mobilization programme for office workers working from home. We considered it important that the online exercise programme we ran be supervised. All sessions were conducted by the same qualified physiotherapist. This allowed him to monitor the positive changes among the members of the intervention group throughout the programme. The physiotherapist was able to supervise whether the participants were performing the exercises he prescribed accurately and correctly, and he was able to answer any questions that arose immediately. The results showed that the exercise programme we had developed improved the core muscle strength and flexibility of the intervention group members, as well as their mental health. The patients participating in the study understood that they could do a lot for their own health by completing our 10-week programme three times a week under supervision, similar to what is described in other literature [20,33]. When developing our programme, we started from the assumption that the programme we designed would improve the strength and flexibility of the participants' core muscles, which would make their office work at home easier and increase their physical activity. We were able to confirm this. as we observed significant improvements among the members of the intervention group using the Kraus-Weber, Core and Kempf tests. We used the Beck questionnaire to assess mental health. We considered it important to examine the impact of our exercise programme on the mental state of the participants. Based on the results, we can say that there was a significant positive change in the intervention group members as a result of our programme. When calculating the results. we were surprised to find that there was also a positive change among the members of the control group. even though we had specified in the initial assessment that they could only continue with their usual habits and activities. However, we can say that the improvement was more significant in the intervention group than in the control group.

Several studies have shown that a decrease in physical activity and an increase in sedentary behaviour result in a decrease in muscle mass, and sitting for 7 h a day also leads to an increase in spinal stiffness [34]. In the initial assessment of mental status, the Beck depression test showed that both the intervention and control groups included individuals suffering from severe depression, moderate depression and mild depression. We believe that months of anxious uncertainty, fear of illness, concern for family members and similarly, isolation, sedentary lifestyle and lack of social contact have significantly contributed to these worrying results. Based on the objectives of our study, we can say that as a result of the online targeted exercise programme, there was an improvement in mental assessment and in muscle strength and stretching tests.

Online exercise programs and the digital sports world have become popular during the pandemic, and many sports clubs and institutions, such as the Debrecen Municipality's Move Debrecen program, have launched sports opportunities in the online space. The digital sports scene has become popular during the pandemic and many sports clubs and institutions, such as the Debrecen Municipality's Move Debrecen programme, have launched sports opportunities in the world of online space. In this period of limited opportunities, online exercise has been shown to prevent mammary diseases caused by sedentary lifestyle and increased physical inactivity, and by creating a sporting community it also plays a role in maintaining mental health as during the sessions the people assessed became more open and encouraged each other while doing the exercises.

Health is determined by the harmonious interaction of physical and mental components. In our modern, fast-paced society, numerous studies have shown that one in four adults worldwide is inactive [32]. Physical inactivity is now called a pandemic that requires urgent action. It is well known that physical activity improves physiological functioning and cognitive, emotional, social and psychosocial functioning [35].

Our research compared several studies to demonstrate the value of remote rehabilitation, an online exercise programme, taking into account the specific needs of patients and healthcare professionals.

The aim of the study by Özel and colleagues was to examine the effect of telerehabilitation-based, remote, supervised or unsupervised structured exercise therapy on pain, disability, and quality of life in patients with chronic nonspecific neck pain. Sixty-six patients with chronic nonspecific neck pain were randomly divided into three groups: remote supervision group (RSG. n = 22), unsupervised group (UG. n = 22), and control group (CG. n = 22). Patients in the remotely supervised and unsupervised groups received a progressive, structured therapy programme four days a week for four weeks. The remotely supervised group was supervised via videoconferencing and text messages. Overall, the improvement in quality of life after treatment was greater in the supervised group than in the unsupervised group. In our study, the results showed that the telerehabilitation-based, remote, supervised exercise programme we had designed had a positive psychological (Beck test) and musculoskeletal (Kraus-Weber test, Core test, Kempf test) effect on the office workers involved [4]. In general, unsupervised exercise programmes have the advantage of being inexpensive and widely available to the public and are particularly popular for the treatment of chronic diseases. However, previous scientific studies suggest that unsupervised exercise programmes are not as effective as supervised programmes in the management of chronic diseases. Experts primarily identify poor adherence as the reason behind the unfavourable results of unsupervised exercise.

Similar to our research, Özül et al. also studied the practical effects of remote rehabilitation among office workers by examining spinal movements. The office workers (n = 120) were randomly divided into three groups participants in group 1 received an interactive telerehabilitation programme (strengthening, motor control and posture correction exercises) for 45 min a day, 3 days a week, for a total of 6 weeks. Participants in group 2 were trained to perform the exercise programme independently at home for 45 min a day, 3 days a week, for a total of 6 weeks. The patients in group 3 only received office ergonomics training. Statistically significant improvements were observed in office workers in terms of pain ($p < 0.001$), range of motion (ROM) ($p < 0.001$), neck disability ($p < 0.001$), functional status ($p < 0.001$) and quality of life ($p < 0.001$). In addition to office ergonomics training, interactive telerehabilitation programmes are the most effective method for treating pain, range of motion and neck disability [36]. In our study, we also achieved positive results with an online exercise programme supervised by a physical therapist in terms of trunk muscle strength and flexibility, using the Core, Kempf and Kraus-Weber tests. Özel et al [4] implemented a 4-day-a-week intervention programme over 4 weeks, while Özül et al. [36] performed a 3-day-a-week programme over 6 weeks. Our telerehabilitation programme performed three times a week for 10 weeks using different tools. The duration and frequency of our intervention not only improved the condition of the musculoskeletal system but also had a positive effect on the mental status of the participants.

Bannell et al. also looked at individuals with sedentary lifestyles similar to our study but with an unsupervised exercise programme. Participants had access to a website that provided virtual resources and instructions for completing a 12-week progressive exercise programme. Although we previously emphasized the advantages of supervised programmes over unsupervised programmes the data suggested that the mobile health (mHealth) technology–supported exercise and physical interventions have a positive impact on the individuals involved in the programme and on their adherence to exercise [37].

Malinska and her colleagues conducted a study among computer workers. They used a questionnaire to collect data on demographics, musculoskeletal disorders, lifestyle characteristics, computer work ergonomics, and psychosocial and physical work characteristics. More than 48% of respondents complained of musculoskeletal disorders in the past year, especially neck and back pain. In their conclusions, they stated that the most effective way to eliminate the risks of musculoskeletal disorders in the workplace is to develop health programmes that promote a healthy lifestyle and to raise employee awareness of workstation ergonomics and work organization, especially among women and older workers [8]. In Debrecen, the importance of community exercise programmes that can encourage the population to adopt a healthier lifestyle is similarly supported.

Numerous studies have been conducted on the online treatment of various diseases. The aim of our study was not to treat existing symptoms or diseases but to create an alternative form of exercise as a preventive measure to interrupt prolonged static sitting.

Yaghoubitajani and his colleagues also wanted to reduce neck and shoulder pain, incorrect posture, sick leave time and reduced work capacity by using an online gymnastics programme. Several parameters (neck and shoulder pain, posture angles, work capacity and muscle activity) related to work performance were improved in this study. This, along with our study, highlights the impact of supervised online intervention programmes on work performance. This scientific evidence can be a useful tool for convincing workplace managers to support the implementation of similar programmes in the future, as improvements in the musculoskeletal and mental health of their employees will also improve productivity [32].

Pérez and colleagues also worked with a small number of patients. including 29 breast cancer patients. As in our pilot study. they carried out the intervention exercise programme in their homes 3 times a week (16 weeks) under online supervision. Their results also showed significant improvement. They found that supervised online exercise in a non-clinical setting is effective. The study supports the role of home-based online exercise programmes in the rehabilitation strategy of breast cancer survivors and confirms that online programmes are justified even in cases of serious chronic diseases [38].

In another area of rehabilitation. Lewis and colleagues studied the implementation of an online exercise programme with 14 individuals with chronic respiratory disease. Similar to our interventions, the programme had a positive effect on patients' functional capacity and mental state (anxiety). They found that there was less dropout in the group of patients treated in telerehabilitation compared to in-person pulmonary rehabilitation It was found that there was less dropout in the group of patients receiving remote rehabilitation than in the group of patients receiving in-person pulmonary rehabilitation. which is another positive argument in favour of remote rehabilitation in the treatment of chronic diseases [39].

Barbosa and colleagues investigated the importance of a six-week telerehabilitation exercise programme in patients with non-specific neck pain through a randomised control trial. Their results suggest that telerehabilitation for patients with non-specific chronic neck pain is effective, easily accessible and low-cost through the use of technology [40].

Vetrovska and colleagues in multiple sclerosis have shown that this type of online exercise programme can help people with MS to remain physically active by providing an online form of exercise programme that is feasible [41].

Limitations

One of the limitations of the study was that many people could not combine the three online training programmes a week with their personal lives or could not find the motivation to complete the 10-week training programme. Another limiting factor was the lack of equipment, as only tools, spare parts and equipment that were available to everyone in their homes were used during the training sessions.

5. Conclusions

Based on the results of our research, we plan to add more mobilisation exercises to the muscle strengthening. In addition, we would like to make our research more complex by examining and treating the cervical spine in more detail, as well as expanding the 10-week training programme by working with more volunteers in person so that we can draw more realistic values and conclusions from more home sampling data. Based on our research, we can say that the complex exercise programme we designed improved the functional status of the spine and the strength and flexibility of the core muscles and mental health.

Author Contributions: Methodology. T.S.; Validation. I.V.-B. and Z.J.; Investigation. B.L., E.K. and A.M.L. All authors have read and agreed to the published version of the manuscript.

Funding: This research received no external funding.

Institutional Review Board Statement: The study was conducted in accordance with the Declaration of Helsinki. and approved by the Research Ethics Committee of the University of Debrecen (6774-2024 on 6 March 2024).

Informed Consent Statement: Informed consent was obtained from all subjects involved in the study.

Data Availability Statement: The original contributions presented in the study are included in the article. further inquiries can be directed to the corresponding author.

Conflicts of Interest: The authors declare no conflict of interest.

References

1. Duchek, D.; McDonough, M.H.; Bridel, W.; McNeely, M.L.; Culos-Reed, S.N. Understanding In-Person and Online Exercise Oncology Programme Delivery: A Mixed-Methods Approach to Participant Perspectives. *Curr. Oncol.* **2023**, *30*, 7366–7383. [CrossRef]
2. Biddle, S.; Mutrie, N. *Psychology of Physical Activity*, 2nd ed.; Routledge: London, UK, 2007. [CrossRef]
3. Salmon, P. Effects of Physical Exercise on Anxiety. Depression. and Sensitivity to Stress: A Unifying Theory. *Clin. Psychol. Rev.* **2001**, *21*, 33–61. [CrossRef]
4. Özel, M.; Ciddi, P.K. The effectiveness of telerehabilitation-based structured exercise therapy for chronic nonspecific neck pain: A randomized controlled trial. *J. Telemed. Telecare* **2024**, *30*, 523–833. [CrossRef] [PubMed]
5. Moretti, A.; Menna, F.; Aulicino, M.; Paoletta, M.; Liguori, S.; Iolascon, G. Characterization of Home Working Population during COVID-19 Emergency: A Cross-Sectional Analysis. *J. Environ. Res. Public. Health* **2020**, *17*, 6284. [CrossRef]
6. Gilbert, A.W.; Jaggi, A.; May, C.R. What is the patient acceptability of real time 1:1 videoconferencing in an orthopaedics setting? A systematic review. *Physiotherapy* **2018**, *104*, 178–186. [CrossRef]
7. Cooper, D.; Ward, K.; Kavanagh, R.; O'cOnnor, S. 'MenoWell': A pilot 6-week novel, online, multimodal exercise and health education programme for women in all stages of menopause living in Laois, Ireland. *J. Phys. Act. Health* **2023**, *7*, 303–318. [CrossRef]
8. Meijer, H.A.; Graafland, M.; Goslings, J.C.; Schijven, M.P. Systematic Review on the Effects of Serious Games and Wearable Technology Used in Rehabilitation of Patients with Traumatic Bone and Soft Tissue Injuries. *Arch. Phys. Med. Rehabil.* **2018**, *99*, 1890–1899. [CrossRef] [PubMed]
9. Malińska, M.; Bugajska, J.; Bartuzi, P. Occupational and non-occupational risk factors for neck and lower back pain among computer workers: A cross-sectional study. *Int. J. Occup. Saf. Ergon.* **2021**, *27*, 1108–1115. [CrossRef]
10. Szalay, P. *Some Psychological and Somatic Consequences of Alcoholism*; Budapest, Hungary, 2005.
11. Shukla, A.; Dogra, D.K.; Bhattacharya, D.; Gulia, S.; Sharma, R. Impact of COVID-19 outbreak on the mental health in sports: A review. *Sport. Sci. Health* **2023**, *19*, 1043–1057. [CrossRef] [PubMed]
12. Tavares, A.I. Telework and health effects review. *Int. J. Healthc.* **2017**, *3*, 30. [CrossRef]
13. Xiao, Y.; Becerik-Gerber, B.; Lucas, G.; Roll, S.C. Impacts of Working from Home During COVID-19 Pandemic on Physical and Mental Well-Being of Office Workstation Users. *J. Occup. Environ. Med.* **2021**, *63*, 181–190. [CrossRef]
14. Samani, S.A. The Impact of Personal Control over Office Workspace on Environmental Satisfaction and Performance. *J. Soc. Sci. Humanit.* **2015**, *1*, 163–175.
15. DeFilippis, E.; Impink, S.M.; Singell, M.; Polzer, J.T.; Sadun, R. Collaborating During Coronavirus: The Impact of COVID-19 on the Nature of Work. *Natl. Bur. Econ. Res.* **2020**. [CrossRef]
16. Majumdar, P.; Biswas, A.; Sahu, S. COVID-19 pandemic and lockdown: Cause of sleep disruption. depression. somatic pain. and increased screen exposure of office workers and students of India. *Chronobiol. Int.* **2020**, *37*, 1191–1200. [CrossRef]
17. Elst, T.V.; Verhoogen, R.; Sercu, M.; Broeck, A.V.D.; Baillien, E.; Godderis, L. Not Extent of Telecommuting. But Job Characteristics as Proximal Predictors of Work-Related Well-Being. *J. Occup. Environ. Med.* **2017**, *59*, e180–e186. [CrossRef]
18. Davis, K.G.; Kotowski, S.E.; Daniel, D.; Gerding, T.; Naylor, J.; Syck, M. The Home Office: Ergonomic Lessons From the "New Normal". *Ergon. Des. Q. Hum. Factors Appl.* **2020**, *28*, 4–10. [CrossRef] [PubMed]
19. Shariat, A.; Bahri, S.; Tamrin, S.B.M.; Arumugam, M. Musculoskeletal disorders and their relationship with physical activities among office workers: A review. *Malaysian J. Public. Health Med.* **2016**, *16*, 62–74.
20. Wu, S.; He, L.; Li, J.; Wang, J.; Wang, S. Visual display terminal use increases the prevalence and risk of work-related musculoskeletal disorders among Chinese office workers: A cross-sectional study. *J. Occup. Health* **2012**, *54*, 34–43. [CrossRef] [PubMed]
21. KSH. *Teleworking Has Gained in Value in the Shadow of Covid19*; KSH: Budapest, Hungary, 2022.
22. Worksafenb. Office Ergonomics Guidelines for Preventing Musculoskeletal Injuries. 2010. Available online: https://docplayer.net/17840851-Office-ergonomics-guidelines-for-preventing-musculoskeletal-injuries.html (accessed on 15 August 2024).
23. GBD 2021 Low Back Pain Collaborators. Global regional and national burden of low back pain, 1990–2020, its attributable risk factors, and projections to 2050: A systematic analysis of the Global Burden of Disease Study 2021. *Lancet Rheumatol.* **2023**, *5*, e316–e329. [CrossRef]

24. GBD 2019: Global Burden of 369 Diseases and Injuries In 204 Countries and Territories, 1990–2019: A Systematic Analysis for the Global Burden of Disease Study 2019. Available online: https://vizhub.healthdata.org/gbd-results/ (accessed on 10 October 2025).
25. Wang, X.-Q.; Zheng, J.-J.; Yu, Z.-W.; Bi, X.; Lou, S.-J.; Liu, J.; Cai, B.; Hua, Y.-H.; Wu, M.; Wei, M.-L.; et al. A meta-analysis of core stability exercise versus general exercise for chronic low back pain. *PLoS ONE* **2012**, *7*, e52082. [CrossRef]
26. Coulombe, B.J.; Games, K.E.; Neil, E.R.; Eberman, L.E. Core Stability Exercise Versus General Exercise for Chronic Low Back Pain. *J. Athl. Train.* **2017**, *52*, 71–72. [CrossRef]
27. Grabovac, B. Organization and Functioning of the Affective Lexicon in Early and Late Serbian Bilinguals. Pécs, Hungary, 2015.
28. Durall, C.J.; Greene, P.F.; Kernozek, T.W. A Comparison of Two Isometric Tests of Trunk Flexor Endurance. *J. Strength Cond. Res.* **2012**, *26*, 1939–1944. [CrossRef] [PubMed]
29. Kovácsné, B.V. Strain Status Assessment and Use of a Strain-Specific Exercise Programme in Dancers. Pécs, Hungary, 2017.
30. Kempf, H.-D. *No More Back Pain*; Novella Publishing: Budapest, Hungary, 1997.
31. Haak, K. *Die Neue Rückenschule*; Falken-Verlag GmbH: Niedernhausen, Germany, 1998.
32. Bartha, É.J. Relaxation and Breathing Exercises. 2019. Available online: https://deac.hu/dokumentumok/EFOP/1.%20Bartha%20%C3%89va%20Judit%20-%20Relax%C3%A1ci%C3%B3s%20%C3%A9s%20l%C3%A9gz%C5%91gyakorlatok_pdf.pdf (accessed on 15 August 2023).
33. Yaghoubitajani, Z.; Gheitasi, M.; Bayattork, M.; Andersen, L.L. Corrective exercises administered online vs at the workplace for pain and function in the office workers with upper crossed syndrome: Randomized controlled trial Randomized Controlled Trial. *Int. Arch. Occup. Env. Health* **2022**, *95*, 1703–1718. [CrossRef]
34. Haileamlak, A. Physical Inactivity: The Major Risk Factor for Non-Communicable Diseases. *Ethiop. J. Health Sci.* **1970**, *29*, 810. [CrossRef]
35. Heneghan, N.R.; Baker, G.; Thomas, K.; Falla, D.; Rushton, A. What is the effect of prolonged sitting and physical activity on thoracic spine mobility? An observational study of young adults in a UK university setting. *BMJ Open* **2018**, *8*, e019371. [CrossRef]
36. Özlü, A.; Ünver, G.; Tuna, H.I.; Erdoğan, A. Effects of Interactive Telerehabilitation Practices in Office Workers with Chronic Nonspecific Neck Pain: Randomized Controlled Study. *Telemed. J. E Health* **2024**, *30*, 238–447. [CrossRef]
37. Lavín-Pérez, A.M.; Collado-Mateo, D.; González, C.H.; Batista, M.; Mayo, X.; Villar, C.R.; Jiménez, A. An online home-based exercise program improves autonomic dysfunction in breast cancer survivors. *Front. Physiol.* **2023**, *14*, 1256644. [CrossRef]
38. Bannell, D.J.; France-Ratcliffe, M.; Buckley, B.J.R.; Crozier, A.; Davies, A.P.; Hesketh, K.L.; Jones, H.; Cocks, M.; Sprung, V.S. Motivate team Adherence to unsupervised exercise in sedentary individuals: A randomised feasibility trial of two mobile health interventions. *Digit. Health* **2023**, *9*, 20552076231183552. [CrossRef]
39. Lewis, A.; Knight, E.; Bland, M.; Middleton, J.; Mitchell, E.; McCrum, K.; Conway, J.; Bevan-Smith, E. Feasibility of an online platform delivery of pulmonary rehabilitation for individuals with chronic respiratory disease. *BMJ Open Respir. Res.* **2021**, *8*, e000880. [CrossRef] [PubMed]
40. Barbosa, J.C.; da Luz, B.V.; da Silva, B.F.P.; Marques, A.P.; Saragiotto, B.T.; Comachio, J.; Magalhaes, M.O. Effectiveness of Telerehabilitation Exercise Programme on Disability and Pain in Patients With Chronic Non-Specific Neck Pain: Randomised Controlled Trial Assessor-Blinded Randomized Controlled Trial. *Musculoskelet. Care* **2025**, *23*, e70119. [CrossRef]
41. Vetrovska, R.; Prochazkova, R.; Havrdova, E.K.; Novotna, K. On-line pilates program as an alternative to conventional rehabilitation therapy for people with multiple sclerosis. pilot study. *J. Bodyw. Mov. Ther.* **2024**, *40*, 689–694. [CrossRef] [PubMed]

applied sciences

MDPI

Article

Comparative Analysis of Different AI Approaches to Stroke Patients' Gait Analysis

Izabela Rojek [1,*], Emilia Mikołajewska [2], Olga Małolepsza [1], Mirosław Kozielski [1] and Dariusz Mikołajewski [1]

1 Faculty of Computer Science, Kazimierz Wielki University in Bydgoszcz, 85-064 Bydgoszcz, Poland; olga.malolepsza@ukw.edu.pl (O.M.); miroslaw.kozielski@ukw.edu.pl (M.K.); dariusz.mikolajewski@ukw.edu.pl (D.M.)

2 Department of Physiotherapy, Faculty of Health Sciences, Ludwik Rydygier Collegium Medicum in Bydgoszcz, Nicolaus Copernicus University in Toruń, 87-100 Toruń, Poland; emiliam@cm.umk.pl

* Correspondence: izabela.rojek@ukw.edu.pl

Abstract: Despite advances in diagnostics, the objective and repeatable assessment of patients with neurological deficits (e.g., stroke) remains a major challenge. Modern methods based on artificial intelligence (AI) are of interest to researchers and clinicians in this area. This study presents a comparative analysis of different AI approaches used to analyze gait of stroke patients using a retrospective dataset of 120 individuals. The main objective is to evaluate the effectiveness, accuracy, and clinical relevance of machine learning (ML) and deep learning (DL) models in identifying gait abnormalities and predicting rehabilitation outcomes. Multiple AI techniques—including support vector machines (SVM), random forests (RF), k-nearest neighbors (k-NN), and convolutional neural networks (CNN)—were trained and tested on time-series gait data with spatiotemporal parameters. Performance metrics such as accuracy, precision, recall, and area under the curve (AUC) were used to compare model results. Initial results indicate that DL models, particularly CNNs, outperform traditional ML methods in capturing complex gait patterns and providing reliable classification. However, simpler models showed advantages in interpretability and computational efficiency. This study highlights the potential and shortcomings of AI-based gait analysis tools in supporting clinical decision-making and planning personalized stroke rehabilitation.

Keywords: artificial intelligence; machine learning; gait parameters; clinical gait assessment

1. Introduction

Despite advances in diagnostics, the objective and repeatable assessment of patients with neurological deficits (e.g., stroke) remains a major challenge. Traditional qualitative gait analysis relies on observational gait testing and is subjective and largely dependent on the observer's clinical experience. Quantitative gait analysis, on the other hand, relies on parameter measurements with good accuracy and repeatability, which translates into more reliable diagnosis and comparative assessment during rehabilitation, even with less experienced diagnosticians/therapists [1]. Modern methods based on artificial intelligence (AI) are of interest to researchers and clinicians in this area [2].

The origins of clinical gait analysis based on artificial intelligence (AI) can be traced back to the long-standing use of biomechanics and motion capture to study human locomotion. Early gait analysis relied on visual observation and simple video recordings, which lacked precision and objectivity [3]. With the development of force plates, marker-based

https://doi.org/10.3390/app152010896

motion capture, and electromyography, clinicians gained access to detailed biomechanical data, although their interpretation remained complex and time-consuming [4–7]. Advances in machine learning have created opportunities to automate data processing and extract clinically relevant patterns [8–10]. Artificial neural networks (ANNs) and support vector machines (SVMs) were first used to classify gait disorders and distinguish between normal and abnormal gait patterns [11]. The advent of deep learning enabled the analysis of raw sensor or video data without the need for extensive manual feature extraction [12]. Wearable sensors, such as inertial measurement units, enabled continuous and real-time gait monitoring, further enhancing the value of AI methods [13]. By combining multimodal data sources, AI models have improved diagnostic accuracy, rehabilitation monitoring, and personalized treatment planning [14]. Wearable technologies and AI (primarily machine learning (ML)) are gaining increasing attention in gait research, but their clinical application is not yet well-established and will not be widely available anytime soon [15]. These tools, combined with telerehabilitation, represent a paradigm shift in the analysis and assessment of post-stroke gait, providing tools for acquiring, storing, and analyzing multivariate, complex gait data with their nonlinear dynamic variability [16]. They also offer significant benefits such as automatic gait classification (e.g., hemiplegic gait), predicting future health status in this area, and analyzing trends and potential for reversal if unfavorable [1]. Recently, explainable AI techniques have been introduced to increase trust and transparency in clinical decision-making [17]. Currently, AI-based gait analysis is emerging as a powerful tool for early disease detection, assessing treatment outcomes, and providing access to healthcare beyond specialized laboratories [18,19].

Artificial intelligence (AI)-based clinical gait analysis in rehabilitation and physical therapy for neurological patients, including those recovering from stroke, requires accurate and reliable motion capture systems to quantify spatiotemporal and kinematic parameters [20]. Integration with wearable sensors, such as inertial measurement units or compression pads, enables continuous monitoring, both in the clinic and at home [21]. The system must handle noisy or incomplete data, as patients may have irregular or compensatory movement patterns [22]. Robust AI algorithms are essential for classifying gait abnormalities, tracking progress, and predicting recovery trajectories [23]. Interoperability with electronic medical records and rehabilitation devices ensures personalized treatment planning and outcome assessment [24]. User-friendly interfaces are essential so that clinicians can easily interpret results without requiring extensive technical training. Real-time feedback capabilities can support motor learning and corrective strategies during therapy sessions [25]. Data privacy and security are crucial given the sensitive nature of patient health information [26]. The system must be adaptable to a wide range of neurological conditions, severity levels, and assistive devices, ensuring inclusiveness [27]. To maintain accuracy, reliability, and clinical utility, validation through clinical trials and continuous model updates is essential [28,29].

AI-based clinical gait analysis in rehabilitation and physical therapy for neurological patients, including stroke patients, faces several research gaps that hinder its widespread clinical implementation. These include

- Limited validation in diverse populations due to any AI models have been developed and validated in controlled settings with small sample sizes, often characterized by a lack of diversity in age, gender, and ethnicity. This limits their generalizability to broader patient populations [30].
- Lack of standardization because a lack of consensus on standardized protocols for data collection, feature extraction, or model evaluation leads to inconsistencies across studies and hinders comparison of results or their integration into clinical practice [31].

- Data privacy concerns because the use of wearable sensors and video recordings poses significant privacy issues, especially for sensitive medical data. Ensuring data security and patient consent is a priority but often insufficiently addressed.
- Interpretability of AI models since many AI systems act as "black boxes," providing predictions without clear explanations. This lack of interpretability can hinder clinician trust and acceptance, as understanding the rationale behind decisions is crucial to decision-making [32].
- Integration with clinical processes because AI tools often operate in isolation and are not seamlessly integrated with existing clinical processes or electronic health record systems, limiting their utility in real-world settings [33].
- Practical application: most AI models are tested in controlled environments, but their effectiveness in real-world settings, such as patients' homes or community clinics, remains understudied [34].
- Challenges with longitudinal monitoring because continuous gait monitoring over extended periods is essential to assessing rehabilitation progress, but current AI systems may lack long-term tracking capabilities or be impractical for daily use [35].
- Multimodal data integration: Integrating data from various sources, such as video, wearable sensors, and clinical assessments, poses challenges in data fusion and analysis but is essential for comprehensive gait analysis [36].
- Personalization of interventions because AI models often provide generalized conclusions without tailoring recommendations to individual patient needs, which is crucial for effective rehabilitation [37].
- Regulatory and ethical issues due to implementing AI in clinical settings raises regulatory concerns about safety, effectiveness, and ethical issues that are still being addressed by policymakers and regulators.

Bridging these gaps is essential for the successful integration of AI-based gait analysis into rehabilitation and physical therapy for neurological patients, ensuring the effectiveness, reliability, and benefits of these tools for patients.

Clinical gait analysis based on artificial intelligence (AI) in rehabilitation and physical therapy for neurological patients, including stroke patients, faces several challenges that limit its full clinical potential. Accurately capturing gait patterns in patients with severe motor impairments is challenging due to compensatory movements and performance variability [38]. Integrating heterogeneous data from wearable sensors, motion capture systems, and video recordings requires advanced preprocessing and synchronization [39,40]. AI models often struggle with small datasets, as collecting large amounts of high-quality patient data is time-consuming and expensive [41]. Ensuring model generalization across different populations and stages of recovery remains a significant hurdle [42,43]. Many AI systems operate as "black boxes," making it difficult for clinicians to interpret and trust their results. Real-time analysis and feedback are computationally intensive, especially when using multiple sensors or high-resolution video recordings [44]. Adherence and usability in home rehabilitation are challenging, as patients may find devices cumbersome or difficult to use. Regulatory approvals and ethical considerations, including data privacy and informed consent, complicate clinical implementation. Seamless integration of AI tools into existing clinical processes and electronic health record systems is often lacking. Continuously updating and validating AI models to reflect evolving patient populations and treatment protocols remains a continuing challenge.

The aim of this article is to provide a comparative diagnostic analysis of different artificial intelligence approaches to gait analysis in stroke patients.

The novelty and contribution of this paper lies in its systematic evaluation of multiple AI techniques applied to a highly complex medical problem. Unlike previous studies that focused on a single algorithm, this work highlights the strengths and weaknesses of different models in capturing specific gait abnormalities caused by stroke. It introduces a direct comparison framework that emphasizes accuracy, interpretability, and clinical relevance, which are crucial in real-world rehabilitation settings. By comparatively analyzing the performance of different datasets and patient profiles, it identifies the most appropriate AI methods for personalized gait assessment. This study contributes to the development of precision rehabilitation by guiding researchers to AI tools that are both technically robust and clinically relevant.

The results presented in this article are only part of a larger study on artificial intelligence-based clinical assessment of gait analysis and the effects of neurorehabilitation in stroke patients, including

- Methods for rapid and easy-to-automate gait analysis;
- New gait parameters, including fuzzy and fractal logic;
- Methods for objectifying the assessment of gait rehabilitation effectiveness.

2. Materials and Methods

This study presents a comparative analysis of different AI approaches used to analyze gait of stroke patients using a retrospective dataset of 120 individuals. The main objective is to evaluate the effectiveness, accuracy, and clinical relevance of machine learning (ML) and deep learning (DL) models in identifying gait abnormalities and predicting rehabilitation outcomes. Multiple AI techniques—including support vector machines (SVM), random forests (RF), k-nearest neighbors (k-NN), and convolutional neural networks (CNN)—were trained and tested on time-series gait data with spatiotemporal parameters. Performance metrics such as accuracy, precision, recall, and area under the curve (AUC) were used to compare model results.

2.1. Dataset

The dataset included historical values of six spatiotemporal gait parameters (vlocity, cadence, stride length, and their values normalized to leg length) in 120 patients (n = 120) from 3 months to 3 years after stroke (convenience sample). Characteristics of the study group are presented in Table 1.

In the study, the dataset included 120 participants ($n = 120$). Although this sample size may be considered limited compared to very large clinical datasets, it is sufficient for the scope and purposes of the analysis. Studies on gait analysis and stroke rehabilitation are often conducted on relatively small cohorts due to the clinical complexity of recruitment and the detailed data collected for each patient. The dataset of 120 subjects provides sufficient statistical power for training and evaluating AI models, ensuring reliable comparisons between different approaches. Furthermore, the sample size is large enough to capture significant variability in gait patterns among stroke patients, strengthening the credibility of the results. Although larger datasets could further improve external validity, the current sample size does not affect the internal validity of the study or the interpretability of the results. Importantly, the primary goal is to compare AI methodologies, not to generalize to the population level. Therefore, the dataset is well suited for assessing the relative performance of the model. With this approach, the study can make a credible contribution to understanding the methodology, even if broader generalizations must be approached with caution.

This study compared different AI approaches for gait analysis in stroke patients using six key spatiotemporal parameters: gait speed, stride length, step length, cadence, and their values normalized to leg length. Data were collected from stroke patients in clinical rehabilitation settings using both video recording and wearable inertial measurement units and pressure sensors during standardized 10 m walking trials. To ensure data consistency, each participant completed multiple walking sessions under identical conditions, and a gold-standard motion capture method was used for a subset of the data to provide reference values. All datasets were anonymized and preprocessed to remove noise, segment gait cycles, and normalize parameters for inter-patient comparison. Model performance was verified by k-fold cross-validation to prevent overfitting and ensure generalization to different patient profiles. Performance metrics such as RMSE, mean absolute error, and sensitivity were calculated for each AI model to assess accuracy and robustness. This detailed data collection and validation protocol improves the transparency, reproducibility, and clinical validity of the comparative analysis.

Table 1. Characteristics of the study group.

Parameter	Study Group (n = 120)
Age [years]	
mean	62.5
SD	8.91
min	40
Q1	59
Median	64
Q3	71
Max	81
Gender	
male	60 (50%)
female	60 (50%)
Time after stroke [weeks]	
mean	56.7
SD	7.98
min	15
Q1	22
Median	61
Q3	102
Max	143
Side affected	
Left	51 (42.50%)
Right	69 (57.50%)

In a study of 120 stroke patients, all individuals in the cohort were survivors of a vascular event. "Normal" cases were defined both by comparison with the healthy population and based on the literature (range of parameter values for healthy individuals).

Table 2 shows a transparent split ratio (70/10/20) prevents information leakage and ensures reliable evaluation. The validation set allows for fine-tuning without contaminating test results. Patient-level splitting preserves independence between training and evaluation. Stratified decomposition ensures that each subset represents different functional levels, which is essential for robust clinical AI models.

Table 2. Data distribution across training, validation, and testing sets.

Subset	Number of Patients	Percentage of Total	Class 1 [%]	Class 2 [%]
Training	84	70	50	50
Validation	12	10	50	50
Testing	24	20	50	50

The presented study handles privacy and security by combining consent, anonymization, encryption, access control, and oversight, while ensuring AI outputs are clinically supervised. In our study on AI approaches to stroke patients' gait analysis, handling data privacy and patient security is a core requirement, especially since gait data can be personally identifying. Typically, such a study addresses it in several ways:

- Ethics and consent: all participants are recruited only after informed consent, explaining how gait data, and health information will be collected, stored, and used, and consent forms specify data-sharing conditions;
- De-identification and anonymization: raw clinical identifiers (name, date of birth, medical record numbers) are replaced with randomized study IDs;
- Data storage and transfer security: data are stored in encrypted servers compliant with GDPR standards, Transfer between sensors and servers uses end-to-end encryption (VPN), Role-based access control ensures only authorized clinicians/researchers access sensitive data;
- Minimization and purpose limitation: only the minimum necessary dataset for analysis is collected (e.g., gait parameters instead of raw continuous video whenever possible).

Data were collected in an MS Excel file (Microsoft, Redmond, WA, USA) converted to the CSV format required by the models.

Figure 1 shows typical skew and variability among post-stroke patients.

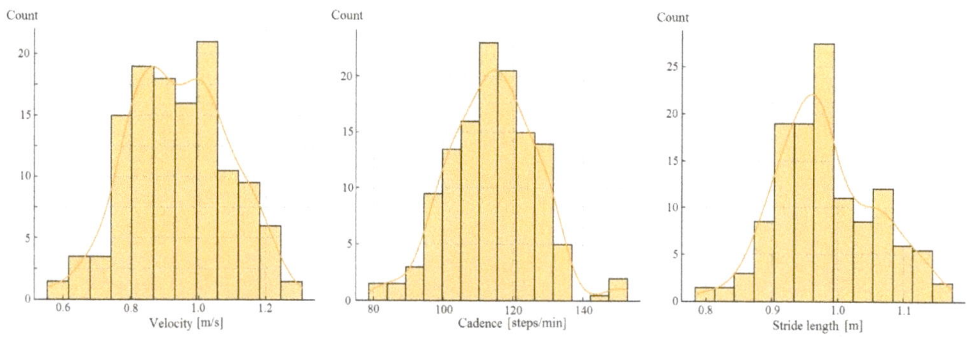

Figure 1. Histograms for velocity, cadence, and stride length.

Higher groups have noticeably higher median velocities, reflecting functional differences (Figure 2).

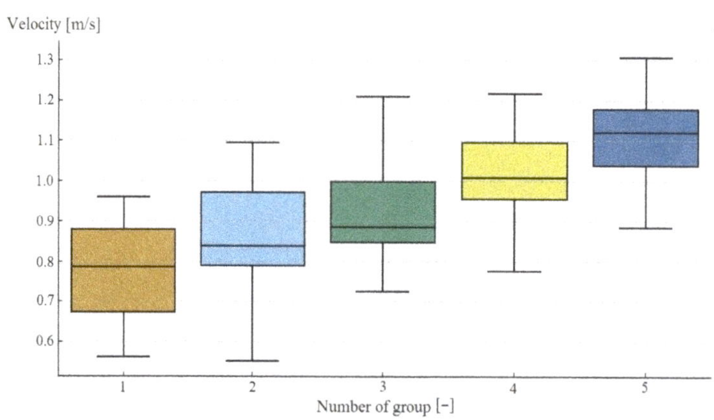

Figure 2. Boxplot of gait velocity by group (sample for five groups—it was possible only in velocity).

Figure 3 shows positive correlation between velocity, cadence and functional level.

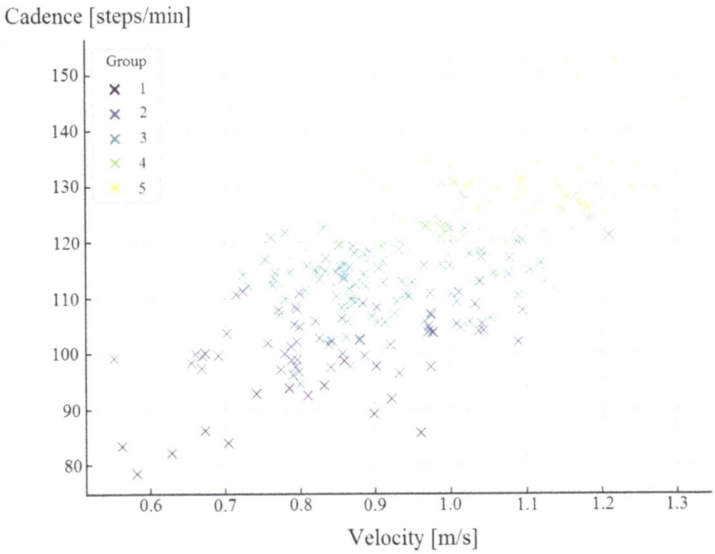

Figure 3. Scatter plot of velocity vs. cadence (colored by group, sample for five groups).

2.2. Statistical Analysis

The arithmetic mean or median was used to determine central tendency. The standard deviation or minimum and maximum values, as well as the lower and upper quartiles, were used to determine dispersion. The Shapiro–Wilk test was used to determine the distribution of variables. Variables with an approximate normal distribution were presented using the arithmetic mean and standard deviation (SD). Variables with a non-normal distribution were presented using the median, minimum, and maximum values, as well as the lower quartile (Q1) and upper quartile (Q3) values (25th and 75th percentiles, respectively). When the results were normally distributed, parametric tests were used; otherwise, nonparametric tests were used. The parametric Student's t-test or its nonparametric equivalent for dependent variables (Wilcoxon signed-rank test) was used to assess the significance of differences within the same group.

2.3. Computational Methods

The selection criteria for AI methods for comparative analysis of stroke gait should primarily consider clinical relevance, ensuring that the selected methods can capture gait variations relevant to rehabilitation outcomes. Accuracy and robustness are essential because the methods must reliably detect subtle gait abnormalities across different patients and stroke severity levels. Interpretability is another key criterion because clinicians must understand and trust the AI's decision-making process to successfully integrate it into practice. The ability to handle data from multimodal studies (such as motion capture, wearable sensors, or video) is crucial given the complexity of gait information. Computational efficiency should also be considered, as real-time or near-real-time feedback is extremely valuable in clinical and home rehabilitation. Furthermore, methods should be evaluated for scalability and adaptability, ensuring their generalizability across different datasets, clinical environments, and patient populations. Ethical and practical considerations must be taken into account in the selection process, including patient data privacy and ease of integration into healthcare systems.

The first stage involved collecting data from a wearable sensor device and a 10 m Walk Test video. After obtaining the data, the initial data preparation process included calculating parameters (velocity, pace, and stride length, and their values normalized to lower limb length to ensure comparability), normalizing and labeling the data (gait time series) to indicate whether a given individual had a hemiplegic gait and on which side, and combining them to prepare for analysis. The data were balanced across classes. In the next stage, the labeled data were divided into training set (70%), test set (20%), and validation set (10%), and the training set was then used to train the model. This led to the construction and optimization of a classification model. Test data were fed into the resulting classification model to generate a binary classification (parameter values within or above the norm (for further diagnosis), normal or hemiplegic gait). Training data helped reduce RMSE and, after training, evaluate the model's performance. The study verified and compared the trained models' ability to correctly classify gait parameter values by calculating metrics (accuracy, precision, sensitivity, and F1 score).

SVM, RF, kNN, and CNN were chosen for the comparative analysis of gait in stroke patients because they represent the spectrum of traditional machine learning methods and modern deep learning methods. SVM is ideal for analyzing gait data due to its ability to handle high-dimensional features and separate complex gait patterns using kernel functions. This is particularly advantageous when the dataset is limited, as it performs well with smaller training sizes, which is often the case in clinical trials. RFs are chosen because they provide robust performance by aggregating multiple decision trees, reducing the risk of overfitting. RFs can capture nonlinear relationships in gait parameters and are interpretable, allowing clinicians to determine which gait features (e.g., stride length, stride asymmetry) have the greatest impact on classification. kNN offers a simple, instance-based learning method that classifies patients' gait by comparing them with similar gait profiles. Its advantage is its intuitive nature, making it easy to understand and apply, particularly in exploratory analysis where transparency is key. However, kNN can be computationally demanding on large datasets, making it a useful benchmark for more scalable methods. CNNs are used due to their state-of-the-art success in pattern recognition, particularly with sequential and spatial data such as gait signals and video-based kinematics. They automatically extract hierarchical features from raw gait data, reducing the need for manual feature engineering. This makes CNNs effective in detecting subtle gait abnormalities that may not be captured by manually constructed features. Comparing neural networks (CNNs) with traditional models such as SVM, RF, and kNN highlights the performance tradeoffs between interpretability, data requirements, and computational cost. This allows for a balanced comparison of models that prioritize accuracy, interpretability, or simplicity. This also reflects real-world clinical needs, where lightweight models may be preferred in resource-constrained settings, while deep learning performs well in advanced laboratories. Therefore, the selection of these four methods provides a comprehensive assessment of the potential of AI in post-stroke gait analysis, bridging traditional and modern paradigms.

Visual Studio Code 1.104.0 (Microsoft, Redmond, WA, USA) was selected as the model execution environment, and Python 3.13.7 (open source) was used as the programming language.

3. Results

In total, over 80 models were developed. All of the tested algorithms, in their best versions, exceeded the assumed accuracy threshold for their usefulness in gait analysis (85%). The CNN model consisted of four convolutional layers, and a rectification linear unit (ReLU) was used as the activation function in each layer. The highest accuracy achieved was 91.88% (RMSE 0.001) (Table 3).

Table 3. The final average results of the performance of the selected models.

Algorithm	Accuracy [%]	RMSE [-]
CNN	91.88	0.001
SVM	89.91	0.01
kNN	87.52	0.02
RF	85.41	0.02

Initial results indicate that DL models, particularly CNNs, outperform traditional ML methods in capturing complex gait patterns and providing reliable classification (Figures 4 and 5). However, simpler models showed advantages in interpretability and computational efficiency.

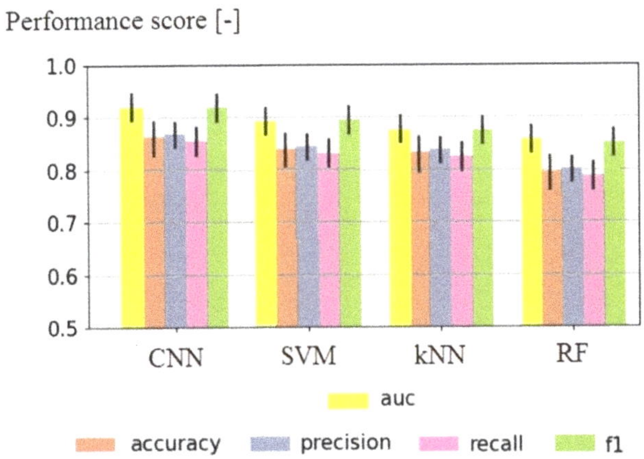

Figure 4. Change in outcomes with model.

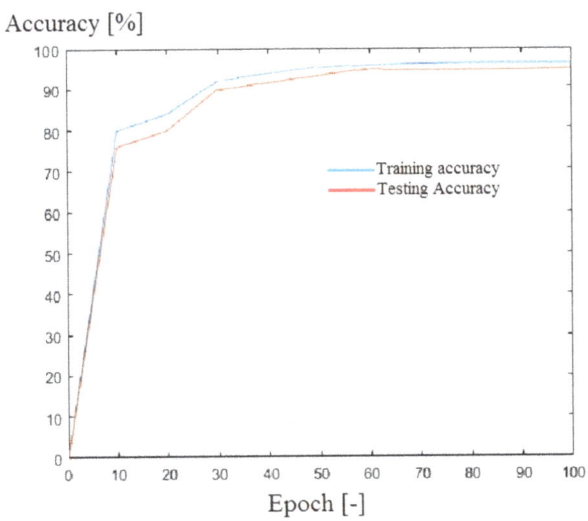

Figure 5. Change in the best model' accuracy with epochs.

Per class metrics reveal strengths and weaknesses in specific patient subgroups. Calibration measures ensure that model results are clinically reliable, not just rank-based. Confidence intervals indicate uncertainty and support fair model comparisons (Table 4).

Table 4. Metrics for classification tasks (best model only).

Class	Precision	Recall	F1	95% CI (F1)
1	91.12	91.12	91.88	0.88–0.94
2	91.02	91.02	91.56	0.88–0.94

Showing the variation in the LOSS curve during training is a clear and generally acceptable way to illustrate convergence, stability, and potential overfitting for any AI model. Below are sample loss curves for our three models, showing typical behavior during training/validation. Figure 6 shows three best models, each with both training and validation loss curves (Figure 6).

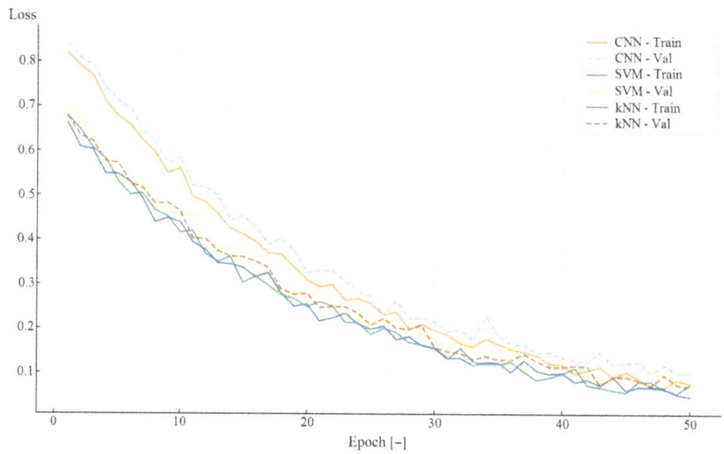

Figure 6. LOSS value curve variation chart.

4. Discussion

Over the last six years (2020–2025), five new articles have been observed on the use of AI in gait analysis of stroke patients, with which our research results were compared. Number of participants was between 100 and 157. A review by Peters et al. [45] found that only 13 (3.67%) of the 354 articles met the inclusion criteria. Significant variation was observed in their quality, reliability, and movement, participants, types of technology used, and recommended gait parameters. Further research is needed to improve the accuracy of devices and software (including AI-based software) for gait assessment among individuals with altered gait after stroke. The use of intelligent insoles allowed us to determine that of the 18 gait parameters studied, 14 were useful for distinguishing patients from healthy individuals, and only 10 (55.55%) were related to stroke severity [46]. The obtained detection rates of the initial and final foot contact with the ground at different sensor positions were over 90% (recall ≥92%, precision ≥97%), but they did not refer to the same practical gait parameters as in our study [47]. New algorithms translate into the connection of exoskeletons for lower limb rehabilitation with rehabilitation assessment, making the rehabilitation assessment process more objective [48]. AI-based assessments were consistent with expert assessments, and expert-system synergy can improve diagnostic accuracy and objectify treatment goals [49].

Literature analysis shows that the threshold for usefulness in gait analysis is 85% accuracy, but solutions that achieve over 90% accuracy in laboratory conditions are becoming increasingly common. Unfortunately, such accuracy is difficult to achieve in clinical settings, where popular markerless methods rely on filming the patient's gait over a short distance, which shortens the examination time and improves access to simple and rapid patient diagnosis.

Previous literature reviews show that despite the advantages of clinical gait analysis, including the portability and accessibility of data acquisition via wearable devices, further studies are needed that monitor gait, including over long periods, among large numbers of participants, and in natural walking conditions (and possibly other patterns, such as running), to analyze actual gait patterns. This would also enable the use of retrospective (historical) data and better planning of prospective studies to more accurately address observed research gaps, including those specific to individuals and their subgroups (including, for example, young stroke patients and elderly stroke survivors). Undoubtedly, the development of a new generation of wearable devices collecting data for gait analysis based on the AI methods presented in this article would be helpful in this regard. Additionally, this could provide data on gait quality that could not previously be determined using traditional gait analysis methods, such as those performed in a gait laboratory. It is necessary to develop prototypes of devices and software, validate them clinically, and establish guidelines for the usability of wearable devices in linear gait analysis and the validity, accuracy, and repeatability of gait quality measurements performed using different devices and software [50].

This study is distinguished by its systematic comparison of different AI approaches used to analyze gait in stroke patients, rather than focusing on a single algorithm. Unlike previous studies, which often emphasize the performance of a single model, the comparative framework provides insight into the strengths and weaknesses of multiple techniques. A key breakthrough is the emphasis on clinically relevant gait parameters, which ensures interpretability of results for rehabilitation professionals. The study also highlights the feasibility of using a standardized 10 m gait test in conjunction with wearable sensors, which increases reproducibility and potential for clinical implementation. Another novelty is the demonstration of how performance varies with the choice of spatiotemporal features, offering guidance for feature selection in future studies. The discussion highlights that while accuracy is important, interpretability and integration into clinical processes are equally crucial, a factor that has been less emphasized in previous work. This study moves the field from the development of isolated models to a broader understanding of how to optimally utilize AI in gait assessment after stroke.

Testing our gait models in real clinical conditions can be the hard part that decides whether a paper prototype becomes a useful clinical tool. We proposed end-to-step plan (design, metrics, analytics, deployment and regulatory checks) to validate and compare multiple AI approaches for stroke-patient gait analysis in a clinical setting:

- Overarching study design: technical/bench testing, prospective observational validation, clinical impact trial;
- Participant selection & sample size (practical guidance): broad range of post-stroke patients (from acute to chronic), varying severity (functional levels), assistive device users vs. independent walkers and comorbidities typical for clinic, sample size guidance (algorithm internal validation, external validation);
- Data collection;
- Outcome measures: clinically meaningful outcomes, technical performance metrics;
- Evaluation metrics;
- Extended statistical analysis (within dataset, between model comparisons, agreement and subgroup analysis;
- Blindig, bias and generalizability,
- Safety of patients and clinical workflow testing;
- Regulatory, ethical and data governance issues;

- Practical logistics;
- Reporting and transparency.

4.1. Limitations of Previous Studies and the Current Study

Current research on AI-based clinical gait analysis in rehabilitation and physical therapy for neurological patients, including stroke patients, has several limitations. Most studies are conducted on small, homogeneous populations, limiting the generalizability of results. Standardized protocols for data collection, feature extraction, and model evaluation are lacking, leading to inconsistent results. Many AI models rely on laboratory-based motion capture, which does not reflect real-world gait patterns. Longitudinal studies are rare, making it difficult to assess the effectiveness of AI-based interventions over time. Integration of multimodal data, such as wearable sensors, clinical assessments, and video analysis, remains limited. Most models focus on classification rather than providing practical insights for personalized therapy. The interpretability of AI results is often insufficient, reducing clinician trust and clinical acceptance. Ethical and privacy issues are often underestimated, particularly with regard to patient data security. Studies often lack robust comparisons with conventional gait assessment methods, leaving their added clinical value uncertain [51]. Studies rarely assess the cost-effectiveness or feasibility of widespread implementation in routine rehabilitation settings [52].

The main limitations of this study include the relatively small sample size, which may limit generalizability. The cohort consisted exclusively of stroke survivors; therefore, there was no healthy control group available to allow for real-world baseline comparisons, and norms were calculated from historical data and the literature. Models were trained on aggregated gait parameters rather than raw sensor or video data, which reduces the richness of the analysis. The cross-sectional design does not account for longitudinal changes in gait recovery. External validation on independent datasets is lacking, limiting conclusions regarding broader applicability.

4.2. Technological Implications

A comparative analysis of different AI approaches to analyzing gait in stroke patients has important technological implications for healthcare and rehabilitation engineering. First, it highlights the differences between AI methods, such as machine learning classifiers, deep learning architectures, and hybrid models, in their ability to capture spatiotemporal gait characteristics. Wearable sensors and motion capture systems generate large, high-dimensional datasets, and the analysis demonstrates that AI can process this complexity far beyond traditional biomechanical techniques. Deep learning models, particularly compound neural networks (CNNs) and randomized neural networks (RNNs), show great potential for automated feature extraction, reducing reliance on manually constructed gait parameters. However, simpler machine learning approaches often offer greater interpretability and lower computational costs, making them suitable for resource-constrained clinical environments. The study suggests that data quality and volume are critical factors: while deep models perform well on large datasets, smaller datasets may favor AI approaches based on ensemble learning or tensor learning. Integration with real-time monitoring systems holds promise, suggesting that AI can enable continuous assessment and adaptive therapy during patient rehabilitation. At the same time, challenges such as model generalization to heterogeneous patient groups and robustness to sensor noise are highlighted as key technological barriers [53]. The comparative analysis also highlights the need for interoperable platforms that can combine gait data from different devices, ensuring scalability and clinical utility [54]. These technological insights pave the way for AI-based gait analysis systems that are not only accurate but also clinically practical, adaptive, and personalized for stroke rehabilitation [55].

4.3. Economical Implications

Comparative analysis of different AI approaches to gait analysis in stroke patients has significant economic implications for healthcare systems and rehabilitation services. By automating gait assessment, AI reduces reliance on time-consuming, manual assessments by clinicians, lowering overall labor costs. Early and accurate detection of gait abnormalities allows for the development of personalized rehabilitation plans, which can shorten recovery times and reduce long-term treatment costs. Wearable AI systems based on sensors are generally more cost-effective than traditional motion capture laboratories, democratizing access to advanced gait analysis technologies. From an economic perspective, the scalability of AI-based platforms means hospitals and rehabilitation centers can serve larger patient volumes without a commensurate increase in staffing and infrastructure costs. At the same time, investing in more advanced deep learning systems can involve high upfront costs for hardware, software, and data management infrastructure. Insurers and healthcare policymakers can see long-term savings through lower readmission rates and fewer complications, improving system-wide cost-effectiveness. However, disparities may emerge between well-funded institutions that can afford advanced AI solutions and smaller, resource-constrained clinics, potentially widening the economic gap in patient care. Comparative analysis suggests that hybrid AI models, balancing accuracy and resource requirements, can provide the best trade-off between cost and benefit in real-world applications. A well-chosen AI approach can transform gait analysis from a specialized, expensive process into a widely available, cost-effective tool that reduces both individual and system burdens in healthcare [56].

4.4. Societal Implications

Comparative analysis of different AI approaches to gait analysis in stroke patients has broad societal implications, influencing the delivery of rehabilitation and long-term care. AI-based gait analysis enables earlier intervention and personalized rehabilitation, improving patients' quality of life and promoting a faster return to daily activities. By facilitating gait assessment outside specialized laboratories, for example, through wearable sensors, AI is helping to expand advanced rehabilitation to rural and underserved communities. This democratization of healthcare technology reduces inequalities in access to effective solutions for stroke recovery. Families and caregivers benefit because AI-based monitoring provides continuous feedback, reducing the emotional and logistical burdens associated with long-term stroke care. From a societal perspective, implementing these systems can reduce pressure on healthcare systems by reducing hospitalization rates and freeing up clinicians' time for more critical cases. At the same time, concerns about privacy and the ethical use of data in continuous gait monitoring underscore the need for responsible implementation. There are also cultural implications, as wider acceptance of AI in rehabilitation could change public perceptions of the technology's role in health and recovery. Potential differences in adoption rates between wealthier and poorer regions underscore the need for policies to ensure equitable access. AI-based gait analysis represents not only a clinical advance but also a societal opportunity to increase the inclusiveness, independence, and dignity of stroke survivors [57].

4.5. Ethical and Legal Implications

Comparative analysis of different AI approaches to gait analysis in stroke patients raises significant ethical and legal implications that shape their clinical application. From an ethical perspective, the use of AI must prioritize patient dignity and autonomy, ensuring that automated gait assessments do not replace, but complement, human clinical judgment. Because gait data are sensitive biometric information, stringent privacy and data protection measures must be implemented to comply with regulations such as GDPR and HIPAA. Legal liability is complex: if an AI system misclassifies gait patterns, leading to incorrect rehabilitation strategies, liability issues arise between clinicians, developers, and healthcare providers. Transparency and explainability of AI models are ethical imperatives, as patients and clinicians must understand how gait decisions are made to maintain trust. There is also a risk of algorithmic bias, where models trained on heterogeneous datasets can lead to unequal rehabilitation outcomes across different populations. Legally, medical AI systems require regulatory approval and certification, and comparative analysis highlights that different AI approaches may face different hurdles depending on their complexity and interpretability. Ethical debates also include ensuring equal access, ensuring that advanced gait analysis systems are not limited to wealthy clinics or regions. The collection and long-term storage of gait data also raises concerns about secondary use and consent, requiring legal safeguards against misuse [58]. Ethical responsibility and legal compliance are as crucial as technical accuracy to ensure that AI-based gait analysis truly benefits stroke patients in a safe, fair, and reliable manner.

4.6. Directions for Further Research

Key venues for future research on AI-based clinical gait analysis in rehabilitation and physical therapy for neurological patients, including stroke patients, focus on improving clinical utility and accessibility. Studies should include larger and more diverse patient populations to increase the generalizability of results across age, gender, ethnicity, and impairment. Establishing standard protocols for data collection, feature extraction, and model evaluation would improve the reproducibility and comparability of studies [59]. Integrating multimodal data from wearable sensors, video recordings, and clinical assessments could provide a more comprehensive understanding of gait patterns [60]. Studies should explore gait monitoring in real-world and home settings to capture daily functioning, not just laboratory settings [61]. Developing understandable AI models will increase clinicians' confidence and facilitate clinical decision-making [62]. Longitudinal studies are needed to assess the impact of AI-assisted interventions on recovery and rehabilitation outcomes over time [63]. Personalizing therapy using AI predictions can optimize rehabilitation plans tailored to individual patient needs [63]. Investigating cost-effectiveness and implementation strategies will support broader clinical adoption [64,65]. Addressing ethical, privacy, and regulatory challenges is essential for safe and responsible implementation. Collaboration between engineers, clinicians, and patients will accelerate innovation and ensure that AI tools meet real-world rehabilitation needs [66,67].

Future research should utilize larger public datasets that allow for more detailed subgroup studies with more diverse sets of spatiotemporal and kinematic parameters, as well as a larger set of neurological conditions. This will enable comparison of gait analysis results in patients with different conditions and computational definition of gait disturbance groups (reflected in parameter values) characteristic for individual conditions (for earlier diagnosis than currently possible).

Future studies could also demonstrate the significance of changes in specific gait parameters with disease phase/progression, to show which ones become more/less diagnostically relevant at different stages. This may require supplementing the dataset with additional data (e.g., joint angles).

The pursuit of transparent and easily human-interpretable ML models has led to the development of a group of techniques generally referred to as eXplainable Artificial Intelligence (XAI). These techniques allow for the assessment of the contribution of individual input features to model outputs, improving their interpretability post hoc. This provides insight into how the trained model uses individual input features to make predictions. This paves the way for further model development and analysis by identifying the influence of context, feature interactions, potential biases, or (in our case) clinically relevant patterns. This could provide insight into the significance of gait parameters in future studies.

5. Conclusions

This study highlights the potential and shortcomings of AI-based gait analysis tools in supporting clinical decision-making and planning personalized stroke rehabilitation. The study results indicate a high probability of developing and implementing more accurate, advanced, and broader solutions for gait analysis in stroke patients based on AI algorithms. The simple, inexpensive, and accurate computational gait analysis tools examined in this article achieved accuracy above 90%, providing a basis for their further development and testing.

This study highlights the potential of ML for classifying gait patterns and identifying clinically relevant gait parameters. Together with XAI, this creates room for the development of more interpretable and personalized diagnostic tools. This approach could become part of preventive medicine, used in healthy patients or at very early stages of clinical evaluation. As part of a periodic health assessment, it could be integrated into screening programs, for example, for populations at particular risk of neurodegenerative diseases or hidden microstrokes. It allows for the development of rapid, automated analysis of gait data in large populations for a timely and scalable decision-making support system, including in telerehabilitation.

Author Contributions: Conceptualization, I.R., E.M., O.M., M.K. and D.M.; methodology, I.R., E.M. and D.M.; software, I.R., E.M., O.M., M.K. and D.M.; validation, I.R., E.M., O.M., M.K. and D.M.; formal analysis, I.R., E.M., O.M., M.K. and D.M.; investigation, I.R., E.M., O.M., M.K. and D.M.; resources, I.R., E.M. and D.M.; data curation, I.R., E.M. and D.M.; writing—original draft preparation, I.R., E.M., O.M., M.K. and D.M.; writing—review and editing, I.R., E.M., O.M., M.K. and D.M.; visualization, I.R., E.M. and D.M.; supervision, I.R. and E.M.; project administration, I.R.; funding acquisition, I.R. All authors have read and agreed to the published version of the manuscript.

Funding: The work presented in this paper was financed by a grant to maintain the research potential of Kazimierz Wielki University.

Institutional Review Board Statement: Approval for the study was given by the Bioethical Committee at the Ludwik Rydygier Collegium Medium in Bydgoszcz, Nicolaus Copernicus University, Torun, Poland (KB 355/2016, 24 May 2016).

Informed Consent Statement: Patients gave their informed consent to take part in the therapy prior to starting the treatment and gave their consent for their results to be used for scientific purposes.

Data Availability Statement: Full anonymized data available from the authors upon request.

Conflicts of Interest: The authors declare no conflicts of interest.

Abbreviations

The following abbreviations are used in this manuscript:

AI	Arificial intelligence
CNN	Convolutional neural network
DL	Deep learning
ML	Machine learning
ReLU	Rectification linear unit
RF	Random forests
SVM	Support vector machine

References

1. Mohan, D.M.; Khandoker, A.H.; Wasti, S.A.; Ismail Ibrahim Ismail Alali, S.; Jelinek, H.F.; Khalaf, K. Assessment Methods of Post-stroke Gait: A Scoping Review of Technology-Driven Approaches to Gait Characterization and Analysis. *Front. Neurol.* **2021**, *12*, 650024. [CrossRef]
2. Pohl, J.; Ryser, A.; Veerbeek, J.M.; Verheyden, G.; Vogt, J.E.; Luft, A.R.; Easthope, C.A. Accuracy of gait and posture classification using movement sensors in individuals with mobility impairment after stroke. *Front. Physiol.* **2022**, *13*, 933987. [CrossRef]
3. Morizio, C.; Compagnat, M.; Boujut, A.; Labbani-Igbida, O.; Billot, M.; Perrochon, A. Immersive Virtual Reality during Robot-Assisted Gait Training: Validation of a New Device in Stroke Rehabilitation. *Medicina* **2022**, *58*, 1805. [CrossRef]
4. Igarashi, T.; Tani, Y.; Takeda, R.; Asakura, T. Minimal detectable change in inertial measurement unit-based trunk acceleration indices during gait in inpatients with subacute stroke. *Sci. Rep.* **2023**, *13*, 19262. [CrossRef]
5. Andrenelli, E.; Ippoliti, E.; Coccia, M.; Millevolte, M.; Cicconi, B.; Latini, L. Features and predictors of activity limitations and participation restriction 2 years after intensive rehabilitation following first-ever stroke. *Eur. J. Phys. Rehabil. Med.* **2015**, *51*, 575–585.
6. Baker, R.; Esquenazi, A.; Benedetti, M.G.; Desloovere, K. Gait analysis: Clinical facts. *Eur. J. Phys. Rehabil. Med.* **2016**, *52*, 560–574. [PubMed]
7. Calabrò, R.S.; Sorrentino, G.; Cassio, A.; Mazzoli, D.; Andrenelli, E.; Bizzarini, E. Robotic-assisted gait rehabilitation following stroke: A systematic review of current guidelines and practical clinical recommendations. *Eur. J. Phys. Rehabil. Med.* **2021**, *57*, 460–471. [CrossRef]
8. Chen, Z.; Li, M.; Cui, H.; Wu, X.; Chen, F.; Li, W. Effects of kinesio taping therapy on gait and surface electromyography in stroke patients with hemiplegia. *Front. Physiol.* **2022**, *13*, 1040278. [CrossRef]
9. Imoto, D.; Hirano, S.; Mukaino, M.; Saitoh, E.; Otaka, Y. A novel gait analysis system for detecting abnormal hemiparetic gait patterns during robot-assisted gait training: A criterion validity study among healthy adults. *Front. Neurorobot.* **2022**, *16*, 1047376. [CrossRef] [PubMed]
10. Nakashima, I.; Imoto, D.; Hirano, S.; Konosu, H.; Otaka, Y. Gait analysis system for assessing abnormal patterns in individuals with hemiparetic stroke during robot-assisted gait training: A criterion-related validity study in healthy adults. *Front Neurorobot.* **2025**, *19*, 1558009. [CrossRef] [PubMed]
11. Goto, K.; Kawasaki, T.; Hamada, H.; Sihan Lim, V.; Shimada, R.; Nakazato, R.; Horimoto, Y.; Ikeda, Y. Changes in Gait Metrics and Motor Strategies Post-Neurocognitive Rehabilitation in Subacute Stroke: A Retrospective Mixed-Methods Study. *Cureus* **2025**, *17*, e86264. [CrossRef]
12. Alammari, B.J.; Schoenwether, B.; Ripic, Z.; Kirk-Sanchez, N.; Eltoukhy, M.; Bishop, L. Validity of AI-Driven Markerless Motion Capture for Spatiotemporal Gait Analysis in Stroke Survivors. *Sensors* **2025**, *25*, 5315. [CrossRef] [PubMed]
13. Scataglini, S.; Abts, E.; Van Bocxlaer, C.; Van den Bussche, M.; Meletani, S.; Truijen, S. Accuracy, Validity, and Reliability of Markerless Camera-Based 3D Motion Capture Systems versus Marker-Based 3D Motion Capture Systems in Gait Analysis: A Systematic Review and Meta-Analysis. *Sensors* **2024**, *24*, 3686. [CrossRef]
14. Lam, W.W.T.; Tang, Y.M.; Fong, K.N.K. A systematic review of the applications of markerless motion capture (MMC) technology for clinical measurement in rehabilitation. *J. Neuroeng. Rehabil.* **2023**, *20*, 57. [CrossRef]
15. Desai, R.; Martelli, D.; Alomar, J.A.; Agrawal, S.; Quinn, L.; Bishop, L. Validity and reliability of inertial measurement units for gait assessment within a post stroke population. *Top. Stroke Rehabil.* **2024**, *31*, 235–243. [CrossRef]
16. Yoma, M.; Herrington, L.; Starbuck, C.; Llurda, L.; Jones, R. Between-Day Reliability of Kinematic Variables Using Markerless Motion Capture for Single-Leg Squat and Single-Leg Landing Tasks. *Int. J. Sports Phys. Ther.* **2025**, *20*, 1160–1175. [CrossRef] [PubMed]
17. Louie, D.R.; Simpson, L.A.; Mortenson, W.B.; Field, T.S.; Yao, J.; Eng, J.J. Prevalence of Walking Limitation After Acute Stroke and Its Impact on Discharge to Home. *Phys. Ther.* **2022**, *102*, pzab246. [CrossRef] [PubMed]
18. Heshmatollah, A.; Darweesh, S.K.L.; Dommershuijsen, L.J.; Koudstaal, P.J.; Ikram, M.A.; Ikram, M.K. Quantitative Gait Impairments in Patients with Stroke or Transient Ischemic Attack: A Population-Based Approach. *Stroke* **2020**, *51*, 2464–2471. [CrossRef]
19. Balaban, B.; Tok, F. Gait Disturbances in Patients with Stroke. *PM&R* **2014**, *6*, 635–642. [CrossRef]
20. Perry, J.; Garrett, M.; Gronley, J.K.; Mulroy, S.J. Classification of walking handicap in the stroke population. *Stroke* **1995**, *26*, 982–989. [CrossRef]
21. Min, Y.S.; Jung, T.D.; Lee, Y.S.; Kwon, Y.; Kim, H.J.; Kim, H.C.; Lee, J.C.; Park, E. Biomechanical Gait Analysis Using a Smartphone-Based Motion Capture System (OpenCap) in Patients with Neurological Disorders. *Bioengineering* **2024**, *11*, 911. [CrossRef]
22. Schmid, A. Improvements in speed-based gait classifications are meaningful. *Stroke* **2007**, *38*, 2096–2100. [CrossRef]
23. Grau-Pellicer, M.; Chamarro-Lusar, A.; Medina-Casanovas, J.; Serda Ferrer, B.C. Walking speed as a predictor of community mobility and quality of life after stroke. *Top. Stroke Rehabil.* **2019**, *26*, 349–358. [CrossRef]

24. Pound, P.; Gompertz, P.; Ebrahim, S. A patient-centred study of the consequences of stroke. *Clin. Rehabil.* **1998**, *12*, 255–264. [CrossRef]
25. Bergamini, E. Multi-sensor assessment of dynamic balance during gait in patients with subacute stroke. *J. Biomech.* **2017**, *61*, 208–215. [CrossRef] [PubMed]
26. Jørgensen, H.S.; Nakayama, H.; Raaschou, H.O.; Olsen, T.S. Recovery of walking function in stroke patients: The Copenhagen stroke study. *Arch. Phys. Med. Rehabil.* **1995**, *76*, 27–32. [CrossRef]
27. Beyaert, C.; Vasa, R.; Frykberg, G.E. Gait post-stroke: Pathophysiology and rehabilitation strategies. *Clin. Neurophysiol.* **2015**, *45*, 335–355. [CrossRef]
28. Price, R.; Choy, N.L. Investigating the Relationship of the Functional Gait Assessment to Spatiotemporal Parameters of Gait and Quality of Life in Individuals with Stroke. *J. Geriatr. Phys. Ther.* **2019**, *42*, 256–264. [CrossRef]
29. Coleman, E.R.; Moudgal, R.; Lang, K.; Hyacinth, H.I.; Awosika, O.O.; Kissela, B.M.; Feng, W. Early Rehabilitation After Stroke: A Narrative Review. *Curr. Atheroscler. Rep.* **2017**, *19*, 59. [CrossRef] [PubMed]
30. Martiš, P.; Košutzká, Z.; Kranzl, A. A Step Forward Understanding Directional Limitations in Markerless Smartphone-Based Gait Analysis: A Pilot Study. *Sensors* **2024**, *24*, 3091. [CrossRef]
31. Tiddia, G.; Mainas, F.; Retico, A.; Oliva, P. Explainable AI Highlights the Most Relevant Gait Features for Neurodegenerative Disease Classification. *Appl. Sci.* **2025**, *15*, 8078. [CrossRef]
32. Rojek, I.; Prokopowicz, P.; Dorożyński, J.; Mikołajewski, D. Novel Methods of AI-Based Gait Analysis in Post-Stroke Patients. *Appl. Sci.* **2023**, *13*, 6258. [CrossRef]
33. Wankhede, V.; Verma, P.; Gahane, S. A Review on the Developments in the Field of AI-Based Gait Analysis. In Proceedings of the 2024 2nd DMIHER International Conference on Artificial Intelligence in Healthcare, Education and Industry (IDICAIEI), Wardha, India, 29–30 November 2024; IEEE: New York, NY, USA, 2024; pp. 1–5.
34. Kim, G.; Kim, H.; Kim, Y.H.; Kim, S.J.; Choi, M.T. Deep Temporal Clustering of Pathological Gait Patterns in Post-Stroke Patients Using Joint Angle Trajectories: A Cross-Sectional Study. *Bioengineering* **2025**, *12*, 55. [CrossRef]
35. Romano, F.; Perpetuini, D.; Cardone, D.; Merla, A. A machine learning framework for gait and EMG analysis for post-stroke motor dysfunctions assessment. In Proceedings of the European Medical and Biological Engineering Conference 2024, Portorož, Slovenia, 9–13 June 2024; SpringerNature: Cham, Switzerland, 2024; pp. 15–22.
36. Teng, T.H.; Kim, G.; Kim, H.; Choi, M.T. Deep Temporal Clustering for Long-Term Gait Recovery Patterns of Post-Stroke Patients using Joint Kinematic Data. In Proceedings of the 2025 11th International Conference on Computing and Artificial Intelligence (ICCAI), Kyoto, Japan, 28–31 March 2025; IEEE: New York, NY, USA, 2025; pp. 654–658.
37. Feng, L.; Duan, Q.; Lai, R.; Liu, W.; Song, X.; Lyu, Y. Development of a three-dimensional muscle-driven lower limb model developed using an improved CFD-FE method. *Comput. Methods Biomech. Biomed. Eng.* **2025**, *28*, 314–325. [CrossRef]
38. Al_Hassani, R.T.; Atilla, D.C. Human Activity Detection Using Smart Wearable Sensing Devices with Feed Forward Neural Networks and PSO. *Appl. Sci.* **2023**, *13*, 3716. [CrossRef]
39. Zia Ur Rehman, M.; Khan, M.; Paul, A. AI-Powered Human Activity Recognition with Attention Networks. In Proceedings of the International Conference on Intelligent Information Hiding and Multimedia Signal Processing, Daegu, Republic of Korea, 5–7 December 2023; Springer Nature: Singapore, 2023; pp. 129–136.
40. Wang, H.; Wang, P.; Wang, C.; Xiao, Z.; Sun, Y. Intelligent Backpack Based on Wireless Mobile Technology. In Proceedings of the International Conference on Artificial Intelligence Security and Privacy, Guangzhou, China, 6–7 December 2024; Springer Nature: Singapore, 2024; pp. 44–58.
41. Chimamiwa, G.; Alirezaie, M.; Pecora, F.; Loutfi, A. Multi-sensor dataset of human activities in a smart home environment. *Data Brief.* **2021**, *34*, 106632. [CrossRef]
42. Kvist, A.; Tinmark, F.; Bezuidenhout, L.; Reimeringer, M.; Conradsson, D.M.; Franzén, E. Validation of algorithms for calculating spatiotemporal gait parameters during continuous turning using lumbar and foot mounted inertial measurement units. *J. Biomech.* **2024**, *162*, 111907. [CrossRef]
43. Fawden, T.; Roberts, I.V.; Goldin, S.; Sharma, Y.; Dunne, H.; Stone, T.; Bance, M. Detection of Gait Events Using Ear-Worn IMUs During Functional Movement Tasks. *Sensors* **2025**, *25*, 3629. [CrossRef] [PubMed]
44. Cordillet, S.; Drapier, S.; Leh, F.; Dumont, A.; Bidet, F.; Bonan, I.; Jamal, K. Detecting Freezing of Gait in Parkinson Disease Using Multiple Wearable Sensors Sets During Various Walking Tasks Relative to Medication Conditions (DetectFoG): Protocol for a Prospective Cohort Study. *JMIR Res. Protoc.* **2025**, *14*, e58612. [CrossRef] [PubMed]
45. Peters, D.M.; O'Brien, E.S.; Kamrud, K.E.; Roberts, S.M.; Rooney, T.A.; Thibodeau, K.P.; Balakrishnan, S.; Gell, N.; Mohapatra, S. Utilization of wearable technology to assess gait and mobility post-stroke: A systematic review. *J. Neuroeng. Rehabil.* **2021**, *18*, 67. [CrossRef] [PubMed]
46. Seo, M.; Shin, M.J.; Park, T.S.; Park, J.H. Clinometric Gait Analysis Using Smart Insoles in Patients With Hemiplegia After Stroke: Pilot Study. *JMIR Mhealth Uhealth* **2020**, *8*, e22208. [CrossRef] [PubMed]

47. Romijnders, R.; Warmerdam, E.; Hansen, C.; Schmidt, G.; Maetzler, W. A Deep Learning Approach for Gait Event Detection from a Single Shank-Worn IMU: Validation in Healthy and Neurological Cohorts. *Sensors* **2022**, *22*, 3859. [CrossRef]
48. Su, D.; Hu, Z.; Wu, J.; Shang, P.; Luo, Z. Review of adaptive control for stroke lower limb exoskeleton rehabilitation robot based on motion intention recognition. *Front. Neurorobot.* **2023**, *17*, 1186175. [CrossRef] [PubMed]
49. Liaw, J.C.; Raab, D.; Weber, M.; Siebler, M.; Hefter, H.; Zietz, D.; Jäger, M.; Kecskeméthy, A.; Geu Flores, F. An Intelligible AI-Driven Decision Support System For Poststroke Mobility Assessment. *J. Rehabil. Med. Clin. Commun.* **2025**, *8*, 42379. [CrossRef]
50. Benson, L.C.; Clermont, C.A.; Bošnjak, E.; Ferber, R. The use of wearable devices for walking and running gait analysis outside of the lab: A systematic review. *Gait Posture.* **2018**, *63*, 124–138. [CrossRef]
51. Bedon, C.; Sciomenta, M.; Mazelli, A. Use of Smartphone-Based Experimental Data for the Calibration of Biodynamic Spring-Mass-Damper (SMD) Pedestrian Models. *Sensors* **2025**, *25*, 1387. [CrossRef]
52. Ma, P.; Bian, Q.; Kim, J.M.; Alsayed, K.; Ding, Z. Measuring Lower-Limb Kinematics in Walking: Wearable Sensors Achieve Comparable Reliability to Motion Capture Systems and Smartphone Cameras. *Sensors* **2025**, *25*, 2899. [CrossRef]
53. Dibbern, K.N.; Krzak, M.G.; Olivas, A.; Albert, M.V.; Krzak, J.J.; Kruger, K.M. Scoping Review of Machine Learning Techniques in Marker-Based Clinical Gait Analysis. *Bioengineering* **2025**, *12*, 591. [CrossRef]
54. Luvizutto, G.J.; Silva, G.F.; Nascimento, M.R.; Sousa Santos, K.C.; Appelt, P.A.; de Moura Neto, E.; de Souza, J.T.; Wincker, F.C.; Miranda, L.A.; Hamamoto Filho, P.T.; et al. Use of artificial intelligence as an instrument of evaluation after stroke: A scoping review based on international classification of functioning, disability and health concept. *Top. Stroke Rehabil.* **2022**, *29*, 331–346. [CrossRef] [PubMed]
55. Senadheera, I.; Hettiarachchi, P.; Haslam, B.; Nawaratne, R.; Sheehan, J.; Lockwood, K.J.; Alahakoon, D.; Carey, L.M. AI Applications in Adult Stroke Recovery and Rehabilitation: A Scoping Review Using AI. *Sensors* **2024**, *24*, 6585. [CrossRef]
56. Daly, J.J.; McCabe, J.P.; Gor-García-Fogeda, M.D.; Nethery, J.C. Update on an Observational, Clinically Useful Gait Coordination Measure: The Gait Assessment and Intervention Tool (G.A.I.T.). *Brain Sci.* **2022**, *12*, 1104. [CrossRef]
57. Adams, J.L.; Dinesh, K.; Snyder, C.W.; Xiong, M.; Tarolli, C.G.; Sharma, S.; Dorsey, E.R.; Sharma, G. A real-world study of wearable sensors in Parkinson's disease. *NPJ Park. Dis.* **2021**, *7*, 106. [CrossRef] [PubMed]
58. Nicora, G.; Pe, S.; Santangelo, G.; Billeci, L.; Aprile, I.G.; Germanotta, M.; Bellazzi, R.; Parimbelli, E.; Quaglini, S. Systematic review of AI/ML applications in multi-domain robotic rehabilitation: Trends, gaps, and future directions. *J. Neuroeng. Rehabil.* **2025**, *22*, 79. [CrossRef] [PubMed]
59. Chang, M.; Kim, T.W.; Beom, J.; Won, S.; Jeon, D. AI Therapist Realizing Expert Verbal Cues for Effective Robot-Assisted Gait Training. *IEEE Trans. Neural Syst. Rehabil. Eng.* **2020**, *28*, 2805–2815. [CrossRef] [PubMed]
60. Mikołajewska, E. The value of the NDT-Bobath method in post-stroke gait training. *Adv. Clin. Exp. Med.* **2013**, *22*, 261–272.
61. Mikołajewska, E. Associations between results of post-stroke NDT-Bobath rehabilitation in gait parameters, ADL and hand functions. *Adv. Clin. Exp. Med.* **2013**, *22*, 731–738.
62. Wu, Y.C.; Huang, Y.J.; Han, C.C.; Cheng, Y.Y.; Chang, C.S. Development of an IMU-Based Post-Stroke Gait Data Acquisition and Analysis System for the Gait Assessment and Intervention Tool. *Sensors* **2025**, *25*, 1994. [CrossRef]
63. Lanotte, F.; Okita, S.; O'Brien, M.K.; Jayaraman, A. Enhanced gait tracking measures for individuals with stroke using leg-worn inertial sensors. *J. Neuroeng. Rehabil.* **2024**, *21*, 219. [CrossRef]
64. Silva, R.S.D.; Silva, S.T.D.; Cardoso, D.C.R.; Quirino, M.A.F.; Silva, M.H.A.; Gomes, L.A.; Fernandes, J.D.; Oliveira, R.A.N.D.S.; Fernandes, A.B.G.S.; Ribeiro, T.S. Psychometric properties of wearable technologies to assess post-stroke gait parameters: A systematic review. *Gait Posture.* **2024**, *113*, 543–552. [CrossRef]
65. Song, Q.; Qin, Q.; Suen, L.K.P.; Liang, G.; Qin, H.; Zhang, L. Effects of wearable device training on upper limb motor function in patients with stroke: A systematic review and meta-analysis. *J. Int. Med. Res.* **2024**, *52*, 3000605241285858. [CrossRef]
66. Rojek, I.; Mikołajewski, D.; Dostatni, E.; Macko, M. AI-Optimized Technological Aspects of the Material Used in 3D Printing Processes for Selected Medical Applications. *Materials* **2020**, *13*, 5437. [CrossRef] [PubMed]
67. Rojek, I.; Mikołajewski, D.; Kopowski, J.; Kotlarz, P.; Piechowiak, M.; Dostatni, E. Reducing Waste in 3D Printing Using a Neural Network Based on an Own Elbow Exoskeleton. *Materials* **2021**, *14*, 5074. [CrossRef] [PubMed]

applied sciences

MDPI

Article

Effect of Taping on Postoperative Recovery Following Saphenectomy

Raquel Michelini Guerero [1], Catarina Clápis Zordão [1], Elisa Helena Subtil Zampieri [2], Andreia Noites [3] and Elaine Caldeira de Oliveira Guirro [1,*]

[1] Postgraduate Program in Rehabilitation and Functional Performance, São Paulo University, Ribeirão Preto 14049-900, São Paulo, Brazil; raquelmichelinig@gmail.com (R.M.G.); catarina.zordao@usp.br (C.C.Z.)

[2] Postgraduate Program of Surgery and Anatomy, São Paulo University, Ribeirão Preto 14049-900, São Paulo, Brazil; elisazampieri@gmail.com

[3] Center for Rehabilitation Research (CIR), ESS, Polytechnic of Porto, Rua Dr. António Bernardino de Almeida, 400, 4200-072 Porto, Portugal; arn@ess.ipp.pt

* Correspondence: ecguirro@fmrp.usp.br; Tel.: +55-16-3315-4584

Abstract: Post-surgical complications are common complications following saphenectomy surgery, and strategies to facilitate its resolution are essential for postoperative recovery. This study evaluated the effects of adhesive elastic taping on edema control in patients undergoing saphenectomy. A randomized controlled clinical trial was conducted with 40 patients of both sexes, divided into two groups: intervention (IG), which received taping immediately after surgery combined with standard compression, and a control group, which received standard treatment with compression stockings (CG). Assessments were performed preoperatively and seven days after surgery, including limb volume (indirect calculation), edema (dielectric constant analysis), Skin Elasticity Assessment (durometer), pain (Visual Analog Scale—VAS), limb functionality (Lower Extremity Functional Scale—LEFS), and ecchymosis area (Image J, version 1.51). Both groups showed a significant increase in edema postoperatively (IG: $p = 0.003$; CG: $p = 0.001$). The intervention group exhibited a trend toward volume reduction ($p = 0.069$), better functionality ($p = 0.006$)—skin elasticity was assessed using a durometer—and fewer ecchymoses ($p = 0.002$). Only the control group showed a significant increase in tissue firmness ($p = 0.012$). No significant difference in pain was observed between groups ($p = 0.203$). The application of taping demonstrated beneficial effects on postoperative functional recovery and ecchymosis control following saphenectomy.

Keywords: functional bandaging; edema; varicose veins; kinesio taping

1. Introduction

The post-surgical recovery of patients undergoing saphenectomy can be significantly affected by complications such as edema and ecchymosis, which delay rehabilitation and impact clinical outcomes [1,2]. Although compression therapy with stockings is considered the standard approach for managing these effects [3], new strategies are being investigated to enhance edema reduction and provide a more efficient postoperative recovery.

Functional elastic bandaging, also known as taping, emerges as an innovative and accessible alternative for surgical edema management, offering advantages in terms of cost-effectiveness, applicability, and patient tolerance. This technique, based on the application of an elastic adhesive cotton tape, is thermoactive and has elasticity similar to the skin, promoting a biomechanical effect that stimulates blood and lymphatic circulation by generating compression and decompression of capillaries during movement [4,5]. Additionally, it may modulate exaggerated inflammatory responses in the tissue, contributing to a more efficient healing process [6,7].

Taping has demonstrated beneficial effects, particularly in recovery control across various surgical contexts, including plastic surgeries [8], third molar extraction [9,10] orthopedic procedures [11], post-mastectomy lymphedema [12,13], and venous insufficiency symptoms [14]. Despite its growing use for different therapeutic purposes, its specific effectiveness in the post-saphenectomy recovery process has yet to be explored.

The objective of this study was to evaluate the impact of taping on post-saphenectomy recovery, analyzing its efficacy compared to conventional methods for postoperative event management. This study aims to expand the use of taping as a complementary resource in vascular disease recovery protocols, providing a practical and non-invasive alternative for patients suffering from chronic venous insufficiency.

2. Materials and Methods

2.1. Ethical Aspects

The study was conducted in accordance with the Declaration of Helsinki and was approved by the Human Research Ethics Committee of the Faculty of Medicine of Ribeirão Preto—São Paulo—Brazil (FMRP/USP) under protocol number 5.596.054. Registered in Clinical Trials under number 12241. Informed consent was obtained from all subjects involved in the study.

2.2. Eligibility Criteria

Individuals of both sexes who underwent unilateral saphenectomy were included in the study, with a mean age of 42 years. Non-inclusion criteria comprised patients diagnosed with allergies, skin lesions, or ulcers in the limb being assessed.

2.3. Description of the Experiment

Participants were randomly allocated into two groups: one receiving standard compression therapy and the other undergoing compression taping immediately after surgery.

The clinical trial was conducted in accordance with the guidelines of the Consolidated Standards of Reporting Trials (CONSORT). One researcher was responsible for determining participant eligibility and conducting assessments using a standardized anamnesis form. A second researcher was in charge of randomizing participants into their respective groups and performing therapeutic procedures, while a third researcher analyzed the data, remaining blinded to group allocation.

Patients were randomly assigned in small blocks of 5:5, meaning that every five participants were allocated to each study group using sealed, opaque envelopes to ensure concealed allocation. Randomization was performed through a lottery draw of the envelopes.

2.4. Conventional Postoperative Treatment

Patients assigned to the conventional treatment group underwent seven days of compressive bandaging following surgery. Subsequently, they were instructed to wear 7/8-length elastic compression stockings with moderate compression (20–30 mmHg) for 24 h per day over a 7-day period. This approach aligns with standard postoperative protocols aimed at minimizing edema, hematoma formation, and improving venous circulation.

2.5. Taping Application

The experimental group received taping applied to the lower limb immediately post-surgery. The technique involved elastic bandages measuring 20 cm and 15 cm, arranged in a fish-scale pattern to optimize lymphatic and blood flow, using AKTive Tape® (Suzhou SunMed, Suzhou, China). The active tension applied in the functional region of the bandage was set at 80%, whereas the base and anchor sections (5 cm) were placed without tension to allow physiological adaptation (Figure 1).

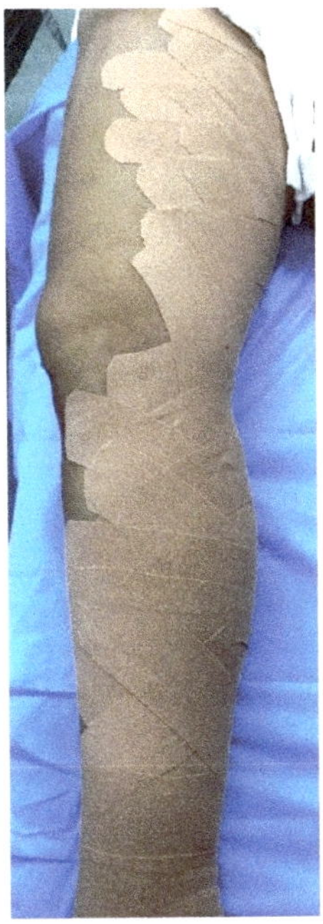

Figure 1. Image of the lower limb in the post-operative period of saphenectomy with taping application.

2.5.1. Evaluation Criteria

Both groups were monitored for postoperative edema, pain levels, and functional recovery using validated assessment scales and imaging techniques. Measurements were taken at baseline (pre-surgery) and at 7-day follow-up to compare the efficacy of both treatments.

2.5.2. Edema Assessment

For edema assessment, a non-invasive method for quantifying water content at different tissue depths was employed using the MoistureMeter D® (Delfin Technologies, Kuopio, Finland). The dielectric constant (DC) of the skin generated a high-frequency electromagnetic (EM) wave at 300 MHz, which was transmitted to the coaxial probe model S15. This probe provides information related to the dermis, with a measurement depth ranging from 1.5 mm to 3 mm [15]. The DC of the tissue is proportional to water content, meaning that higher readings indicate greater hydration levels [15,16].

Participants remained at rest for 5 min in a supine position on an examination table in a temperature-controlled room (24 °C) before the procedure. The probe was positioned perpendicularly to the skin with stable pressure, taking measurements at five points along the medial region of the lower limbs.

Lower limb perimetry was performed by the same evaluator, using a graduated measuring tape in centimeters, with the participant in a supine position [17]. Measurements were taken at the dorsum of the foot, ankle, and from this joint, five additional measurements were recorded every 10 cm up to the inguinal region.

The limb volume was calculated based on perimetry data using an indirect method, summing the approximate volume of six truncated cones according to the following formula [18]: volume = (h) (C1 $\times$ C2 + C12 + C22)/12π

2.5.3. Skin Elasticity Assessment

Skin elasticity was assessed using the ElastiMeter® (Delfin Technologies, Kuopio, Finland), a device that enables instantaneous measurements of elastic properties and biophysical changes in the skin. For each evaluation point, five consecutive measurements were performed, and the mean value of these readings was used to determine the Elastic Support Index (ESI), expressed in N/m. Assessments were conducted at five distinct points on the medial region of the lower limbs, with 10 cm spacing between them.

2.5.4. Pain Assessment

Pain was quantified using the Numerical Visual Scale (NVS), where a score of 0 represented no pain and a score of 10 indicated the worst imaginable pain [19]. The scale was applied on the seventh postoperative day.

2.5.5. Lower Limb Functionality Assessment

The functionality of the lower limbs was assessed using the Lower Extremity Functional Scale—LEFS [20], which measures musculoskeletal dysfunctions in this region. The scale consists of 20 items, each scored on a scale from zero to four, with a maximum possible score of 80. Patients completed the questionnaire, and the total score was obtained by summing the responses to all 20 items. A minimum score of 0 indicates a reduced functional level, while a maximum score of 80 represents high functionality [20,21]. The scale was applied on the seventh postoperative day.

2.5.6. Ecchymosis Area Analysis

The ecchymosis area was assessed using Image J (Wayne Rasband at the National Institutes of Health—NIH, Bethesda, MD, USA), version 1.51, 2018, an advanced image analysis software that enables precise quantification of lesion dimensions. The tool allowed for accurate measurement of the affected region by analyzing digital images and detecting the boundaries of the ecchymoses. This methodology ensured objective evaluation, minimizing potential biases and enhancing the reliability of the data collected.

2.5.7. Statistical Analysis

The sample size calculation was processed using SPSS software, version 20 (IBM, Armonk, NY, USA). The sample size was determined based on the study conducted by Aguilar-Ferrándiz et al. [14], considering the effect of taping on edema as the primary outcome variable. A statistical power of 80%, a significance level of 5%, and potential sample losses were accounted for, resulting in a required sample size of 20 volunteers per group.

Statistical analysis was performed using SPSS software (version 20), applying non-parametric statistical tests due to the nature of the data and the observed distribution. For comparisons between more than two groups, the Kruskal-Walli's test was applied. For paired comparisons, the Wilcoxon test was used, while the Mann-Whitney test was employed for comparisons between independent groups. The adopted significance level was 5% ($p < 0.05$).

3. Results

The following flow chart (Figure 2) presents the stages of the clinical trial, detailing the process of participant selection, intervention, and analysis.

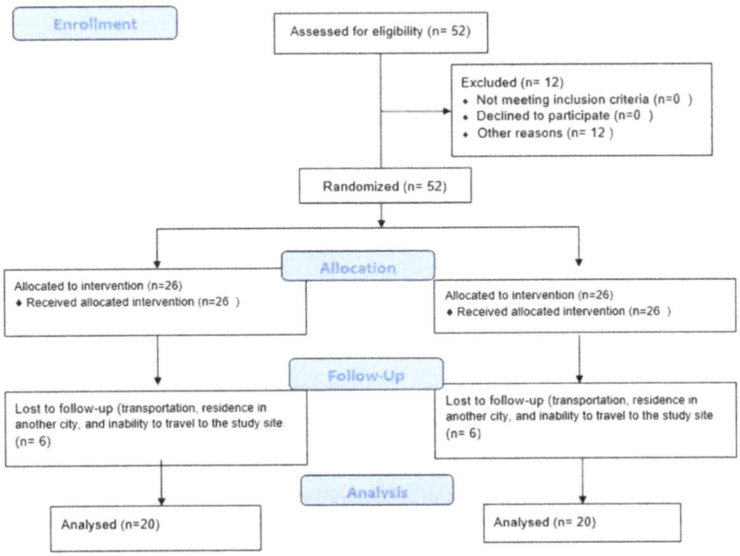

Figure 2. Study flow chart.

In Table 1, the analysis of tissue hydration in the lower limbs (MoistureMeter), shows no statistically significant difference was observed between the Intervention group (IG) and Control group (CG) at any of the evaluated time points (pre- and post-operative), both in the limb where saphenectomy was performed ($p = 0.387$ and $p = 0.285$, respectively) and in the limbs without saphenectomy ($p = 0.745$ and $p = 0.839$).

Table 1. Comparison of lower limb tissue hydration between the Intervention group (IG) and Control group (CG) at pre- and post-operative time points.

MOMENT	Group	Mean	Medien	SD	Mín	Máx	*p* (Kruskal Wallis)
SAF PRE-OP	IG	29.42	29.76	5.04	17.70	37.91	0.387
	CG	27.64	28.32	5.04	16.11	34.50	
SAF POST-OP	IG	32.90	34.26	5.79	17.13	44.29	0.285
	CG	31.16	30.49	5.80	20.50	40.96	
NO SAF PRE-OP	IG	28.08	27.32	4.15	19.51	34.40	0.745
	CG	27.10	29.05	5.01	16.16	34.19	
NO SAF POST-OP	IG	28.44	27.95	4.70	17.03	44.29	0.839
	CG	28.40	31.65	5.38	18.41	38.49	

SAF PRE-OP: Limb undergoing saphenectomy in the preoperative period. SAF POST-OP: Limb undergoing saphenectomy in the postoperative period. NO SAF PRE-OP: Control limb in the preoperative period. NO SAF POST-OP: Control limb in the postoperative period. SD: Standard deviation. IG: Intervention group. CG: Control group. $p < 0.05$ indicates statistical significance. Unit of measurement for volume: cm^3.

Similarly, the comparison of lower limb volume in Table 2 also showed no significant differences between the groups at any of the analyzed time points ($p > 0.05$).

Table 2. Lower limb volume—comparisons between groups in the preoperative and postoperative periods.

MOMENT	Group	Mean	Medien	SD	Min	Max	p (Kruskal-Wallis)
SAF PRE-OP	IG	9316.77	9233.85	1393.81	6954.77	12,422.69	0.185
	CG	9932.40	9924.84	1452.12	7511.37	12,787.42	
SAF POST-OP	IG	9393.12	9492.86	1279.74	7198.66	12,211.65	0.204
	CG	10,004.80	9805.97	1629.31	7510.83	12,935.08	
NO SAF PRE-OP	IG	9217.15	9144.54	1249.96	6921.18	11,932.57	0.160
	CG	9838.56	9791.84	1454.21	7469.07	12,759.05	
NO SAF POST-OP	IG	9234.61	9119.94	1172.23	7312.98	11,557.13	0.291
	CG	9675.42	9885.49	1438.56	7383.17	12,094.63	

SAF PRE-OP: Limb undergoing saphenectomy in the preoperative period. SAF POST-OP: Limb undergoing saphenectomy in the postoperative period. NO SAF PRE-OP: Control limb in the preoperative period. NO SAF POST-OP: Control limb in the postoperative period. SD: Standard deviation. IG: Intervention group. CG: Control group. $p < 0.05$ indicates statistical significance. Unit of measurement for volume: cm^3.

Table 2 showed a significant increase in postoperative edema in both groups (Intervention Group: $p = 0.003$; Control Group: $p = 0.001$), with no significant differences between them in the preoperative ($p = 0.387$) and postoperative periods ($p = 0.285$). Regarding lower limb volumetry, as presented in Table 3, there was a trend toward a reduction in lower limb volume in the intervention group (IG) between the pre- and post-intervention periods ($p = 0.069$), although without statistical significance. In the remaining analyses, no significant intragroup differences were observed. Intergroup analyses also did not reveal relevant statistical differences between the pre- and post-intervention periods ($p > 0.05$).

Table 3. Instantaneous sin elasticity between GT and GC group at preoperative and postoperative time points.

MOMENT	Group	Mean	Medien	SD	Min	Max	p (Krushal-Wallis)
SAF PRE-OP	IG	67.48	68.21	13.32	38.14	89.71	0.117
	CG	61.45	57.14	12.71	45.29	86.00	
SAF POST-OP	IG	74.85	67.71	22.50	34.86	114.14	0.387
	CG	71.59	60.92	25.59	39.86	144.43	
NO SAF PRE-OP	IG	65.96	66.57	15.08	37.71	99.29	0.417
	CG	61.46	61.50	10.82	44.57	76.00	
NO SAF POST-OP	IG	70.25	70.85	15.67	38.57	92.43	0.262
	CG	66.26	62.92	16.38	39.86	105.57	

SAF PRE-OP: Limb undergoing saphenectomy in the preoperative period. SAF POST-OP: Limb undergoing saphenectomy in the postoperative period. NO SAF PRE-OP: Control limb in the preoperative period. NO SAF POST-OP: Control limb in the postoperative period. SD: Standard deviation. IG: Intervention group. CG: Control group. $p < 0.05$ indicates statistical significance. Unit of measurement for volume: cm^3.

The analysis of instantaneous skin elasticity using the Elastimeter (Table 3) showed no significant differences between the Intervention group (IG) and Control group (CG) in the preoperative and postoperative periods, both in limbs that underwent saphenectomy ($p = 0.117$ and $p = 0.387$, respectively) and in those that did not ($p = 0.417$ and $p = 0.262$).

Table 4 presents the intergroup analysis of the variables ecchymosis, visual analog scale (VAS) for pain, and lower extremity functional scale (LEFS). Regarding the presence of postoperative ecchymosis in the limb subjected to saphenectomy, a significant difference was observed between groups (p = 002), with the intervention group (IG) showing a lower mean ecchymosis value (0.085) compared to the control group (CG)—0.174.

Table 4. Comparison of the presence of ecchymosis, pain scale (VAS), and power extremity functional scale (LEFS) between the intervention (IG) and control (CG) groups in the postoperative period of saphenectomy.

Analyzed Variables	Moment	Group	Mean	Medien	SD	Min	Max	p (Kruskal-Wallis)
ECCHYMOSIS	SAF POST-OP	IG	0.085	0.050	0.109	0.006	0.510	0.002 *
		CG	0.174	0.105	0.133	0.019	0.428	
VAS	SAF POST-OP	IG	1.25	1.00	1.61	0	6	0.203
		CG	1.65	1.00	1.34	0	4	
LEFS	SAF POST-OP	IG	66.20	73.00	18.63	17	80	0.006 *
		CG	58.60	59.00	11.81	32	79	

IG: Intervention group. CG: Control group; SAF POST-OP: Limb undergoing saphenectomy in the postoperative period. Significance level $p < 0.05$ *.

For pain assessment, the IG also exhibited a lower mean score (1.25) than the CG (1.65), although this difference was not statistically significant (p = 0.203).

Finally, functional assessment of the lower limbs demonstrated a significant difference between groups (p = 0.006), with the IG achieving a higher mean functional score (66.20) compared to the CG (58.60), suggesting better functional recovery in patients who received taping intervention.

In the comparison between the IG and CG groups, as shown in Table 5, no significant difference was observed in pain intensity (VAS) between the groups (p = 0.203), although the IG group showed a slightly lower mean (1.25 ± 1.62) compared to the CG group (1.65 ± 1.35). On the other hand, lower extremity functionality, as measured by the Lower Extremity Functional Scale (LEFS), was significantly higher in the IG group (66.20 ± 18.64) than in the CG group (58.60 ± 11.82), with p = 0.006.

Table 5. Comparison between IG and CG groups on VAS and LEFS scales in the postoperative period.

Safenectomy	Varable	Moment	Mean	Medien	SD	Min	Max	p (Wilxocon)
IG	MoistureMeter	PRE-OP	29.42	29.76	5.04	17.70	37.91	0.001 *
		POST-OP	32.90	34.26	5.79	17.13	44.29	
CG	MoistureMeter	PRE-OP	27.64	28.32	5.04	16.11	34.50	0.003 *
		POST-OP	31.16	30.49	5.80	20.50	40.96	
IG	Volumetria	PRE-OP	9316.77	9233.85	1393.81	6954.77	12,422.69	0.411
		POST-OP	9393.12	9492.86	1279.74	7198.66	12,211.65	
CG	Volumetria	PRE-OP	9932.40	9924.84	1452.12	7511.37	12,787.42	0.881
		POST-OP	10,004.80	9805.97	1629.31	7510.83	12,935.08	
IG	Elastimeter	PRE-OP	67.48	68.21	13.32	38.14	89.71	0.079
		POST-OP	74.85	67.71	22.50	34.86	114.14	
CG	Elastimeter	PRE-OP	61.45	57.14	12.71	45.29	86.00	0.012 *
		POST-OP	71.59	60.92	25.59	39.86	144.43	

IG: Intervention group. CG: Control group; VAS = Visual Analog Scale for pain; LEFS = Lower Extremity Functional Scale; PRE-OP: preoperative period. POST-OP: postoperative period. SD = Standard Deviation; P25–P75 = 25th and 75th Percentiles. Mann–Whitney test used for between-group comparison. * Significance level $p < 0.05$.

4. Discussion

This study aimed to evaluate the effect of taping on the recovery of lower limbs following unilateral saphenectomy. Studies have indicated that taping may assist in the reorganization of tissue pressures and improvement of lymphatic circulation, thereby supporting the management of postoperative complications [8,22–25]. However, its specific effects in the postoperative period of saphenectomy are still not well established.

No significant differences were observed between the groups undergoing unilateral saphenectomy in terms of edema control. This finding is likely due to the fact that both groups received standard compression therapy, which is widely recognized for its efficacy in managing postoperative edema [26,27]. Unlike compression stockings, taping primarily exerts elastic traction forces rather than circumferential pressure or direct compression.

The absence of significant differences between groups regarding edema control with the use of elastic adhesive bandaging was an expected finding, given the well-established evidence supporting the superiority of conventional compression therapy. Compression garments exert continuous circumferential pressure, which facilitates venous return, reduces capillary filtration, and promotes lymphatic drainage, mechanisms widely documented in the literature as effective in managing postoperative edema [28]. A clinical trial [29] comparing elastic adhesive bandaging with compression stockings in patients with chronic venous disease demonstrated that compression was significantly more effective in reducing edema, whereas taping showed more modest effects. Furthermore, a systematic review on the use of taping for edema control found that studies reporting positive outcomes are few, methodologically weak, and present notable limitations.

Although elastic adhesive bandaging is often perceived as a supportive intervention for edema control due to its elastic traction, its mechanism of action differs substantially from that of compression stockings [11]. Unlike the latter, bandaging does not provide uniform circumferential pressure and instead relies on localized tension, which may be insufficient to counteract the hemodynamic forces driving edema formation, particularly in the acute postoperative phase

The analysis of instantaneous skin elasticity, measured by the Elastimeter, also did not reveal significant differences between the intervention group (IG) and the control group (CG), either in the operated or non-operated limbs, across the pre- and postoperative periods. However, although statistical results did not reach the conventional significance threshold ($p < 0.05$), the p-value (0.052) suggests a trend toward significance that warrants further investigation. A significant difference was observed after seven days ($p = 0.012$).

The relatively small sample size (20 participants per group) may have limited the ability to detect moderate effects, as suggested by near-significant p-values like 0.052 for skin elasticity. This raises the possibility of a type II error, underscoring the need for future studies with larger cohorts to confirm the clinical benefits of taping after saphenectomy.

The presence of edema seven days after saphenectomy remains a common and clinically relevant occurrence. Even after the initial recovery period, interstitial fluid accumulation may persist due to lymphatic injury, transient venous dysfunction, and local inflammatory responses triggered by surgery [30]. This edema, even when subclinical, can alter tissue biomechanics by increasing interstitial volume and compromising the integrity of the extracellular matrix (ECM), thereby reducing its density and elasticity [31].

It is also important to consider additional hypotheses, such as the influence of individual factors like pre-existing vascular conditions. Chronic venous insufficiency is one of the leading causes of lower limb edema, resulting from increased capillary hydrostatic pressure and vascular permeability, which facilitate fluid buildup in the interstitial space. The severity of edema may vary according to individual clinical profiles, with venous hypertension being a key determinant [32,33].

In the present study, a significant reduction in ecchymosis was observed seven days after surgery, consistent with previous findings in other surgical contexts that reported positive outcomes related to ecchymosis modulation through taping [34]. These findings reinforce the potential of taping in the reabsorption of blood extravasation, even though its effects appear more peripheral than central in the affected area, corroborating the clinical trial results.

Although this study did not show statistically significant results regarding pain reduction, Aguillar-Ferrandiz et al. [14] suggested that taping may exert a placebo effect on pain. Therefore, taping is considered a promising therapeutic strategy in postoperative rehabilitation, particularly for pain relief and improved functionality both observed in the current study.

The improvements in LEFS scores and subcutaneous hemorrhage suggest clinically meaningful effects of taping. The LEFS changes approach the established MCID of nine points [20], and prior studies support taping's role in reducing postoperative edema and ecchymosis [8,35].

Another study also reported reductions in pain and increases in knee flexion range in patients undergoing total knee arthroplasty, although the evidence quality remains low [36]. A similar effect was seen in post-thoracotomy patients [37], where taping significantly reduced pain, improved respiratory function, and decreased analgesic consumption compared to the control group. It was also effective in reducing pain and enhancing physical tolerance in postoperative laparoscopic abdominal surgery, further supporting its role as an adjunct in functional recovery. The timing of the postoperative period may also influence the outcomes of this study. Significant improvements in ecchymosis, edema, and pain were observed in liposuction patients at 7, 14, and 21 days postoperatively [8].

The findings of this study are limited to the acute phase of the inflammatory and edematous process, which prevents assessment of tissue response progression in later repair stages, suggesting the need for further research.

Evaluating therapeutic effects within the first seven days after saphenectomy carries strong clinical relevance, as this period marks the peak of the acute inflammatory response and postoperative edema formation. During this window, hemodynamic changes, increased vascular permeability, and interstitial fluid accumulation can directly affect limb mobility and raise the risk of secondary complications such as wound dehiscence and infection. Investigating interventions capable of modulating this process early on is therefore strategically valuable not only to accelerate functional recovery but also to reduce postoperative care costs. Further studies are needed to evaluate the progression of effects during later stages of tissue repair, as well as to investigate how comorbidities and age-related physiological differences may influence the effectiveness of taping in postoperative recovery.

A limitation of this study was the inability to blind participants and therapists due to the nature of the intervention, which may have affected subjective outcomes such as pain. To minimize this bias, we employed concealed allocation through randomization and standardized procedures, along with validated instruments for outcome assessment.

5. Conclusions

The study demonstrated significant effects of taping following saphenectomy, including reduced ecchymosis and improved skin elasticity and lower limb functionality. However, no statistically significant result was observed in relation to edema control.

Author Contributions: Conceptualization, E.C.d.O.G. and R.M.G.; methodology, E.C.d.O.G. and R.M.G., software, R.M.G.; validation, E.C.d.O.G. and R.M.G.; formal analysis, R.M.G. and C.C.Z.; investigation, R.M.G.; resources, E.C.d.O.G.; data curation, R.M.G. and C.C.Z.; writing—original draft preparation, R.M.G.; writing—review and editing, R.M.G., E.C.d.O.G. and A.N.; visualization, R.M.G., E.C.d.O.G., E.H.S.Z., C.C.Z. and A.N.; supervision, E.C.d.O.G. All authors have read and agreed to the published version of the manuscript.

Funding: This study was financed in part by the Coordination for the Improvement of Higher Education Personnel (CAPES), Brazil–Finance Code 001.

Institutional Review Board Statement: The study was conducted in accordance with the Declaration of Helsinki, and the protocol was approved by the Ethics Committee Research Ethics Committee of the University Hospital, Ribeirão Preto Medical School–University of São Paulo (CAAE: 59928122.2.0000.5440) on 22 August 2022.

Informed Consent Statement: Written informed consent has been obtained from the patient(s) to publish this paper.

Data Availability Statement: The original contributions presented in this study are included in the article. Further inquiries can be directed to the corresponding author.

Conflicts of Interest: The authors declare no conflicts of interest. Furthermore, this study has not been submitted to any other journal.

References

1. Mik, S.M.L.; Stubenrouch, F.E.; Dink, A.; Legemate, D.A.; Balm, R.; Ubbink, D.T. Treatment of varicose veins, international consensus on which major complications to discuss with the patient: A Delphi study. *Phlebology* **2019**, *34*, 201–207. [CrossRef]
2. Da Silva, E.M.; Hochman, B.; Ferreira, L.M. Perioperative corticosteroids for preventing complications following facial plastic surgery. *Cochrane Database Syst. Rev.* **2014**, *6*, CD009697. [CrossRef] [PubMed]
3. Mariani, F.; Marone, E.M.; Gasbarro, G.; Bucalossi, M.; Spelta, S.; Amsler, F.; Agnati, M.; Chiesa, R. Multicenter randomized trial comparing compression with elastic stocking versus bandage after surgery for varicose veins. *J. Vasc. Surg.* **2011**, *53*, 115–122. [CrossRef] [PubMed]
4. Shim, J.Y.; Lee, H.R.; Lee, D.C. The use of elastic adhesive tape to promote lymphatic flow in the rabbit hind leg. *Yonsei Med. J.* **2003**, *44*, 1045–1052. [CrossRef] [PubMed]
5. Ristow, O.; Pautke, C.; Kehl, V.; Koerdt, S.; Hahnefeld, L.; Hohlweg-Majert, B. Kinesiologic taping reduces morbidity after oral and maxillofacial surgery: A pooled analysis. *Physiother. Theory Pract.* **2014**, *30*, 390–398. [CrossRef]
6. Beidler, S.K.; Douillet, C.D.; Berndt, D.F.; Keagy, B.A.; Rich, P.B.; Marston, W.A. Inflammatory cytokine levels in chronic venous insufficiency ulcer tissue before and after compression therapy. *J. Vasc. Surg.* **2009**, *49*, 1013–1020. [CrossRef]
7. Frangogiannis, N.G. The inflammatory response in myocardial injury, repair, and remodeling. *Nat. Rev. Cardiol.* **2017**, *14*, 254–272. [CrossRef]
8. Zingaretti, N.; Albanese, R.; Pisano, G.; Isola, M.; Giusti, A.; De Martino, M.; De Francesco, F.; Riccio, M.; Parodi, P.C. Evaluation of Kinesio Taping for Edema, Ecchymosis, and Pain After Liposuction: A Prospective Pilot Study. *Aesthet. Surg. J.* **2023**, *43*, NP787–NP796. [CrossRef]
9. Jarón, A.; Jedliński, M.; Grzywacz, E.; Mazur, M.; Trybek, G. Kinesiology Taping as an Innovative Measure against Post-Operative Complications after Third Molar Extraction—Systematic Review. *J. Clin. Med.* **2020**, *9*, 3988. [CrossRef]
10. Tatli, U.; Benlidayi, I.C.; Salimov, F.; Guzel, R. Effectiveness of kinesio taping on postoperative morbidity after impacted mandibular third molar surgery: A prospective, randomized, placebo-controlled clinical study. *J. Appl. Oral Sci.* **2020**, *28*, e20200159. [CrossRef]
11. Hörmann, J.; Vach, W.; Jakob, M.; Seghers, S.; Saxer, F. Kinesiotaping for postoperative oedema—What is the evidence? A systematic review. *BMC Sports Sci. Med. Rehabil.* **2020**, *12*, 14. [CrossRef]
12. Pop, T.B.; Karczmarek-Borowska, B.; Tymczak, M.; Hałas, I.; Banaś, J. The influence of Kinesiology Taping on the reduction of lymphoedema among women after mastectomy-preliminary study. *Contemp. Oncol.* **2014**, *18*, 124–129. [CrossRef] [PubMed]
13. Torres-Lacomba, M.; Navarro-Brazález, B.; Prieto-Gómez, V.; Ferrandez, J.C.; Bouchet, J.Y.; Romay-Barrero, H. Effectiveness of four types of bandages and kinesio-tape for treating breast-cancer-related lymphoedema: A randomized, single-blind, clinical trial. *Clin. Rehabil.* **2020**, *34*, 1230–1241. [CrossRef] [PubMed]
14. Aguilar-Ferrándiz, M.E.; Castro-Sánchez, A.M.; Matarán-Peñarrocha, G.A.; Guisado-Barrilao, R.; García-Ríos, M.C.; Moreno-Lorenzo, C.A. A randomized controlled trial of a mixed Kinesio taping–compression technique on venous symptoms, pain,

peripheral venous flow, clinical severity and overall health status in postmenopausal women with chronic venous insufficiency. *Clin. Rehabil.* **2014**, *28*, 69–81. [CrossRef] [PubMed]
15. Miettinen, M.; Mönkkönen, J.; Lahtinen, M.R.; Nuutinen, J.; Lahtinen, T. Measurement of oedema in irritant-exposed skin by a dielectric technique. *Skin Res. Technol.* **2006**, *12*, 235–240.
16. Alanen, E.; Nuutinen, J.; Nicklén, K.; Lahtinen, T.; Mönkkönen, J. Measurement of hydration in the stratum corneum with the MoistureMeter and comparison with the Corneometer. *Skin Res. Technol.* **2004**, *10*, 32–37. [CrossRef]
17. Bertelli, D.F.; de Oliveira, P.; Gimenes, A.S.; Moreno, M.A. Postural drainage and manual lymphatic drainage for lower limb edema in women with morbid obesity after bariatric surgery: A randomized controlled trial. *Am. J. Phys. Med. Rehabil.* **2013**, *92*, 697–703. [CrossRef]
18. Brorson, H.; Svensson, B.; Ohlin, K. Volume measurements and follow-up. In *Lymphedema: Presentation, Diagnosis, and Treatment*; Greene, A.K., Slavin, S.A., Brorson, H., Eds.; Springer: Cham, Switzerland, 2015; pp. 115–122. [CrossRef]
19. Heller, G.; Manuguerra, M.; Chow, R. How to analyze the Visual Analogue Scale: Myths, truths and clinical relevance. *Scand. J. Pain* **2016**, *13*, 67–75. [CrossRef]
20. Binkley, J.M.; Stratford, P.W.; Lott, S.A.; Riddle, D.L. The Lower Extremity Functional Scale (LEFS): Scale development, measurement properties, and clinical application. *Phys. Ther.* **1999**, *79*, 371–383. [CrossRef]
21. Pereira, L.M.; Dias, J.M.; Mazuquin, B.F.; Castanhas, L.G.; Menacho, M.O.; Cardoso, J.R. Translation, cross-cultural adaptation and analysis of the psychometric properties of the lower extremity functional scale (LEFS): LEFS-BRAZIL. *Braz. J. Phys. Ther.* **2013**, *17*, 272–280. [CrossRef]
22. Rezende, V.J.; Carvalho, L.C.; Yanagihara, G.; Rocha, C.B.J.; Maia, P.R.; Marino, L.d.S.; Iunes, D.H. Effect of kinesio-taping on the acute phase of the post-operative reconstruction of the anterior cruciate ligament: A randomized controlled trial. *J. Bodyw. Mov. Ther.* **2023**, *35*, 320–325. [CrossRef]
23. Labianca, L.; Andreozzi, V.; Princi, G.; Princi, A.A.; Calderaro, C.; Guzzini, M.; Ferretti, A. The effectiveness of Kinesio Taping in improving pain and edema during early rehabilitation after Anterior Cruciate Ligament Reconstruction: A prospective, randomized, controlled study. *Acta Biomed.* **2022**, *92*, e2021336. [CrossRef]
24. Cakmak, M.; Cigdem-Karacay, B. The effect of kinesio taping on edema, pain, and functionality after total knee arthroplasty: A randomised sham-controlled double blinded clinical study. *J. Orthop. Sci.* **2023**, *29*, 983–989. [CrossRef] [PubMed]
25. Sobiech, M.; Czępińska, A.; Zieliński, G.; Zawadka, M.; Gawda, P. Does application of lymphatic drainage with kinesiology taping have any effect on the extent of edema and range of motion in early postoperative recovery following primary endoprosthetics of the knee joint? *J. Clin. Med.* **2022**, *11*, 3456. [CrossRef] [PubMed]
26. Belczak, C.E.Q.; Godoy, J.M.P.; Ramos, R.N.; Belczak, S.Q.; Caffaro, R.A. Is the wearing of elastic stockings for half a day as effective as wearing them for the entire day? *Phlebology* **2009**, *24*, 21–25. [CrossRef]
27. Gloviczki, P.; Comerota, A.J.; Dalsing, M.C.; Eklof, B.G.; Gillespie, D.L.; Gloviczki, M.L.; Lohr, J.M.; McLafferty, R.B.; Meissner, M.H.; Murad, M.H.; et al. The care of patients with varicose veins and associated chronic venous diseases: Clinical practice guidelines of the Society for Vascular Surgery and the American Venous Forum. *J. Vasc. Surg.* **2011**, *53*, 2S–48S. [CrossRef]
28. De Moraes Silva, M.A.; Nelson, A.; Bell-Syer, S.E.; Jesus-Silva, S.G.; Miranda, F., Jr. Compression for preventing recurrence of venous ulcers. *Cochrane Database Syst. Rev.* **2024**, *3*, CD002303. [CrossRef]
29. Baha, N.; Ozyilmaz, S.; Aygutalp, N.; Demir, R.; Baltaci, G.; Yigit, Z. Effects of Kinesio Taping and compression stockings on pain, edema, functional capacity and quality of life in patients with chronic venous disease: A randomized controlled trial. *Clin. Rehabil.* **2020**, *34*, 783–793. [CrossRef]
30. Terada, Y.; Fukuda, S.; Tohda, E.; Kigawa, I.; Wanibuchi, Y.; Mitsui, T. Venous function and delayed leg swelling following saphenectomy in coronary artery bypass grafting. *Jpn. J. Thorac. Cardiovasc. Surg.* **1999**, *47*, 559–562. [CrossRef]
31. Subramani, R.; Izquierdo-Alvarez, A.; Bhattacharya, P.; Meerts, M.; Moldenaers, P.; Ramon, H.; Van Oosterwyck, H. The influence of swelling on elastic properties of polyacrylamide hydrogels. *Front. Mater.* **2020**, *7*, 212. [CrossRef]
32. Santa Cruz, M.; Singh, N.; Poltiyelova, E.; Marks, N.; Ascher, E.; Hingorani, A. Seasonal variation in swelling of lower extremity edema in patients with chronic venous insufficiency. *J. Vasc. Surg. Venous Lymphat. Disord.* **2023**, *11*, 498–501. [CrossRef]
33. Ortega, M.A.; Fraile-Martínez, O.; García-Montero, C.; Álvarez-Mon, M.A.; Chaowen, C.; Ruiz-Grande, F.; Pekarek, L.; Monserrat, J.; Asúnsolo, A.; García-Honduvilla, N.; et al. Understanding Chronic Venous Disease: A Critical Overview of Its Pathophysiology and Medical Management. *J. Clin. Med.* **2021**, *10*, 3239. [CrossRef]
34. Vercelli, S.; Colombo, C.; Tolosa, F.; Moriondo, A.; Bravini, E.; Ferriero, G.; Francesco, S. The effects of kinesio taping on the color intensity of superficial skin hematomas: A pilot study. *Phys. Ther. Sport* **2017**, *25*, 24–29. [CrossRef]
35. Bruno, A.; Vigliano, T.; Foti, R. The role of taping in early recovery after abdominoplasty: A comprehensive review and proposal of a standardized protocol. *Aesthetic Plast. Surg.* **2025**, *49*, 112–120. [CrossRef]

36. Ickert, E.C.; Griswold, D.; Ross, O.; Dudash, S.; Duchon, C.; Learman, K. Effects of kinesiotaping during early post-operative rehabilitation in individuals who underwent a total knee arthroplasty: A systematic review and meta-analysis of randomized control trials. *Clin. Rehabil.* **2024**, *38*, 345–356. [CrossRef]
37. Aksu, N.T.; Erdoğan, M.; Erdoğan, A. The effect of kinesio taping on pain, respiratory function, and muscle strength after thoracotomy. *Turk. Gogus Kalp Damar Cerrahisi Derg.* **2023**, *31*, 507–516. [CrossRef] [PubMed]

applied sciences

MDPI

Article

Perspectives of Physiotherapists on Immune Functioning in Oncological Rehabilitation in the Netherlands: Insights from a Qualitative Study

Anne M. S. de Hoop [1,2,3,*], Karin Jäger [1], Jaap J. Dronkers [1], Cindy Veenhof [1,3], Jelle P. Ruurda [4], Cyrille A. M. Krul [2], Raymond H. H. Pieters [2,5] and Karin Valkenet [1,3]

1 Research Group Innovation of Human Movement Care, Research Center for Healthy and Sustainable Living, HU University of Applied Sciences Utrecht, 3584 CE Utrecht, The Netherlands; jaap@dronkers.nu (J.J.D.); cindy.veenhof@hu.nl (C.V.); k.valkenet@umcutrecht.nl (K.V.)
2 Research Group Innovative Testing in Life Sciences and Chemistry, Research Center for Healthy and Sustainable Living, HU University of Applied Sciences Utrecht, 3584 CE Utrecht, The Netherlands; cyrille.krul@hu.nl (C.A.M.K.); raymond.pieters@hu.nl (R.H.H.P.)
3 Department of Rehabilitation, Physiotherapy Science & Sport, UMC Utrecht Brain Center, University Medical Center Utrecht, Utrecht University, 3584 CX Utrecht, The Netherlands
4 Department of Surgery, University Medical Center Utrecht, University of Utrecht, 3584 CX Utrecht, The Netherlands; j.p.ruurda@umcutrecht.nl
5 IRAS-Toxicology, Population Health Sciences, Faculty of Veterinary Sciences, Utrecht University, 3584 CS Utrecht, The Netherlands
* Correspondence: anne.dehoop@hu.nl

Abstract: Oncology physiotherapists frequently provide care for patients experiencing severe immunosuppression. Exercise immunology, the science that studies the effects of exercise on the immune system, is a rapidly evolving field with direct relevance to oncology physiotherapists. Understanding oncology physiotherapists' perspectives on the subject of immune functioning is essential to explore its possible integration into clinical reasoning. This study aimed to assess the perspectives of oncology physiotherapists concerning immune functioning in oncology physiotherapy. For this qualitative research, semi-structured interviews were performed with Dutch oncology physiotherapists. Results were analyzed via inductive thematic analysis, followed by a validation step with participants. Fifteen interviews were performed. Participants' ages ranged from 30 to 63 years. Emerging themes were (1) the construct 'immune functioning' (definition, and associations with this construct in oncology physiotherapy), (2) characteristics related to decreased immune functioning (in oncology physiotherapy), (3) negative and positive influences on immune functioning (in oncology physiotherapy), (4) tailored physiotherapy treatment, (5) treatment outcomes in oncology physiotherapy, (6) the oncology physiotherapist within cancer care, and (7) measurement and interpretation of immune functioning. In conclusion, oncology physiotherapists play an important role in the personalized and comprehensive care of patients with cancer. They are eager to learn more about immune functioning with the goal of better informing patients about the health effects of exercise and to tailor their training better. Future exercise-immunology research should clarify the effects of different exercise modalities on immune functioning, and how physiotherapists could evaluate these effects.

Keywords: physiotherapy; cancer; exercise-immunology; clinical reasoning

1. Introduction

Exercise immunology, the science that studies the effects of exercise on the immune system, is an emerging field that is gaining increasing attention in the field of oncology physiotherapy [1,2]. Historically, physiotherapists working with patients with cancer have primarily focused on physical functioning. Over time, the role of the physiotherapist has broadened towards a movement specialist who also has a role in improving general health. Research has shown that exercise can have significant immunological effects, which can be an underlying mechanism for many health outcomes, although this area of research remains relatively underexplored [2–4].

The immunological effects of exercise are dependent on the type and workload of the training [2]. This implies that physiotherapists can differentially influence immune functioning (IF) by adjusting the training's frequency, intensity, timing, and type (also known as FITT parameters). This is particularly relevant for physiotherapists working in oncological rehabilitation, i.e., the multidisciplinary approach aimed at helping individuals with cancer to manage and recover from the impairments caused by the disease and its treatments [5]. Since cancer treatments often have immunosuppressive effects, the IF of these patients can be more vulnerable. This means that oncology physiotherapists face the challenge of finding the right balance regarding physically challenging patients without overreaching. Knowledge regarding the optimal FITT parameters to support IF would therefore contribute to optimization of physiotherapy treatment.

With a growing field of research regarding exercise immunology in oncology, the question arises if and how this knowledge can be incorporated into the oncology physiotherapy practice. Current guidelines do not provide guidance on this yet [6,7]. This lack of direction presents clinical challenges, such as uncertainty in how to tailor exercise interventions to support immune function. To integrate this subject into the clinical reasoning of oncology physiotherapists, it is essential to understand the current role of IF in oncology physiotherapy. Additionally, physiotherapists' thoughts on this subject are crucial. Here, IF is defined by the authors as the dynamic activity of the immune system, encompassing both innate and adaptive mechanisms. It consists of a complex network of cells, tissues and organs that protect the body from harmful internal and external influences, with the aim of maintaining homeostasis and promote recovery [8,9]. This study aimed to assess the perspectives of oncology physiotherapists on IF in oncology physiotherapy.

2. Materials and Methods

2.1. Design and Setting

A generic qualitative research design was adopted to explore participants' experiences, opinions, and ideas [10,11]. We did not aim to generate theory (as in grounded theory), nor to explore lived experience in the phenomenological sense, but rather to identify patterns of meaning across participants' answers. The generic qualitative design offers a flexible and pragmatic approach, which was suitable for our exploratory study on the perspectives of oncology physiotherapists.

This study adheres to the Standards for Reporting Qualitative Research (SRQR) [12]. Semi-structured interviews were performed with primary- and secondary care oncology physiotherapists in the Netherlands. Interviews took place in the work- or home-setting of the physiotherapist, or via an online video meeting, between September 2021 and March 2022.

2.2. Participants

Physiotherapists were recruited via purposeful sampling to ensure heterogeneity in age, sex, and years of working experience. Inclusion criteria were (1) registered as oncology physiotherapist in the Dutch quality register, (2) treating patients with cancer or cancer survivors on a monthly basis, (3) able to understand and speak the Dutch language.

An invitation for participation was spread within the professional network of the researchers via social media channels, e-mail, and phone. Also, oncology physiotherapists were invited via participants in the study (i.e., the snowball method). Eligible persons who were interested in participating received an information letter about the study at least one week before the interview took place. Before the onset of the interview, participants signed informed consent. The ethical committee of the University of Applied Sciences Utrecht approved this study protocol and waived further medical ethical review and approval because The Medical Research Involving Humans Act (Dutch: WMO) did not apply to this study (reference number 151-000-2021).

2.3. Data Collection

A topic list was developed by the research team (Supplementary Materials). Prior to the first interview, the topic list, interview strategies and coding strategies were discussed with an independent external researcher with expertise on qualitative research (peer debriefing).

Semi-structured interviews were conducted by AH (MSc, researcher, female) and KJ (MSc, oncology physiotherapist, female). AH and KJ had educational training for qualitative research, but no prior experience. Therefore, one pilot interview was performed by KJ, which was not included in the analyses. The first five interviews were conducted with an observer who was allowed to add questions, and discussed directly afterwards, to improve interview techniques and reflexivity [13].

Before and during the interviews, as minimal information regarding current knowledge on exercise immunology as possible was given, to prevent biasing the outcomes. An interview started with a mutual introduction and explanation of the goal of the interview. To avoid social desirable answering, it was emphasized that the objective was to explore a wide range of ideas and viewpoints, and there were no right or wrong answers. Initial questions focused on the work setting and work experience of the participant. Next, participants were asked to share their associations with IF. The rest of the interview included questions such as the examples below, which were further elaborated on via in-depth questions:

- 'Which parameters do you take into account when tailoring training?'
- 'Does IF of the patient play a role in your clinical reasoning, and if so, how?'

2.4. Analysis

Face-to-face interviews were audio-recorded and online meetings were screen + audio recorded. The recordings were verbatim transcribed manually by KJ and AH. Texts were analyzed via inductive thematic analysis, involving open coding, axial coding, and selective coding [14]. Coding was performed in Atlas.ti (version 23.0.6) [15].

The first three interviews were coded independently by the researchers KJ and AH. Next, codes were compared and discussed until consensus was reached. All other interviews were coded by either KJ or AH. In an iterative process parallel to data collection, codes were discussed and grouped (constant comparison) [16]. Codes of one interview were reviewed by KV and RP to reach final consensus on the coding and to check for personal biases in interpretation of the raw data. During two final sessions, themes were identified based on the code groups. This process involved exploring the meaning and context of codes within each group, by examining how codes were used across interviews, and identifying relationships and patterns between codes. The naming and definition of themes was based on a discussion to understand the underlying concepts represented by each group of codes.

Data collection stopped when theoretical saturation was reached, which was defined as little to no new information (i.e., codes, themes, and links between codes) emerging from newly collected data [17]. A summary of the outcomes was written, including anonymized quotes from the interviews. This summary was sent per e-mail to all participants for a member check; the participants were asked whether they recognized their own thoughts, and/or they wanted to add or change anything to the summary. This input was incorporated in the final analysis.

3. Results

In total, 25 oncology physiotherapists were approached to participate. Seven were unable to participate because they lacked time and three did not respond. Eventually, fifteen interviews were conducted. After thirteen interviews, the researchers observed that no new codes were emerging anymore, and the data began to show strong consistency across participants. To ensure theoretical saturation, two additional interviews were conducted, which aligned with earlier findings and did not yield new insights. Based on this, it was concluded that theoretical saturation was reached. Eleven participants responded to the member check.

Demographic characteristics of the participants are shown in Table 1. The age of the physiotherapists ranged between 30 and 63 years, and the working experience of the oncology physiotherapists ranged between one and fourteen years. Twelve participants were female and three were male, which is a good reflection of the Dutch oncology physiotherapy. The majority (11/15) worked in primary care, the others (4/15) in secondary care in a hospital. The duration of the interviews varied between 35 and 60 min. The results are described below per theme, with exemplary quotes including the participant's number in squared brackets. Figure 1 provides an overview of the themes, and examples of codes.

Table 1. Demographic characteristics of the participants.

Participant ID	Age (Years)	Sex	Years of Experience as Oncology Physiotherapist	Years of Experience as General Physiotherapist	Setting
1	44	F	12	21	1
2	38	F	14	15	1
3	63	F	7	42	1
4	50	F	8	32	1
5	63	M	7	41	1
6	37	F	11	15	1
7	36	F	10	15	1
8	33	F	4	10	1
9	35	F	10	12	2
10	36	M	7	10	2
11	32	M	4	10	1
12	30	F	5	7	2
13	51	F	13	26	2
14	36	F	6	15	1
15	47	F	1	25	1

Note. F: female, M: male, Setting 1: primary care, Setting 2: secondary care.

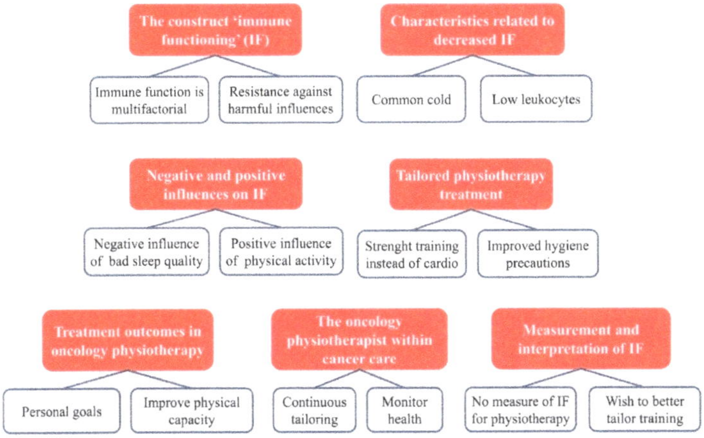

Figure 1. Overview of the identified themes and exemplary codes. Note. Codes are exemplary and were selected because they occurred frequently and meaningfully across the dataset.

3.1. The Construct 'Immune Functioning'

Participants associated IF with resistance against harmful influences on the human body. Most of the participants described external influences such as pathogens, and some also described harmful internal processes such as the development of cancer. Participants also associated the immune system with loading capacity of the patient and the ability to recover.

- "I see the immune system as the system that makes the body resistant against external influences that may negatively influence health" [1]
- "...the immune system actually is one of the most important links in the whole [treatment] trajectory. Related to recovery, but also to tumor activity." [8]

IF and the immune system were described as multifactorial and dependent on many different lifestyle factors. Participants put emphasis on different lifestyle factors.

- "I believe that the immune system is complex, and not to be described by one factor. I think that many factors play a role, including nutrition, sleep and physical activity." [9]

Two relationships were mentioned in the context of cancer and cancer treatment. First, the development of cancer was related to the inability of the immune system to eliminate cancer cells. Second, cancer treatment severely challenges the immune system, and during cancer treatment the immune system is often suppressed.

3.2. Characteristics Related to Decreased IF

Participants linked decreased IF to developing diseases. Symptoms indicating this were, amongst others, general malaise, fever, flu, and common cold symptoms. Participants described slower recovery, more fatigue, decreased loading capacity, and decreased physical fitness as characteristics of patients with decreased IF.

- "... and also, when people start to have a harder time performing their prescribed exercises, or harder than usual" [14]

3.3. Negative and Positive Influences on IF

Many factors were mentioned for having a negative or positive influence on IF. These included nutrition, sleep, stress, medical treatment, (physical) overload and underuse. Participants each put emphasis on different factors, and whether or not they incorporated this into their treatment plan.

- "We often discuss the patients' sleep quality. Bad sleep quality also weakens the immune system." [6]
- "When the training intensity is too high, for example what you see in elite sports, then there is a chance that your immune functioning decreases." [6]

Most participants believed that physical activity positively influences IF, and many emphasized that they could not explain the mechanism behind this. Several participants said that the right intensity of physical activity can positively influence the immune system while physical overload may have a negative impact.

- "Studies have shown that physical activity can improve your immune system, so physical activity is important, but how should you be physically active? Which advices should we give our patients?" [3]

Several participants mentioned that physical activity can influence cancer development, and recovery from cancer treatment. For example, physical training could positively influence negative effects of treatment.

- "<immunosuppressive effect of treatments> are often an inevitable consequence of treatment, but we can influence it. We cannot prevent it, but physical activity can positively influence these consequences." (....) [4]
- "... The better the physical fitness before surgery, the better the chances that you have some reserve capacity to improve your recovery. I think the immune system is associated with this." [9]

3.4. Tailored Physiotherapy Treatment

All participants emphasized the large heterogeneity between patients with cancer, and the need to continuously tailor training parameters to the individual. Patients should be tightly monitored and adaptations in the training schedule are often necessary because of changing circumstances.

Various symptoms were mentioned as reason to adapt the training parameters or training setting. Some symptoms were a reason to refrain from a training session: fever, diarrhea or headache in case of a risk of brain metastasis. General malaise and fatigue could be a reason to 'take it easier', which generally meant decreasing intensity of training or choosing another type of training such as strength training instead of endurance training. All of just mentioned symptoms were specifically associated with reduced IF by several participants.

- "For example, general malaise, slight fever, fatigue that is unrelated to the exertion, those are things that I notice, and tell me that I have to be careful, because the immune system might be vulnerable." [3]

Other symptoms that prompted to downgrade intensity of training or take a small break included pain, dizziness, dyspnea, heavy transpiration, a deviating heart rate, increased resting heart rate, changes in blood pressure or saturation, and prolonged time to recovery after a training session. The participants described their 'gut feeling' as one of the most important reasons to adjust training. Explanations of what was meant by this 'gut feeling' or what determined it, were for example "the way someone enters the room" or "a persons' glance".

Participants associated chemotherapy, immunotherapy, or stem cell transplantation with suppressed IF. In these cases, more hygiene measures were taken during the training to reduce infection risk. Some participants also decreased the intensity of the training, or would choose strength training instead of endurance training. Some physiotherapists would inform themselves about blood markers such as leukocyte numbers during these medical treatments. Low leukocyte numbers were interpreted as suppressed IF and a higher risk of infections, meaning that there was a need for more hygiene measures.

3.5. Treatment Outcomes in Oncology Physiotherapy

Goals of the treatment were dependent on the patients' personal context and wishes. Treatment goals were generally aimed at improving physical functioning in daily life and quality of life. IF was generally not seen as a treatment outcome, meaning that the oncology physiotherapists did not aim to improve IF with their training.

- "<treatment goals> differ between individuals <. . .>. One person may have the goal to be able to babysit its grandchild, a second wants to be able to do on hour of cycling each day and a third is happy when able to take care of himself in and around the house." [14]
- "it is not the first thing that comes to mind, like, I would like to optimize the immune system." [6]

3.6. The Oncology Physiotherapist Within Cancer Care

The oncology physiotherapist was seen as someone who treats patients with complex health problems, who continuously tailors exercise interventions to the individual patients' needs and circumstances. Multidisciplinary collaboration was always seen as desirable for oncology physiotherapists and all participants actively participated in a monodisciplinary or multidisciplinary network for knowledge exchange and collaboration.

- ". . .you have to work together with other primary care specialists, get involved in the networks, then you can treat your patient best." [3]

Many participants emphasized that they had relatively frequent and close contact with the patient. Because of this, monitoring the patients' well-being was seen as an important responsibility, as well as being a consultant for the patient.

- "We are of course also consultants for many issues that they experience. Sometimes even more than their nurse because we see people two times per week." [2]

This was especially the case for primary care physiotherapists. A hospital-based physiotherapist emphasized in the response to the member check that this was not the case in their work setting, since they spent less time with their patients.

3.7. Measurement and Interpretation of IF

Oncology physiotherapists saw potential value of assessing IF in order to better inform their patients about the health benefits of exercise. Furthermore, they wished to further tailor training parameters (type, frequency, intensity, duration, timing) such as knowing the best timing or intensity to train during a chemotherapy treatment. They thought that incorporating knowledge regarding exercise immunology would help them do so.

- "I would like to have some indicators, like, when should you challenge someone with cardio or with strength training, or when is perhaps relaxation therapy the right therapy for that moment [1]"

Participants were generally not familiar with measurement instruments for IF and most lacked knowledge of interpretation of measures such as biomarkers in blood. If a measurement instrument for IF would exist for the physiotherapist, they would need knowledge to interpret the outcome and guidelines to tailor their training based on this outcome. Lastly, a measurement instrument for IF should have additional value for the patient in order to be valuable for oncology physiotherapy.

4. Discussion

This study assessed the perspectives of oncology physiotherapists on exercise immunology in oncology physiotherapy. Seven emerging themes were identified: the construct 'immune functioning', symptoms related to decreased IF, influences on IF, tailored physiotherapy treatment, treatment outcomes in oncology physiotherapy, the oncology physiotherapists within cancer are and measurement and interpretation of IF.

Participants' definitions of and associations with the construct IF generally aligned with current literature [8]. Interestingly, diverse opinions existed regarding the most influential factors affecting IF and the physiotherapist's role in addressing them: While exercise, nutrition, stress, and sleep were commonly linked to IF, variations existed in how these elements were integrated into clinical reasoning. Some participants discussed these factors with patients, actively incorporating them into treatment plans. Notably, these participants had often followed courses or did personal research on exercise immunology. Others did not discuss the topic of IF with their patient since they did not feel confident regarding their knowledge of the topic. This can be explained by the novelty of this subject within physiotherapy. Existing Dutch and international oncology physiotherapy guidelines lack guidance on integrating the immunological effects of exercise [6,7].

Previous research showed that the effects of exercise on the immune system interact with other lifestyle factors [18–20]. This highlights the need for a multidisciplinary approach for treatment, as well as a better fundamental understanding of the interaction of these effects. Interestingly, a recurring theme in this study was collaboration of oncology physiotherapists in multidisciplinary networks and teams including, for example, nurses, dieticians, and psychologists. This aligns with a growing recognition of the importance of a comprehensive approach in healthcare for patients with cancer, addressing not only physical function but also overall well-being and quality of life [21–23].

The findings also underscore the importance of a personalized approach in oncology physiotherapy. The heterogeneity of patients with cancer undergoing cancer treatment necessitates tailored exercise interventions that account for each patient's unique circumstances and needs [7]. This approach reflects the core tenets of person-centered care, which emphasize respect for patients' values, preferences, and active participation in their care [22,24]. Continuous tailoring is crucial due to significant variations in patients' physical conditions between training sessions. Oncology physiotherapists play a pivotal role in monitoring patients, adjusting training parameters (type, frequency, intensity, timing, and duration), and making real-time decisions to optimize patient outcomes. Participants recognized the potential value of incorporating exercise immunology into their clinical reasoning to further enhance decision-making—such as optimizing the timing of training sessions during chemotherapy—thereby reinforcing a dynamic, evidence-informed, and person-centered model of care.

This study showed that oncology physiotherapists tailor their training based on various factors, including patient-reported symptoms such as dizziness and pain, observations such as changes in heart rate or color of the skin, as well as their 'gut feeling'. Some common symptoms such as fever, fatigue, and pain were specifically associated with decreased IF by participants. Importantly, IF was rarely utilized as parameter or outcome on itself. Participants often emphasized that they were not familiar with a measurement instrument for physiotherapists to assess and interpret IF. Future research should therefore explore whether a non-invasive measure of IF could be a facilitator for oncology physiotherapists to incorporate the subject IF in their clinical reasoning.

Study Limitations

A potential limitation of this study is the risk of selection bias. It is likely that people with an interest in the topic of exercise immunology were more inclined to participate in this study. To mitigate this, we used diverse recruitment channels and emphasized in the invitation that prior knowledge or interest in IF was not required. Nonetheless, it is possible that the participants may not be fully representative of all oncology physiotherapists and their perspectives may reflect a more favorable view of the integration of IF into their practice. Also, the sample included only four physiotherapists from the secondary care setting, which may have limited the inclusion of context-specific views relevant to hospital-based oncology physiotherapy. Future research should aim to include a more diverse sample, potentially through a mixed-methods approach—using surveys to reach a wider audience, followed by interviews to explore perspectives in depth. The perspectives on this subject could also be enriched by including a group with a broader spectrum of knowledge and engagement with this topic.

5. Conclusions

In conclusion, this study provides valuable insights into the perspectives of oncology physiotherapists on IF and the potential integration of this subject in their clinical reasoning. The emphasis on personalized and comprehensive care for patients with cancer underscores the crucial role of oncology physiotherapists. On the one hand, IF is currently not a treatment outcome or training parameter, and some participants did not feel confident (yet) to discuss the topic with their patient. On the other hand, this study highlights an interest among physiotherapists in incorporating this subject into their practice. Oncology physiotherapists recognized potential value of incorporating knowledge about exercise immunology to better tailor training and better inform patients about the health benefits of exercise, and they were eager to learn more about the topic. Future research should explore the possibility to measure and interpret IF in oncology physiotherapy. Second, guidelines and educational modules on IF in oncology patients and how to tailor training parameters based on IF should be developed.

Supplementary Materials: The following supporting information can be downloaded at: https://www.mdpi.com/article/10.3390/app15158673/s1, Supplementary data S1 Topiclist.

Author Contributions: Conceptualization, A.M.S.d.H., K.J., J.J.D., C.V., J.P.R., C.A.M.K., R.H.H.P. and K.V.; Data curation, A.M.S.d.H. and K.J.; Formal analysis, A.M.S.d.H. and K.J.; Investigation, A.M.S.d.H., J.J.D., R.H.H.P. and K.V.; Methodology, A.M.S.d.H.; Project administration, J.J.D. and R.H.H.P.; Supervision, J.J.D., C.V., J.P.R., C.A.M.K. and R.H.H.P.; Visualization, A.M.S.d.H.; Writing—original draft, A.M.S.d.H. and K.J.; Writing—review and editing, A.M.S.d.H., K.J., J.J.D., C.V., J.P.R., C.A.M.K., R.H.H.P. and K.V. All authors have read and agreed to the published version of the manuscript.

Funding: This research received no external funding.

Institutional Review Board Statement: Ethical review and approval were waived for this study by the Ethical Committee of the University of Applied Sciences Utrecht, because the Medical Research Involving Humans Act (Dutch: WMO) did not apply to this study.

Informed Consent Statement: Informed consent was obtained from all subjects involved in the study.

Data Availability Statement: Data available on request: a pseudonymized version of the data presented in this study are available on request from the corresponding author. The data are not publicly available due to privacy of the participants.

Conflicts of Interest: The authors declare no conflicts of interest.

Abbreviations

The following abbreviations are used in this manuscript:

IF	Immune functioning

References

1. Nieman, D.C.; Pence, B.D. Exercise immunology: Future directions. *J. Sport Health Sci.* **2020**, *9*, 432–445. [CrossRef]
2. Nieman, D.C.; Wentz, L.M. The compelling link between physical activity and the body's defense system. *J. Sport Health Sci.* **2019**, *8*, 201–217. [CrossRef]
3. Turner, J.E.; Brum, P.C. Does Regular Exercise Counter T Cell Immunosenescence Reducing the Risk of Developing Cancer and Promoting Successful Treatment of Malignancies? *Oxid Med. Cell Longev.* **2017**. *Epub ahead of print.* [CrossRef]
4. Arem, H.; Moore, S.C.; Patel, A.; Hartge, P.; De Gonzalez, A.B.; Visvanathan, K.; Campbell, P.T.; Freedman, M.; Weiderpass, E.; Adami, H.O.; et al. Leisure time physical activity and mortality: A detailed pooled analysis of the dose-response relationship. *JAMA Intern. Med.* **2015**, *175*, 959–967. [CrossRef]
5. Mayer, R.S.; Engle, J. Rehabilitation of Individuals with Cancer. *Ann. Rehabil. Med.* **2022**, *46*, 60–70. [CrossRef]
6. Campbell, K.L.; Winters-Stone, K.M.; Wiskemann, J.; May, A.M.; Schwartz, A.L.; Courneya, K.S.; Zucker, D.S.; Matthews, C.E.; Ligibel, J.A.; Gerber, L.H.; et al. Exercise Guidelines for Cancer Survivors: Consensus Statement from International Multidisciplinary Roundtable. *Med. Sci. Sports Exerc.* **2019**, *51*, 2375–2390. [CrossRef] [PubMed]
7. Sweegers, M.G.; van Doormaal, M.C.M.; Conijn, D.; Stuiver, M.M. *KNGF-Richtlijn Oncologie*; KNGF/VvOCM: Utrecht, The Netherlands, 2022.
8. Institute for Quality and Efficiency in Health Care. How Does the Immune System Work? 2020. Available online: https://www.ncbi.nlm.nih.gov/books/NBK279364/ (accessed on 19 January 2024).
9. Su, B.; Ng, L.G. Immunological modulation in health and disease. *Cell Mol. Immunol.* **2023**, *20*, 981–982. [CrossRef] [PubMed]
10. Merriam, S.B. Basic interpretive qualitative research. In *Qualitative Research in Practice*; Jossey-Bass: San Francisco, CA, USA, 2002; pp. 37–39.
11. Kahlke, R.M. Generic Qualitative Approaches: Pitfalls and Benefits of Methodological Mixology. *Int. J. Qual. Methods* **2014**, *13*, 37–52. [CrossRef]
12. O'Brien, B.C.; Harris, I.B.; Beckman, T.J.; Reed, D.A.; Cook, D.A. Standards for Reporting Qualitative Research. *Acad. Med.* **2014**, *89*, 1245–1251. [CrossRef] [PubMed]
13. Dowling, M. Approaches to reflexivity in qualitative research. *Nurse Res.* **2006**, *13*, 7–21. [CrossRef] [PubMed]
14. Braun, V.; Clarke, V. Using thematic analysis in psychology. *Qual. Res. Psychol.* **2006**, *3*, 77–101. [CrossRef]
15. *ATLAS.ti 22 Windows*; ATLAS: Berlin, Germany, 2022.
16. Dierckx de Casterlé, B.; De Vliegher, K.; Gastmans, C.; Mertens, E. Complex Qualitative Data Analysis: Lessons Learned from the Experiences with the Qualitative Analysis Guide of Leuven. *Qual. Health Res.* **2021**, *31*, 1083–1093. [CrossRef] [PubMed]
17. Rahimi, S.; Khatooni, M. Saturation in qualitative research: An evolutionary concept analysis. *Int. J. Nurs. Stud. Adv.* **2024**, *6*, 100174. [CrossRef] [PubMed]
18. Shao, T.; Verma, H.K.; Pande, B.; Costanzo, V.; Ye, W.; Cai, Y.; Bhaskar, L.V.K.S. Physical Activity and Nutritional Influence on Immune Function: An Important Strategy to Improve Immunity and Health Status. *Front. Physiol.* **2021**, *12*, 751374. [CrossRef] [PubMed]
19. Aleksandrova, K.; Pischon, T.; Jenab, M.; Bueno-De-Mesquita, H.B.; Fedirko, V.; Norat, T.; Romaguera, D.; Knüppel, S.; Boutron-Ruault, M.-C.; Dossus, L.; et al. Combined impact of healthy lifestyle factors on colorectal cancer: A large European cohort study. *BMC Med.* **2014**, *12*, 168. [CrossRef] [PubMed]
20. de Frel, D.L.; Atsma, D.E.; Pijl, H.; Seidell, J.C.; Leenen, P.J.M.; Dik, W.A.; van Rossum, E.F.C. The Impact of Obesity and Lifestyle on the Immune System and Susceptibility to Infections Such as COVID-19. *Front. Nutr.* **2020**, *7*, 597600. [CrossRef] [PubMed]
21. Loiselle, C.G.; Howell, D.; Nicoll, I.; Fitch, M. Toward the development of a comprehensive cancer experience measurement framework. *Support. Care Cancer* **2019**, *27*, 2579–2589. [CrossRef] [PubMed]
22. Stout, N.L.; Brown, J.C.; Schwartz, A.L.; Marshall, T.F.; Campbell, A.M.; Nekhlyudov, L.; Zucker, D.S.; Basen-Engquist, K.M.; Campbell, G.; Meyerhardt, J.; et al. An exercise oncology clinical pathway: Screening and referral for personalized interventions. *Cancer* **2020**, *126*, 2750–2758. [CrossRef] [PubMed]
23. O'Brien, K.; Ried, K.; Binjemain, T.; Sali, A. Integrative Approaches to the Treatment of Cancer. *Cancers* **2022**, *14*, 5933. [CrossRef] [PubMed]
24. McCormack, B.; McCance, T. *Person-Centred Nursing: Theory and Practice*; Wiley: Hoboken, NJ, USA, 2010. [CrossRef]

applied sciences

MDPI

Article

Effectiveness of an Intervention Program for Informal Carers of Children Admitted to a Rehabilitation Centre

Sónia Morais [1], Rui Esteves Pimenta [2,3,4,*], Carminda Morais [2,5,6], Rui Macedo [7], Inês Ribeiro [2] and Pedro Lopes Ferreira [2,3]

1 North Rehabilitation Center, 4405-565 Valadares, Portugal
2 Centre for Health Studies and Research, University of Coimbra, 3004-512 Coimbra, Portugal
3 Centre for Innovative Biomedicine and Biotechnology, 3004-512 Coimbra, Portugal
4 REQUIMTE/LAQV, Health School, Polytechnic of Porto, 4200-072 Porto, Portugal
5 Health School, Polytechnic of Viana do Castelo, 4900-314 Viana do Castelo, Portugal
6 Health Sciences Research Unit: Nursing, UICISA: E, 3046-851 Coimbra, Portugal
7 Rehabilitation Research Center, Health School, Polytechnic of Porto, 4200-072 Porto, Portugal
* Correspondence: rep@ess.ipp.pt

Abstract: Health literacy improves informal caregivers' knowledge and ability to provide care. The aim of this study is to analyze the impact of a group-based intervention on the health literacy of informal caregivers of children with special health needs (SHNs). The intervention focuses on movement, hydrotherapy, walking, and relaxation, with three evaluation stages. Participants included 34 informal caregivers of children with SHNs, recruited at a Physical Medicine and Rehabilitation Service. We collected sociodemographic data of the participants and measured their health literacy through the short-form version of the European Health Literacy Survey Questionnaire (HLS-EU-Q16). Around 70% of the participants were female, with a mean age of 41.06 ± 4.98 years, and nearly 85% were married or in a de facto union. About 26% were unemployed, and more than 65% had completed secondary education or higher. The mean scores of the HLS-EU-Q16 were statistically significantly higher throughout the intervention, with differences over the evaluation stages [$F(2,56) = 75.55$; $p < 0.05$]. A structured, dynamic, and group-based intervention plan showed improvements in the health literacy of the participants, with an increase in the percentage of participants with sufficient and excellent levels of health literacy at the end of the intervention.

Keywords: informal caregivers; health literacy; physical activity

1. Introduction

According to UNICEF [1], children with special health needs (SHNs) are those with health problems whose condition may include: (a) those who were born with a pathology that affects their development; (b) those who have suffered a lifelong injury, infection or nutritional deficiency that contributes to long-term dysfunction; (c) those who have been exposed to environmental toxins that result in developmental delays; and (d) those who have developed anxiety or depression as a result of limiting events. The increasing prevalence of these situations has led to a rise in dependency cases within this population, often requiring the dedicated support of informal caregivers [2].

In Portugal, the Statute of the Informal Caregiver (Law no. 100/2019, 6 September) defines an informal caregiver as "the spouse, de facto partner or relative up to the fourth degree of the person being cared for, who permanently accompanies and cares for him/her, lives in a shared residence, and does not receive any payment for professional activity or the care provided to the person" [3].

The demanding nature of caregiving often leads to significant personal sacrifice. Informal caregivers frequently experience feelings of isolation and neglect their own well-being and basic needs, which can result in considerable mental and physical health problems [4,5]. For these caregivers, enhanced knowledge and health literacy are crucial. Better-informed caregivers are more capable of making conscious and well-considered decisions, ultimately improving the quality of life for the individuals they care for [6]. Indeed, health literacy is a topic of growing importance in public health and healthcare, recognizing its role in empowering individuals to manage their health effectively [7,8].

In 2012, the European Health Literacy Consortium introduced a comprehensive definition of health literacy that states: "Health literacy is linked to literacy and entails people's knowledge, motivation and competences to access, understand, appraise, and apply health information in order to make judgments and take decisions in everyday life concerning healthcare, disease prevention and health promotion to maintain or improve quality of life during the life course" [8]. Individuals with low health literacy often experience more hospitalizations, increased emergency care usage, underutilization of preventive services, difficulty interpreting health information, poorer health, higher mortality, and increased healthcare expenditures [9,10].

Informal caregivers of children with SHN face multiple challenges that require special knowledge to fulfill their role effectively. Providing them with social recognition and appreciation is important to ensure they receive the necessary support for more effective and sustainable care. It is essential to invest in their training and provide professional, legal, and financial assistance [2].

Despite efforts to integrate informal caregivers into support systems and improve the care they provide, significant gaps remain. A study on informal caregivers revealed that 65% cited restrictions on their personal and social lives as a major challenge, followed by health limitations (42%), lack of time (39%), and insufficient personal or technical support (38%). Additionally, 33% reported financial difficulties, and 30% experienced challenges in their relationship with the dependent person. Over 40% identified in-home support, technical aids, and free time as their most pressing needs [11]. These unmet needs underscore the necessity of developing targeted plans and strategies to address them [12,13].

In the literature, most studies focus on caregivers of the adult population, and there is a distinct lack of specific activity programs designed for informal caregivers of children with SHNs. To bridge this gap and align with Portuguese informal caregiver guidelines, a survey was conducted to identify the perceived needs and difficulties within this specific population. This research directly informed the development of the intervention project 'Cuide+ de Si' (Take better care of yourself).

Through this program, it is expected that the informal caregiver will acquire knowledge that promotes the improvement of their health literacy, benefiting both themselves and their dependents. The intervention was designed to help caregivers develop greater awareness of their physical and emotional state. By participating in activities, the connection between physical exercise and the production of neuromediators (such as endorphins, serotonin, and dopamine [14]) promoted their overall well-being [15]. We aimed to encourage intentional adoption of practical self-care strategies (e.g., energy conservation techniques, proper postures) and strengthen their ability to understand, apply, and share health information, given the group dynamic. The intervention consists of a group-based plan with thematic interventions—movement, hydrotherapy, walking, and relaxation. According to Mahindru et al. [15], physical activity has significant positive effects on mental health, including reducing anxiety and depression, boosting self-esteem, and improving mood and general well-being. These factors create more favorable conditions for acquiring new knowledge and developing the cognitive and emotional skills essential for health literacy. Thus, by promoting caregivers' emotional balance and well-being through physical activity and relaxation, the intervention indirectly but significantly enhances their capacity to comprehend and utilize health information more autonomously and consciously.

The primary objective of this study is to assess the impact of the intervention on the health literacy of informal caregivers of children with SHN. Secondary objectives include the characterization of informal caregivers based on their sociodemographic characteristics and examination of the relationship between these characteristics and health literacy levels.

2. Materials and Methods

2.1. Type of Study and Data Source

The study is characterized as quasi-experimental, utilizing a single intervention group to facilitate causal inference. Descriptive analysis was performed on the health literacy levels of the study population, in addition to an exploration of their correlation with sociodemographic data. Data collection occurred in 2024 within the Pediatrics Department of the Physical Medicine and Rehabilitation Service at the North Rehabilitation Center, which is part of the Local Health Unit of Vila Nova de Gaia/Espinho.

2.2. Study Population

The study population consisted of all informal caregivers of children admitted to this unit. It included all informal caregivers of children admitted to the unit who agreed to participate, ensuring everyone had an equal probability of inclusion. The inclusion criteria were: the person cared for was aged 17 years or younger, and participants did not present any physical activity-limiting pathology. The exclusion criteria were: the cared-for person was aged 18 years or older, or the participant had any physical activity-limiting pathology. Informed consent was obtained from all participants.

2.3. Procedures and Intervention Design

The design of the intervention was centered on four key areas:

i. Physical exercise (movement) that promotes changes in body structure and function, preventing various pathologies (e.g., diabetes, cardiovascular diseases), and improving physical health, mental health, and quality of life [16–19];
ii. Hydrotherapy through physical exercise in water induces physiological changes in body systems due to hydrostatic and hydrodynamic pressure. Benefits include improved balance, motor coordination, and increased muscle strength [20,21];
iii. Beyond its sustainable component, walking improves the cardiovascular system and increases flexibility, as noted by Schmidt et al. [21];
iv. Relaxation through less impactful, calmer interventions promotes whole-body work focused on resistance, flexibility, and mental strength [22].

For the intervention's implementation, approval was secured from the Clinical Board of the North Rehabilitation Center and the Ethics Committee of the Local Health Unit of Vila Nova de Gaia/Espinho (approval document no. CES 32/2024-1, dated 4 June 2024).

The questionnaires were administered in three stages: initial (S0), after 6 weeks of the intervention (S1), and at the end of the program (S2), which was determined by the dependent's hospital discharge. At S0, participants also completed a sociodemographic characterization form and informed consent.

The intervention was led by physiotherapists, with more than ten years of experience working with children and their caregivers, and involved bi-weekly sessions organized into thematic cycles: one week of movement (physical exercise), one of hydrotherapy, one of walking, and one of relaxation.

2.4. Measurement Instruments

To achieve the study's objective, a questionnaire was developed to collect sociodemographic data from informal caregivers, including age, sex, employment status, education, and marital status. It also gathered information about the cared-for person's age, sex, and relationship to the caregiver.

The European Health Literacy Survey Questionnaire (HLS-EU) was originally created in 2012 [23]. A short-form version was developed in 2016 [7], the HLS-EU-Q16 [24], with good internal consistency (Cronbach's alpha = 0.89). We selected the HLS-EU-Q16 as the health literacy assessment instrument for this population.

The HLS-EU-Q16 measures health literacy levels based on an individual's self-perception across three domains: healthcare, health prevention, and health promotion. The classification of overall health literacy and the descriptive analysis of its dimensions followed the methodology proposed by Sørensen et al. [23]. These domains are assessed through 16 items [7], each with a response range from 'very difficult' to 'very easy,' indicating the perceived difficulty of performing a specific task. The obtained values are standardized according to Pelikan et al. [25], making the instrument scores range from 0 to 50 with the following levels, for total and for dimensions: inadequate from 0 to 25; problematic from 25.1 to 33; sufficient from 33.1 to 42; and excellent from 42.1 to 50.

2.5. Data Analysis

After collecting the completed questionnaires, data were analyzed using IBM SPSS® Statistics 29, with a significance level of 5% considered.

We performed a quantitative analysis of the responses. For both the sociodemographic data and the questionnaires at each time point (S0, S1, S2), a descriptive analysis was conducted, presenting frequencies, means, maximums, minimums, and standard deviations.

Regarding internal consistency, Cronbach's alpha values between 0.70 and 0.79 were considered acceptable, and values higher than 0.80 were deemed adequate [26,27].

Before conducting any parametric tests, we ensured that the necessary assumptions were met. Normality was assessed using the Kolmogorov–Smirnov test for sample sizes (N) of 30 or greater, and the Shapiro–Wilk test for N less than 30. Homogeneity of population variances was verified using Levene's test.

To compare the three evaluation stages (S0, S1, S2), we used the mean values of the total scores. We initially verified the assumptions of normality and sphericity, then performed an ANOVA for repeated measures. If statistically significant differences were found between stages, a Sidak post hoc test was conducted.

Finally, to compare the results from each scale with sociodemographic data, we used the Pearson correlation test for total scores (S0, S1, and S2) and age. The chi-square test was applied to compare the categorized total scores (S0, S1, and S2) with the education level.

3. Results

3.1. Sample Size

During the collection stage, out of 38 informal caregivers with children hospitalized, 34 agreed to participate in the program. Of the 34 informal caregivers, all completed the first two evaluation stages (S0 and S1), with 29 completing the third stage (S2).

On average, caregivers participated in the project for 14 weeks, with a maximum duration of 29 weeks.

3.2. Sample Characterization

Table 1 presents the sociodemographic data of the study population. Around 70% (N = 24) of the sample were female, aged between 25 and 59 (41.06 ± 4.98) years, and 85.3% were married or in a de facto union. Regarding employment status, 73.5% were employed. All participants had some level of education, with 70.6% having completed secondary education or higher.

Table 1. Sociodemographic data of informal caregivers.

Variables		N (%)
Sex	Female	24 (70.6%)
	Male	10 (29.4%)
Age	Mean ± standard deviation	41.06 ± 4.98
	Min–Max	25–59
Marital status	De facto union	5 (14.7%)
	Married	24 (70.6%)
	Divorced	3 (8.8%)
	Widowed	2 (5.9%)
Employment status	Employed	25 (73.5%)
	Unemployed	9 (26.5%)
Education	Lower Secondary Education	3 (8.8%)
	Upper Secondary Education	7 (20.6%)
	Secondary Education	12 (35.3%)
	Bachelor's Degree	7 (20.6%)
	Postgraduate	2 (5.9%)
	Master's Degree	2 (5.9%)
	Doctoral Degree	1 (2.9%)

3.3. Results of the HLS-EU-Q16

The HLS-EU-Q16 presented a good internal reliability (Cronbach's α = 0.759).

As demonstrated in Table 2, over the three evaluation stages, the scale scores showed an increasing trend: S0 with an average score of 30.36, S1 with a value of 37.19, and S2 with a value of 43.14.

Table 2. HLS-EU-Q16 scores at three evaluation stages.

	Minimum	Maximum	Mean	Standard Deviation
Stage 0	16.67	44.79	30.36	6.93
Stage 1	29.17	50.00	37.19	4.98
Stage 2	32.29	50.00	43.14	4.98

Stage 0: beginning of the intervention; Stage 1: 6 weeks after the beginning of the intervention; Stage 2: end of the intervention.

Analyzing the health literacy levels, presented in Table 3, there was a continuous increase over the three evaluation stages. Stage 0 was the only one where an inadequate level was found, and there were mostly problematic levels of health literacy. Stage 1 predominantly presented 'sufficient' levels, and Stage 2 presented 'excellent' levels in more than half of the sample.

Table 3. Health literacy levels at three evaluation stages.

Health Literacy Levels	S0 N (%)	S1 N (%)	S2 N (%)
Inadequate	7 (20.6%)	-	-
Problematic	15 (44.1%)	7 (20.6%)	1 (3.4%)
Sufficient	10 (29.4%)	23 (67.6%)	11 (37.9%)
Excellent	2 (5.9%)	4 (11.8%)	17(58.6%)

S0: beginning of the intervention; S1: 6 weeks after the beginning of the intervention; S2: end of the intervention.

Regarding the three domains, it was observed that the healthcare domain consistently exhibited the lowest health literacy levels across all stages (problematic at S0 and sufficient at S1 and S2) when compared to the other domains. At Stages 0 and 2, the disease prevention domain showed the highest average levels (sufficient and excellent, respectively), whereas at Stage 1, the health promotion domain had the highest average (sufficient). These results are presented in Table 4.

Table 4. Mean scores of health literacy for each domain, at three evaluation stages.

Health Literacy Domain	S0 Mean (sd)	S1 Mean (sd)	S2 Mean (sd)
Global health literacy	30.36 (6.93)	37.19 (4.98)	43.14 (4.98)
Healthcare	25.77 (9.80)	34.10 (8.51)	40.80 (6.26)
Disease prevention	34.51 (8.16)	38.73 (7.29)	45.06 (5.95)
Health promotion	33.21 (10.23)	40.69 (6.89)	44.83 (6.17)

sd: standard deviation; S0: beginning of the intervention; S1: 6 weeks after the beginning of the intervention; S2: end of the intervention.

A repeated measures ANOVA test was performed, which revealed statistically significant differences in health literacy levels between the evaluation stages [$F(2,56) = 75.55$; $p < 0.05$]. The Sidak post hoc test further indicated statistically significant differences ($p < 0.001$) between all stages. These statistical test results are presented in Table 5.

Table 5. Comparison of HLS-EU-Q16 scores between the three evaluation stages (based on estimated marginal means).

Literacy (I)	Literacy (J)	Mean Difference (I − J)	Standard Deviation	Sig.
S0	S1	−6.36	1.028	<0.001
	S2	−12.21	1.092	<0.001
S1	S0	6.36	1.028	<0.001
	S2	−5.86	0.845	<0.001
S2	S0	12.21	1.092	<0.001
	S1	5.86	0.845	<0.001

S0: beginning of the intervention; S1: 6 weeks after the beginning of the intervention; S2: end of the intervention.

The remaining sociodemographic data did not correlate with the scales, indicating that the characteristics of the present sample did not influence the scores obtained on each scale. Furthermore, program duration also showed no correlation with the scale scores.

4. Discussion

The sociodemographic profile of informal caregivers in this study largely aligns with the existing literature, particularly concerning sex, age, employment status, education level, and their relationship with the dependent child.

This study found that 71% of informal caregivers were women, consistent with the literature indicating that caregiving is a highly gendered activity [28–33]. Despite universal care needs and various gender equality initiatives, the distribution of caregiving tasks remains disproportionate. Society continues to assign and associate the caregiving role primarily with women [28,33]. This disparity acts as a barrier to gender equality and negatively impacts women's employment opportunities, especially at younger ages. Notably, 50% of women in this study remained employed while providing informal care, mirroring findings in both Portuguese and other European populations [28,31–33].

While most reviewed studies report caregiver age in broad ranges (typically between 35 and 64 years [29–33]), the mean age of participants in this study was 41.06 ± 4.98 years, falling in this range. Participants also showed a similar distribution between secondary and higher education levels, consistent with data from the Caring for Informal Caregivers Movement study [32]. We observed that the dependent individuals were minors with an average age of 9 years, predominantly cared for by their parents, an observation consistent with other studies [30,32].

Our intervention was designed to foster greater awareness of physical and emotional well-being among caregivers through their participation in activities. They learn practical self-care strategies (such as energy conservation techniques and proper posture) and strengthen their ability to understand, apply, and share health information within a group dynamic. This hands-on experience directly empowers caregivers to make informed decisions for themselves and in the context of the care they provide.

As Mahindru et al. [15] highlight, physical activity significantly impacts mental health by reducing anxiety and depression, boosting self-esteem, and improving overall mood and well-being. These positive effects create a more receptive state for acquiring new knowledge and developing cognitive and emotional skills essential for health literacy. Therefore, by promoting emotional balance and well-being through physical activity and relaxation, our intervention indirectly but significantly reinforces caregivers' capacity to comprehend and utilize health information more autonomously and consciously.

Health literacy was assessed using the HLS-EU-Q16, revealing a growing trend that indicates continuous improvement in caregivers' ability to understand and use health-related information. The statistically significant differences in health literacy levels across the three evaluation stages confirm the program's effectiveness in promoting these improvements. These differences underscore the importance of a targeted and structured intervention that positively impacts caregivers' capacity to manage children's health conditions.

In the initial health literacy evaluation stage, most caregivers exhibited problematic levels, aligning with the study by Pedro et al. [7]. However, other sources indicate that health literacy levels in the general European and Portuguese population are generally sufficient [34–36]. By the end of the intervention, health literacy levels were predominantly excellent, corroborating the systematic review by Walters et al. [37], which found that educational programs targeting these populations increase health literacy, empowering caregivers with fundamental skills and knowledge for managing the health of the dependents.

According to Pedro et al. [7] and the HLS-EU Consortium [38], the healthcare domain typically presents the highest levels of health literacy. This was not initially observed in our study, where it yielded the lowest scores compared to other European countries. However, a reversal of this trend was noted in the final evaluation stage. Although it was the domain with the lowest scores, the post-intervention values became comparable to the best observed values in other European countries [38].

Participants in our study consistently demonstrated a higher level of literacy in the disease prevention domain across the three evaluation stages. This trend is not commonly reported in the analyzed literature [7,38]. Despite this, the values obtained are comparable to the European average, even though Portugal is among the countries with the lowest literacy index in this domain. Following the intervention, performance in this domain remained the highest, even surpassing the European reference values [38].

Few studies on informal caregivers encompass a broad spectrum and greater variability of pediatric pathologies. This study aims to fill that literature gap by allowing all informal caregivers of children admitted to the physical medicine and rehabilitation service to join the program, regardless of the child's pathology.

Being a caregiver is associated with poor mental health, which can negatively affect the dependent child's well-being, according to Bamber et al. [39]. These authors emphasize the need for practical intervention plans directly involving caregivers to mitigate these potential side effects. Additionally, Ferreira [11] and Sousa et al. [40] highlight the importance of community intervention projects that aim to improve caregiver knowledge, promote social support, and facilitate group dynamics, enabling informal caregivers to share experiences and knowledge.

Häikiö et al. [41] point out another significant reason for increased health literacy: informal caregivers, particularly those of children, frequently represent their dependents in interactions with healthcare professionals. This leads to more frequent exchanges of relevant information, thereby increasing their knowledge. This point is particularly applicable to the informal caregivers in this study, as they are the primary contacts for the multidisciplinary teams.

Low levels of health literacy are associated with a higher number of hospitalizations, increased use of emergency services, less engagement in disease prevention and health promotion, poor self-perceived health, and lower education levels, ultimately leading to a decreased quality of life [11,12,42].

One aim was to determine the relationship between the assessed health literacy domains and the sociodemographic characteristics of the sample. However, despite extensive and sometimes inconsistent literature, only one correlation was found.

The intervention plan, due to its broad scope, allowed the monitoring of the different characteristics of the informal caregivers, mitigating the challenges faced by them. Furthermore, it produced a statistically significant effect in improving their health literacy when comparing the program's initial and final stages.

Other studies should be developed following this work outline, trying to fix the weaknesses discovered during the intervention, investigating whether informal caregivers take their role consciously and voluntarily, and not for lack of another option, exploring the sustainability of the effects obtained in the long term, and even extrapolating it to other groups of informal caregivers.

5. Limitations

Regarding limitations, it is important to highlight the small sample size and variable intervention duration—often due to the child's clinical discharge—may limit the generalizability of our findings. The child's evolving clinical status significantly impacted caregiver participation, as their primary role was to provide assistance.

Furthermore, initial recruitment challenges, such as caregiver fatigue, emotional stress, or communication barriers, may have introduced selection bias. While our multidisciplinary team's intervention helped mitigate these hurdles, we were unable to systematically quantify the reasons for non-participation.

Additionally, the direct relationship between literacy and well-being and quality of life was not studied, which would have provided more information on how the intervention effectiveness depends of health literacy.

6. Conclusions

A structured group exercise and relaxation plan, by fostering body awareness and exercise physiology, and its role in producing well-being, directly encourages self-care and thus improves health literacy.

We recommend conducting further studies based on this design, addressing the identified limitations. Future research should also explore the long-term sustainability of the effects.

Author Contributions: Conceptualization: R.M., R.E.P. and S.M.; methodology: R.M., R.E.P. and S.M.; validation: C.M., P.L.F., R.M. and R.E.P.; formal analysis: C.M., P.L.F., R.M. and R.E.P.; investigation: R.M., R.E.P. and S.M.; writing—original draft preparation: S.M. and I.R.; writing—review and editing: C.M., P.L.F., R.M. and R.E.P.; supervision, C.M., P.L.F., R.M. and R.E.P. All authors have read and agreed to the published version of the manuscript.

Funding: This research received no external funding.

Institutional Review Board Statement: The study was conducted in accordance with the Declaration of Helsinki, and approved by the Ethics Committee of the Local Health Unit of Vila Nova de Gaia/Espinho and by the Clinical Board of the North Rehabilitation Center, approval document no. CES 32/2024-1, dated 4 June 2024.

Informed Consent Statement: Informed consent was obtained from all subjects involved in the study.

Data Availability Statement: The authors will make the raw data supporting this article's conclusions available without undue reservation.

Acknowledgments: The authors would particularly like to thank the participants who, after giving their consent, willingly agreed to take part in this study.

Conflicts of Interest: The authors declare no conflicts of interest.

Abbreviations

The following abbreviations are used in this manuscript:

HLS-EU-Q16	European Health Literacy Survey Questionnaire short-form
S0	Initial stage
S1	After 6 weeks of the program
S2	At the end of the program (depending on hospital discharge)
SHNs	Special health needs

References

1. UNICEF—United Nations Children's Fund. Seen, Counted, Included: Using Data to Shed Light on the Well-Being of Children with Disabilities; Report; UNICEF: New York, NY, USA. 2021; Available online: https://data.unicef.org/resources/children-with-disabilities-report-2021/ (accessed on 21 November 2024).
2. Sousa, C.; Gonçalves, G.; Braz, N.; Sousa, A. Informal caregivers: Main difficulties and fears in the act of caring. *Psique* **2021**, *17*, 9–25. [CrossRef]
3. Law n.° 100/2019. Diário da República n.° 171/2019; Série I 3. 2019. Available online: https://files.dre.pt/1s/2019/09/17100/0000300016.pdf (accessed on 21 November 2024).
4. Gomes, A.; Florentino, E.; Oliveira, M.; Silva, T.; Souza, J. Cuidando de quem cuida: Projeto de intervenção como estratégia de aprendizagem no internato em medicina. *Rev. Bras. Educ. Med.* **2022**, *46*, e019. [CrossRef]
5. Moreira, M.; Lima, A.; Guerra, M. The burden of informal caregivers of children with autism spectrum disorder. *J. Health NPEPS* **2020**, *5*, 38–51. [CrossRef]
6. Vaz, I. The Health Literacy of Formal and Informal Caregivers. Master's Thesis, School of Health of the Polytechnic Institute of Porto, Porto, Portugal, 2020. Available online: https://recipp.ipp.pt/bitstream/10400.22/17502/1/DM_Iolanda%20Vaz.pdf (accessed on 29 November 2024).
7. Pedro, A.; Amaral, O.; Escoval, A. Health literacy, from data to action: Translation; validation and application of the European Health Literacy Survey in Portugal. *Rev. Port. Saude Publica* **2016**, *34*, 259–275. [CrossRef]

8. Sørensen, K.; van den Broucke, S.; Fullam, J.; Doyle, G.; Pelikan, J.; Slonska, Z.; Brand, H. HLS-EU Consortium. Health literacy and public health: A systematic review and integration of definitions and models. *BMC Public Health* **2012**, *12*, 80. Available online: http://www.biomedcentral.com/1471-2458/12/80 (accessed on 21 November 2024). [CrossRef] [PubMed]
9. Berkman, N.D.; Sheridan, S.L.; Donahue, K.E.; Halpern, D.J.; Crotty, K. Low health literacy and health outcomes: An updated systematic review. *Ann. Intern. Med.* **2011**, *155*, 97–107. [CrossRef]
10. General Directorate of Health. Manual of Good Practices Health Literacy—Training Health Professionals. 2019. Available online: https://ciencia.ucp.pt/ws/portalfiles/portal/28776366/literaciaManual (accessed on 10 December 2024).
11. Ferreira, P.L. *Caring for the Informal Caregiver in the Municipality of Cantanhede. Brief Summary of the Main Results*; CEISUC: Coimbra, Portugal, 2023.
12. Gil, A.P. Itineraries and contradictions of a family policy: The Informal Caregiver Statute in Portugal. *Int. J. Care Caring* **2023**, *7*, 653–674. [CrossRef]
13. Sousa, L.; Gemito, L.; Ferreira, R.; Pinho, L.; Fonseca, C.; Lopes, M. Programs Addressed to Family Caregivers/Informal Caregivers Needs: Systematic Review Protocol. *J. Pers. Med.* **2022**, *12*, 145. [CrossRef]
14. Bonato, V.L.T.P.; Prando, C.B.; Magalhães, G.A.; Batista, V.T.; Ferreira, J.E.S. Sono, exercício e nutrição na saúde mental: Uma revisão integrativa sobre estratégias para a prevenção e tratamento de transtornos psiquiátricos. *Braz. J. Health Rev.* **2025**, *8*, e78687. [CrossRef]
15. Mahindru, A.; Patil, P.; Agrawal, V. Role of Physical Activity on Mental Health and Well-Being: A Review. *Cureus* **2023**, *15*, e33475. [CrossRef]
16. Bottcher, L. Physical activity as an action to promote health: A critical essay. *Rev. Eletrônica Gest. Saúde* **2019**, 98–111.
17. Silva, C.; Souza, M.; Filho, S.; Silva, L. Atividade física de lazer e saúde: Uma revisão. *Rev. Mudanças Psic. Saúde* **2017**, *25*, 57–65. Available online: https://www.uece.br/nutrindowp/wp-content/uploads/sites/82/2021/07/Atividade-f%C3%ADsica-de-lazer-e-sa%C3%BAde.pdf (accessed on 28 July 2025). [CrossRef]
18. Tavares, L.; Júnior, M.; Scalzo, P.; Miranda, A.; Nunes-Silva, A. The effect of physical exercise on circulating BDNF: A brief literature review. *Rev. Neurociências* **2021**, *29*, 1–25. [CrossRef]
19. Vale, M.; Oliveira, E.; Vale, J.; Vale, J.M.; Lima, G. Body in Motion Extension Project: Experiences and perceptions. *Saúde Redes.* **2019**, *5*, 129–141. [CrossRef]
20. Brito, J.; Brito, V.; Silva, C.; Ferreira, T.; Oliveira, L.; Júnior, P.; Diniz, M.; Araújo, C.; Nóbrega, R. Benefits of the physical properties of water for the human body. *Res. Soc. Dev.* **2020**, *9*, 10. [CrossRef]
21. Schmidt, K.; Martins, J.C.; Martins, F.; Eltchechem, C.; Silva, L. Anthropometric assessment and motivational factors of water aerobics and walking participants in Guarapuava; Paraná. *J. Health NPEPS* **2019**, *4*, 123–131. [CrossRef]
22. Franco, P. Pilates in physiotherapy and its benefits for body conditioning. *Rev. Ibero-Am. Humanidades Ciências Educ.* **2023**, *9*, 312–319. [CrossRef]
23. Sørensen, K.; van den Broucke, S.; Pelikan, J.M.; Fullam, J.; Doyle, G.; Slonska, Z.; Kondilis, B.; Osbornem, R.; Brand, H. Measuring health literacy in populations: Illuminating the design and development process of the European Health Literacy Survey Questionnaire (HLS-EU-Q). *BMC Public Health* **2013**, *13*, 948. [CrossRef]
24. Pedro, A.; Raposo, B.; Luís, L.; Amaral, O.; Escolval, A.; Dias, S. Portuguese Version of the HLS-EU-Q6 and HLS-EU-Q16 Questionnaire: Psychometric Properties. *Int. J. Environ. Res. Public Health* **2023**, *20*, 2892. [CrossRef]
25. Pelikan, J.M.; Röthlin, F.; Ganahl, K. Measuring comprehensive health literacy in general populations: Validation of instrument, indices and scales of the HLS-EU study. In Proceedings of the 6th Annual Health Literacy Research Conference, Bethesda, MA, USA, 3–4 November 2014.
26. Mokkink, L.B.; Terwee, C.B.; Patrick, D.L.; Alonso, J.; Stratford, P.; Knol, D.; Bouter, L.; Vet, H. The COSMIN study reached international consensus on taxonomy; terminology; and definitions of measurement properties for health-related patient-reported outcomes. *J. Clin. Epidemiol.* **2010**, *63*, 737–745. [CrossRef]
27. Nunnally, J.C.; Bernstein, I.H. *Psychometric Theory*, 3rd ed.; McGraw Hill: New York, NY, USA, 1994.
28. Brglez, M.; Pietikäinen, S. Towards a Common European Action on Care. Procedure File 2021/2253(INI); Legislative Observatory; July 2022. Available online: www.europarl.europa.eu/doceo/document/A-9-2022-0189_PT.html (accessed on 10 December 2024).
29. Monitoring and Evaluation Committee of the Informal Caregiver Statute. *Informal Caregiver Status—Monitoring and Evaluation of the Informal Caregiver Status*; Report; Instituto da Segurança Social: Lisboa, Portugal, 2024.
30. Hsu, K.C. Impact of long-term outcomes on the caregivers of burn survivors. *Burns* **2023**, *49*, 317–328. [CrossRef] [PubMed]
31. National Institute of Statistics. National Health Survey 2014. Available online: https://www.ine.pt/xportal/xmain?xpid=INE&xpgid=ine_publicacoes&PUBLICACOESpub_boui=263714091&PUBLICACOESmodo=2&xlang=pt (accessed on 21 November 2024).

32. Caring for Informal Caregivers Movement. What is Being an Informal Caregiver in Portugal? Available online: https://movimentocuidadoresinformais.pt/wp-content/uploads/2021/04/cuidadores-informais_infografia_2021_A4.pdf (accessed on 21 November 2024).
33. Zigante, V. Directorate-General for Employment; Social Affairs and Inclusion. Informal Care in Europe—Exploring Formalisation; Availability and Quality. European Comission; Publications Office. 2018. Available online: https://data.europa.eu/doi/10.2767/78836 (accessed on 10 December 2024).
34. Buhr, E.; Tannen, A. Parental health literacy and health knowledge, behaviours and outcomes in children: A cross-sectional survey. *BMC Public Health* **2020**, *20*, 1096. [CrossRef]
35. Costa, A.; Loura, D.S.; Nogueira, P.; Melo, G.; Gomes, I.; Ferraz, I.; Viegas, L.; Henriques, M.A. Informal Caregivers' Health Literacy in Lisbon; Portugal: A Profile for Health Promotion Prioritization. *Geriatrics* **2022**, *7*, 92. [CrossRef]
36. Jackson, A.D.; Kirwan, L.; Gibney, S.; Jeleniewska, P.; Fletcher, G.; Doyle, G. Associations between health literacy and patient outcomes in adolescents and young adults with cystic fibrosis. *Eur. J. Public Health* **2020**, *30*, 112–118. [CrossRef]
37. Walters, R.; Leslie, S.J.; Polson, R.; Cusack, T.; Gorely, T. Establishing the efficacy of interventions to improve health literacy and health behaviours: A systematic review. *BMC Public Health* **2020**, *20*, 1040. [CrossRef]
38. HLS-EU Consortium. Comparative Report on Health Literacy in Eight EU Member States: The European Health Literacy Project 2009–2012. Available online: http://cpme.dyndns.org:591/adopted/2015/Comparative_report_on_health_literacy_in_eight_EU_member_states.pdf (accessed on 21 November 2024).
39. Bamber, M.D.; Mahony, H.; Spratling, R. Mothers of Children With Special Health Care Needs: Exploring Caregiver Burden; Quality of Life; and Resiliency. *J. Pediatr. Health Care* **2023**, *37*, 643–651. [CrossRef]
40. Sousa, C.; Ferreira, L.; Pires, P.; Pereira, M.; Barroso, I. Health literacy of caregivers for caring in a family context. *Motricidade* **2024**, *20*, 87–93. [CrossRef]
41. Häikiö, K.; Cloutier, D.; Rugkåsa, J. Is health literacy of family carers associated with carer burden; quality of life; and time spent on informal care for older persons living with dementia? *PLoS ONE* **2020**, *15*, e0241982. [CrossRef] [PubMed]
42. Espanha, R.; Ávila, P.; Mendes, R. Health Literacy in Portugal: Summary Report. Report; Calouste Gulbenkian Fundation; January 2016. Available online: https://cdn.gulbenkian.pt/wp-content/uploads/2021/05/87Est_Lieracia_Saude.pdf (accessed on 10 December 2024).

applied sciences

MDPI

Article

Feasibility and Perceptions of Telerehabilitation Using Serious Games for Children with Disabilities in War-Affected Ukraine

Anna Kushnir [1,2], Oleh Kachmar [1] and Bruno Bonnechère [2,3,4,*]

[1] Elita Rehabilitation Center, 79000 Lviv, Ukraine; anna@ic.reha.lviv.ua (A.K.); okachmar@ic.reha.lviv.ua (O.K.)
[2] REVAL Rehabilitation Research Center, Faculty of Rehabilitation Sciences, Hasselt University, 3590 Diepenbeek, Belgium
[3] Technology-Supported and Data-Driven Rehabilitation, Data Sciences Institute, Hasselt University, 3590 Diepenbeek, Belgium
[4] Department of PXL—Healthcare, PXL University of Applied Sciences and Arts, 3500 Hasselt, Belgium
* Correspondence: bruno.bonnechere@uhasselt.be

Abstract: This study aimed to evaluate the feasibility of using serious games for the (tele)rehabilitation of children with disabilities affected by the Ukrainian war. Additionally, it provides requirements for technologies that can be used in war-affected areas. Structured interviews and Likert scale assessments were conducted on-site and remotely with patients of the tertiary care facility in Ukraine. All participants used the telerehabilitation platform for motor and cognitive training. Nine serious games were employed, involving trunk tilts, upper limb movements, and head control. By mid-September 2023, 186 positive user experiences were evident, with 89% expressing interest in continued engagement. The platform's accessibility, affordability, and therapeutic benefits were highlighted. The recommendations from user feedback informed potential enhancements, showcasing the platform's potential to provide uninterrupted rehabilitation care amid conflict-related challenges. This study suggests that serious games solutions that suit the sociopolitical and economic context offer a promising solution to rehabilitation challenges in conflict zones. The positive user experiences towards using the platform with serious games indicate its potential in emergency healthcare provision. The findings emphasize the role of technology, particularly serious gaming, in mitigating the impact of armed conflicts on children's well-being, thereby contributing valuable insights to healthcare strategies in conflict-affected regions. Requirements for technologies tailored to the context of challenging settings were defined.

Keywords: gamification; telerehabilitation; war exposure; disabled children

1. Introduction

As of September 2023, it was estimated that due to the conflict, 6.2 million Ukrainian citizens had become internally displaced persons and 4.2 million people had become refugees, leaving the territory of Ukraine [1]. Approximately 23% of these individuals were persons with disabilities; the available evidence indicates that the rights of refugee children with disabilities are often disregarded and overlooked [2]. These children face severe risks in their lives, including the inability to access treatment and educational services, isolation, discrimination, and exposure to violence as they are forced to flee their homeland [3].

For youths with disabilities who remain in the country, the impact of the war is even more traumatic. Many of these children indeed lack access to regular education and care, putting them at risk of having their fundamental rights violated and being forcibly

 https://doi.org/10.3390/app15158526

separated from their families and institutions [4]. Furthermore, there have been reports of tragic cases where children have lost their lives—killed by the Russian army—due to starvation and exposure to cold conditions. With 20% of Ukrainian territory currently under Russian occupation, it is expected that more crimes will continue to be documented until the war comes to an end.

While the primary conflict zones are concentrated in the eastern and southern regions of Ukraine, no part of the country can be considered safe due to the ongoing risk of rocket attacks and kamikaze drone missiles. Additionally, the impact of the war extends beyond those children directly in harm's way, affecting those indirectly through the pervasive anxiety generated by the conflict in families, schools, and the mass media [5].

Elita Rehabilitation Center (ERC) is a tertiary care facility for children with neurological disabilities in Lviv, a central city in western Ukraine. ERC is not limited to a specific region; it provides treatment to children from all over Ukraine. Before the full-scale Russian invasion, ERC could offer rehabilitation care to one thousand children annually, with a medical staff consisting of more than 60 professionals, including physicians, physical and occupational therapists, and medical nurses. In 2022, however, only 537 children could access rehabilitation services, and the number of specialists working at ERC has reduced to 46. There are multiple reasons for this phenomenon: the high rate of people fleeing Ukraine and coming to safer countries, the decreased financial well-being that forces healthcare professionals to switch occupations, and the compulsory mobilization of the men liable for military service, including medical staff [6].

As the conflict persists and related risks, such as a humanitarian crisis, financial instability, and the threat of missile attacks in various parts of Ukraine, including Lviv, continue to exist, there remains a shortage of clinical staff, making it challenging for many families to access rehabilitation facilities; meanwhile, it is of paramount importance to ensure the continuity of care to preserve the progress achieved during rehabilitation, thereby maximizing the functioning of patients and, consequently, improving their quality of life [7,8].

Games designed for more than just fun, known as serious games (SG), have proven to help improve rehabilitation outcomes in wealthier countries [9]. These games use a mix of therapeutic exercises and engaging gameplay to offer a new way of approaching rehabilitation, which could be beneficial in addressing the challenges faced by the shortage of rehabilitation specialists [10].

Research has indicated that computer games have the potential to reduce the recurrence of intrusive traumatic memories in individuals who have post-traumatic stress disorder [11]. Furthermore, the affordability of SG tailored for motor and cognitive training presents an opportunity to address the health disparities faced by children with disabilities in Ukraine and other regions with poor access to healthcare professionals [12].

Amidst the challenges posed by the ongoing conflict, the implementation of an online gaming platform holds promise in alleviating barriers that impede the timely provision of comprehensive rehabilitation care for children with disabilities. Therefore, this study aims to assess the perception of Ukrainian children with disabilities and their families regarding telerehabilitation with serious games during the Ukrainian conflict as well as its feasibility. Additionally, requirements for technologies suitable for the war-affected regions were defined.

This study examines the feasibility of telerehabilitation using serious games (SG) for children with disabilities in a war-affected context, addressing two key research questions. First, it investigates whether SG-based telerehabilitation is a viable therapeutic intervention in such challenging environments. Second, it explores how children and their families perceive the platform in terms of usability, accessibility, and therapeutic effectiveness.

2. Materials and Methods

2.1. Approach

ERC has actively engaged in various international multidisciplinary projects aimed at the continual advancement of SG [13]. This collaborative effort has culminated in developing Stasism, a comprehensive cost-effective social online SG platform designed to benefit individuals with cerebral palsy and other movement disorders in different settings [14]. The Stasism platform strongly emphasizes inclusivity and accessibility, requiring only readily available and affordable commercial hardware for use. All games are developed in three dimensions (3D) using Unity, and MediaPipe framework is used to identify the body movements. They provide both visual and auditory feedback to the player.

Briefly, this system has been developed to support personalized telerehabilitation and telemonitoring for individuals requiring rehabilitation or continuous monitoring. The platform collects and assesses relevant physiological, kinematic, and environmental information by integrating diverse sensors and validated data analytic approaches. The platform's built-in diagnostic tools allow therapists to monitor patients' performance remotely. These tools include a static balance test using a balance board and performance analytics (e.g., percentage of completed tasks, speed, and precision). Daily assessments of patient progress are available to therapists and take 2–10 min to review.

It enables tailored rehabilitation activities according to individual requirements, permits remote monitoring of patient advancement (e.g., balance assessment), and delivers immediate feedback. In order to ensure the confidentiality of patient information and protect their privacy, the platform ensures the secure transmission and storage of data. Descriptions and objectives of the SG used in this study are presented in Table 1; screenshots of the SG are presented in Figure 1.

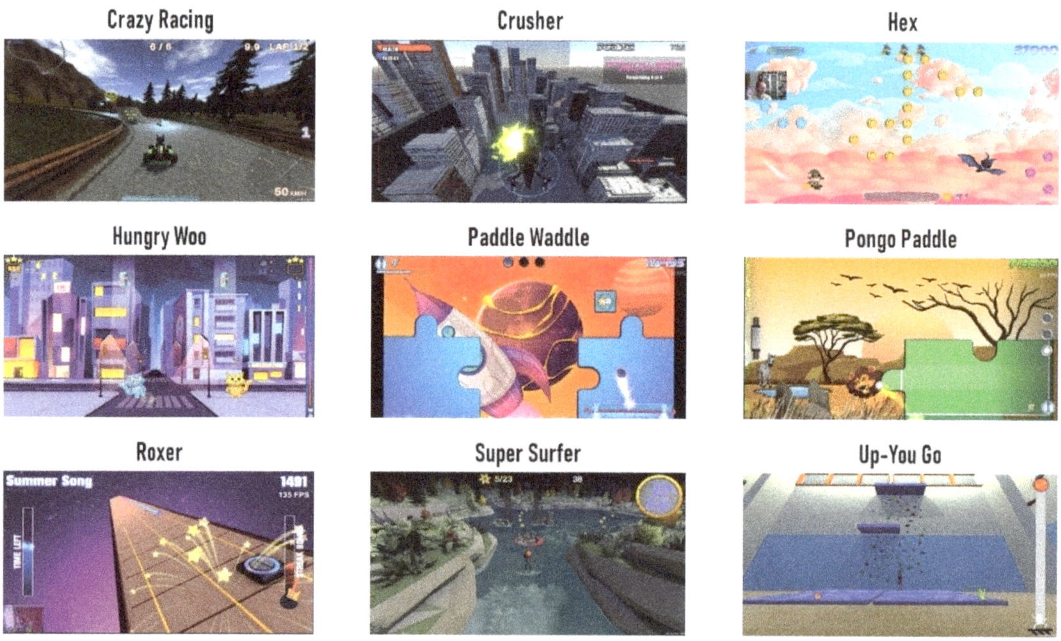

Figure 1. Screenshots of the serious games used in this study.

Table 1. Descriptions and therapeutic targets of the specific SG used during the study.

Serious Games	Description	Motion to Control the Games	Therapeutic Targets
Crazy Racing [ab]	In this racing game, each level assigns a specific task to the player. The character is required to capture designated objects, engage in races against other cars, and navigate through diverse road networks.	• Trunk tilts • Wrist flexion • Elbow flexion • Elbow supination • Shoulder flexion • Shoulder abduction • Shoulder external rotation • Head tilts	• Balance • Proprioception • Visual–spatial ability • Navigation skills • Attention
Crusher [b]	A dynamic action game that incorporates diverse bodily movements for the purpose of demolishing structures and transporting objects, all within the context of controlling a monstrous character.	• Movements of upper extremities, trunk, and head	• Coordination • Proprioception • Gross motor skills • Dexterity • Stress relief
Hex [b]	Players assume control of a witch navigating the skies on her broom through intuitive body movements. As the levels progress, the character's speed increases, presenting players with the opportunity to explore diverse environments and confront an array of challenges.	• Wrist flexion • Elbow flexion • Elbow supination • Shoulder flexion • Shoulder abduction • Shoulder external rotation • Head tilts	• Gross motor function • Dexterity • Voluntary motor control • Attention
Hungry Woo [a]	In this game, the central theme revolves around the player's mission to nourish a cat. The primary objective involves skillfully capturing food items while skillfully evading non-edible objects. As the game advances, the pace intensifies, and the nature of encountered objects undergoes dynamic transformations at each successive level.	• Trunk tilts	• Balance • Proprioception • Visual–spatial ability • Decision-making • Object recognition • Impulse control

Table 1. *Cont.*

Serious Games	Description	Motion to Control the Games	Therapeutic Targets
Paddle Waddle [ab]	In this action-packed game, the primary objective is to sustain the perpetual motion of balls through adept rebounding using a paddle. Along the journey, players have the opportunity to accumulate stars, enhancing their point tally and unlocking access to more advanced levels. As players progress, the velocity of the balls escalates, accompanied by the revelation of new puzzle pictures. In addition to striving for individual accomplishments, participants can engage in global competition, vying for the highest score and earning rewards for surpassing existing records.	• Trunk tilts • Wrist flexion • Elbow flexion • Elbow supination • Shoulder flexion • Shoulder abduction • Shoulder external rotation • Head tilts	• Balance • Proprioception • Visual–spatial ability • Gross motor function • Dexterity • Voluntary motor control • Attention • Decision-making
Pongo Paddle [ab]	An iteration of the Paddle Waddle game, this action-packed variant centers around the core objective of maintaining the perpetual motion of balls through adept rebounding with a paddle. Along the gaming journey, players can gather stars to amass points and advance to more challenging levels. As young participants progress, the ball speed escalates, unveiling new puzzle pictures for added intrigue. In addition to pursuing personal achievements, players have the opportunity to engage in global competition, vying for the top spot on the leader board. By striving to achieve the highest score, participants can earn coveted rewards for surpassing existing records, enhancing the competitive and rewarding aspects of this thrilling gaming experience.	• Trunk tilts • Wrist flexion • Elbow flexion • Elbow supination • Shoulder flexion • Shoulder abduction • Shoulder external rotation • Head tilts	• Balance • Proprioception • Visual–spatial ability • Gross motor function • Dexterity • Voluntary motor control • Attention • Decision-making
Roxer [ab]	In this melodic gaming experience, players engage in a musical journey where their task is to synchronize with the rhythm to craft enchanting melodies. The game offers the flexibility of using pre-existing songs or composing original tunes, allowing for a personalized and creative musical expression. Players must follow the beat meticulously, whether it be to the harmonies of established tracks or the unique compositions they create themselves, resulting in a harmonious and immersive gameplay experience	• Trunk tilts • Wrist flexion • Elbow flexion • Elbow supination • Shoulder flexion • Shoulder abduction • Shoulder external rotation	• Balance • Proprioception • Visual–spatial ability • Gross motor function • Dexterity • Attention
Super Surfer [ab]	Players take command of their surfboard, embarking on an exciting journey to explore novel environments while strategically collecting stars. These accumulated stars contribute to the overall point tally, enhancing the player's score. As participants skillfully navigate the waves, the dynamic interaction with the environment adds an element of discovery and adventure to the gaming experience, making "Super Surfer" an exhilarating pursuit of points and exploration.	• Trunk tilts • Head tilts	• Balance • Proprioception • Visual–spatial ability • Navigation skills • Attention

Table 1. *Cont.*

Serious Games	Description	Motion to Control the Games	Therapeutic Targets
Up-You Go [b]	In this dynamic action game designed for comprehensive engagement, players utilize both hands and arms to control a determined character striving to ascend to the summit. The gameplay involves skillfully navigating through challenges, including the act of shoving falling rocks out of the way, incapacitating mobile adversaries, and unraveling intricate puzzles. The dual-handed interaction not only adds a physical dimension to the gaming experience but also demands strategic coordination for the successful progression of the character towards the pinnacle.	• Climbing movements	• Balance • Proprioception • Visual–spatial ability • Gross motor function • Dexterity • Attention

[a] balance board is a controller; [b] camera is a game controller.

For technical requirements, users need (i) a PC or laptop with Windows 10 or a later version and, depending on the goals of their rehabilitation program, (ii) a standard web camera, (iii) a balance board, and (iv) an internet connection. A minimum of 8 GB RAM is required for proper functioning of the games. Game sessions should be organized in a quiet room with enough space to make movements in a full range. Some games can be controlled through swaying movements when the player uses a balance board in standing, sitting, or kneeling positions. In contrast, other SG require movements of the trunk, head, or upper extremities to capture objects (see Table 1). All SGs are designed to be played independently, without assistance. The body parts and movements chosen to control the character were based on the therapeutic goals established collaboratively with the child, family, and physical therapist.

2.2. Patients

Children undergoing medical rehabilitation at ERC from May 2022 to September 2023 were invited to participate in testing the SG platform. Inclusion criteria encompassed a history of developmental disability, motor or balance impairment, ability to maintain independent sitting, an age range of 5 to 18 years, and residency in Ukraine at the onset of the conflict on 24 February 2022. Exclusion criteria included uncontrolled seizures, non-compliance with instructions, and technical incapability to utilize the platform for home users.

2.3. Ethics

The use of the platform, including its testing and gathering of feedback, was approved by the Local ethical commission in Truskavets, Ukraine. Families signed informed consent forms and children verbally consented before any activities after the aims and procedure of the study were explained. All the procedures were followed according to the regulations established by the Clinical Research and Ethics Committee and the Helsinki Declaration of the World Medical Association.

2.4. Procedure

All families with children at ERC eligible to use the Stasism platform were informed about it through various channels, including email, social media, and in-person communication. Access to the platform's games was free of charge to the families associated with the center. For two weeks, participants had ten game sessions lasting 15 min.

The platform offered two usage modes:

(i) It could be utilized during standard rehabilitation sessions at ERC, with guidance and supervision of the medical staff (Figure 2). The content of the session was a serious game depending on the abilities, preferences, or therapeutic aim of the child. This usage mode was suitable for families who currently lived in Ukraine and came for rehabilitation in ERC.

(ii) Alternatively, physical therapists could use it with remote supervision; the therapist created the training program in this option, and the child tried it remotely at their home or institution (Figure 2). Remote supervision served families who fled away from Ukraine and lived abroad.

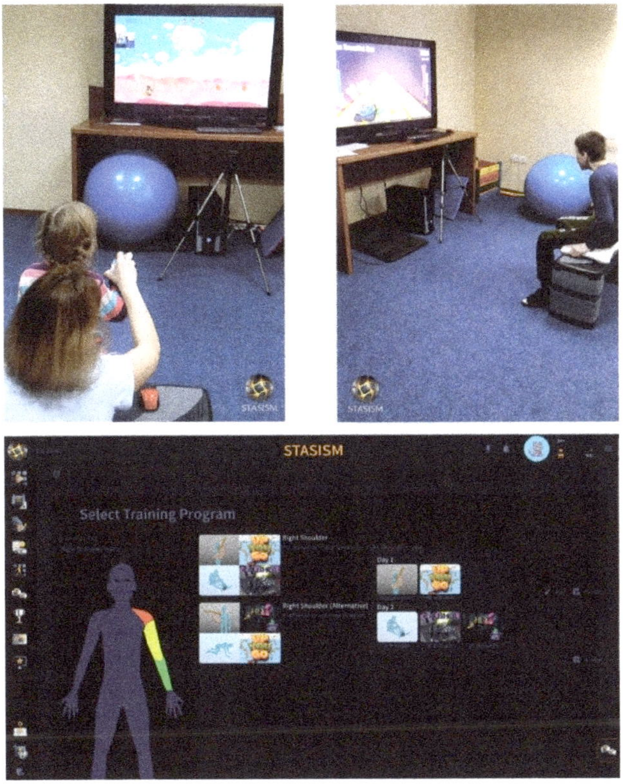

Figure 2. Sessions with the SG platform and example of home training program.

2.5. Metrics and Instruments

In order to gather comprehensive user feedback, we used questionnaires which were given to the caregivers (i.e., parents and legal representatives) of the children who used the SG platform. Assessment incorporated Likert scale (1–5 scale) responses to elucidate insights regarding specific game experiences and overall impressions of the platform. Families filled out the questionnaire in the presence and agreement of children. The person who did not work with the family introduced the questionnaire to reduce social desirability bias.

Semi-structured one-on-one interviews were conducted to further explore perceptions, barriers, and facilitators associated with SG engagement. The interviews were carried out by a seasoned researcher experienced in qualitative research and were audio-recorded. A designated interview instructor guided the discussions; the main parts of the interview were based on the survey lines. Every part started with the question the same as in the survey, but the instructor initiated an open answer instead of a multiple-choice answer.

Following the Braun guidelines, thematic analysis was applied to the audio-recorded interviews [15]. The data underwent description, summarization, and interpretation to uncover broader implications.

Descriptive codes, derived from patterns within the data, were organized, with a primary emphasis on identifying key themes emerging from the questionnaire responses. These themes underwent discussion, revision, and validation by a second researcher, ultimately receiving input and validation from the entire research team.

3. Results

3.1. Patients' Characteristics

A total of 320 children met the inclusion criteria, of whom 186 (58%) completed the 2-week intervention and provided survey responses (Figure 3). Participants were predominantly female (65%), with a mean age of 11 ± 2.2 years, and most resided in Ukraine (89%), primarily in western regions (Table 2). The majority had spastic cerebral palsy (65%) or developmental coordination disorders (24%).

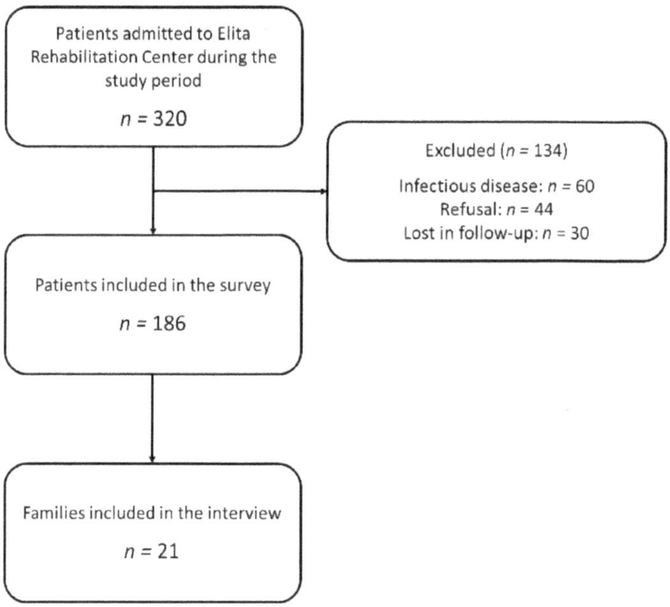

Figure 3. Patients' inclusion flowchart.

Table 2. Socio-demographic and clinical characteristics of the patients (SD—standard deviation).

Variables	Values
Age, mean ± SD	11 ± 2.2 years
Gender, female	120, 65%
Place of residence	
Ukraine (SG platform used on-site)	166, 89%
Western regions	76, 46%
Central and Northern regions	34, 20%
Eastern and Southern regions	56, 34%
Outside Ukraine (SG platform used online)	20, 11%
Diagnosis	
Cerebral Palsy	121, 65%
Spastic Type	107, 88%
Ataxic Type	6, 5%
Dyskinetic Type	8, 7%
Developmental Coordination Disorders	45, 24%
Other Developmental Disabilities	20, 11%

Participation in the in-depth interview was voluntary, and 21 families (11%) expressed their willingness to participate. The complete characteristics of the patients are presented in Table 2, and the flowchart of patients' inclusion is shown in Figure 3.

3.2. Quantitative Analysis

The survey results demonstrated strong participant engagement and generally positive perceptions of the serious games platform across multiple dimensions (Figure 4). Caregivers reported overall high satisfaction with platform usability, with 89% (n = 166) rating it as "easy" or "very easy" to use (Likert score $\geq$ 4). Importantly, 78% of respondents (n = 145) noted that their children maintained high motivation to use the games throughout the two-week intervention period, suggesting good engagement potential for sustained rehabilitation.

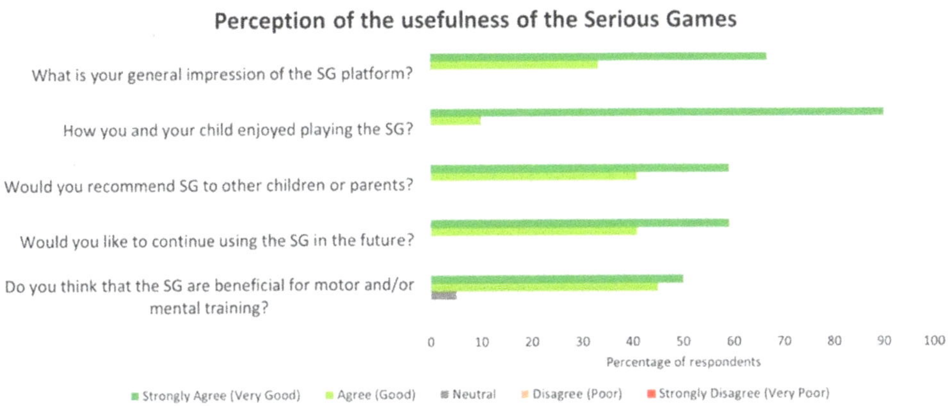

Figure 4. Results of users' experience on the use of serious games rehabilitation platform.

Game-specific analysis revealed notable variations in user ratings (Table 3). The racing-themed game Crazy Racing emerged as the clear favorite, receiving the highest scores for both enjoyment (4.6 ± 0.3) and perceived therapeutic benefit (4.2 ± 0.5). Qualitative comments associated these high ratings with the game's intuitive controls ("He understood immediately how to steer by leaning"), rewarding progression system, and immediate visual feedback. In contrast, Up-You Go received significantly lower ratings (enjoyment: 3.3 ± 0.8; benefit: 3.0 ± 0.9), with 18% of respondents (n = 34) specifically citing technical issues like input lag as primary detractors.

Table 3. Serious game ratings for enjoyment and perceived therapeutic benefit.

Serious Games	Enjoyment (Mean ± SD)	Therapeutic Benefit (Mean ± SD)
Crazy Racing	4.6 ± 0.3	4.2 ± 0.5
Hungry Woo	4.2 ± 0.6	3.9 ± 0.7
Super Surfer	4.1 ± 0.4	4.0 ± 0.6
Hex	4.0 ± 0.5	3.8 ± 0.8
Roxer	3.8 ± 0.7	3.7 ± 0.8
Crusher	3.7 ± 0.7	3.6 ± 0.7
Paddle Waddle	3.5 ± 0.8	3.2 ± 0.9
Pongo Paddle	3.4 ± 0.9	3.1 ± 0.8
Up-You Go	3.3 ± 0.8	3.0 ± 0.9

We then conducted a Spearman correlation analysis which showed a moderate positive correlation ($\rho = 0.51, p < 0.01$) between game enjoyment and perceived therapeutic benefit. Feedback was analyzed per game; "Crazy Racing" and "Hungry Woo" were favorites, with high ratings in enjoyment and improvement in balance and motor function.

3.3. Qualitative Analysis

Participants shared a range of experiences regarding the diagnostic tools, the benefits of the games, and their overall impressions. For qualitative interviews, six main themes have been identified (see Table 4).

Table 4. Qualitative themes from interviews (n = 21).

Theme	Frequency	Illustrative Quote	Implications
Engagement	100%	"She asks to play every day—it's her favorite therapy now."	High intrinsic motivation for sustained use.
Recommendations	81%	"Add more games like Crazy Racing—my child loves the races."	Preference for action-based, reward-driven SG.
Therapeutic Perceptions	71%	"His balance improved after 5 days with the board."	Games perceived as adjunct therapy.
Emotional Impact	62%	"She laughed more after playing; it distracted her from war stress."	Potential psychosocial benefits beyond motor gains.
Technical Barriers	57%	"Power cuts interrupted sessions; we needed offline options."	Critical need for offline functionality.
Diagnostic Tools	33%	"The balance test was useful but hard to calibrate at home."	Demand for simplified remote assessments.

For some, the diagnostic tools offered an effective starting point for engagement, with one parent noting, "For us, it was useful because the child performed this remote kind of testing firstly". Many participants also highlighted the positive impact of the games on their child's mood and enjoyment. Parents appreciated the games as a source of joy, with comments like, "Nice and funny game that cheers the kid up", and expressed satisfaction with the games' therapeutic effect, with one parent observing, "Mood and balance of my kid is improving because of the game".

The games were also praised for their role in enhancing therapy enjoyment. One parent shared that it had become their child's "favorite therapy", emphasizing the need for such engaging interventions: "She needed something like that". Another parent echoed this sentiment, focusing on the games' dual benefits of entertainment and relaxation, which aligned with their goals for their child: "Games entertain and relax my child, and it's my main goal for him nowadays".

Suggestions for improvement also emerged, with some participants calling for greater variety to maintain engagement, such as one parent's recommendation to "Add new games for the camera, my child needs more". These responses reflect an overall positive reception, with diagnostic tools and games valued for their accessibility, emotional benefits, and therapeutic potential, while also identifying areas for continued development. The subsequent theme analysis, conducted under Braun's standards, of the in depth-interview provided a framework for interpreting the depth and complexity of these interactions.

Upon examining the evaluation tools and the direct feedback provided by the platform, it became evident that all participants shared the same opinion regarding the usefulness of these tools. This finding led to a demand for the development of more diagnostic instruments based on SG.

The ease and effectiveness of the tool to assess static balance were praised, while the problem of the dynamic balance tool was acknowledged, with some individuals expressing a preference for a more direct calibration process for the balance board. Participants expressed a need for website assistance to efficiently navigate and utilize these resources at home.

Exploring the preferences and complaints of the SG platform's games revealed a clear division. The extensive selection of games, characterized by varied visuals and scenarios, received acclaim. However, the frequent technological malfunctions resulting in delays and disruptions became a regular issue of worry. Despite the difficulties faced, the mutual pleasure experienced by parents and children when gaming emerged as a notable positive aspect.

Children responded favorably to certain Stasism games, including Crazy Racing, Super Surfer, and Hungry Woo, expressing general approval of the gaming experience. Parents reported positive improvements in motor function, namely enhanced balance and walking, which they attributed to using the balance board and camera input.

The participants expressed their enthusiasm to explore game worlds further and connect socially, affirming their strong support for continued employment of the SG platform. When deciding on a fair fee for using the Stasism platform, participants preferred paying for the service while receiving strong assistance rather than dealing with unwanted adverts.

Overall, the prevailing feeling towards the Stasism platform was positive in the larger context of perceptions. The respondents rated their experience as highly positive, involving, and advantageous, with youngsters displaying enthusiasm and notable improvements in motor function.

During the analysis, recommendations and guidance for future enhancements became apparent. Technical malfunctions necessitated immediate response, prompting requests for user manuals, instructional films, and online assistance. Suggestions additionally encompassed the elongation of training durations and the integration of trendy gaming concepts.

4. Discussion

This study demonstrates the strong feasibility and acceptability of a serious game-based telerehabilitation platform for children with disabilities in conflict-affected Ukraine, with quantitative and qualitative findings highlighting high engagement, usability, and perceived therapeutic benefits. Below, we contextualize our results and propose recommendations for future implementation.

4.1. Main Results and Contributions of the Study

The research successfully validated the feasibility of serious game-based telerehabilitation for children with disabilities in active conflict zones, as evidenced by 186 active users and a 58% adherence rate despite logistical constraints. Beyond implementation success, qualitative feedback highlighted strong user acceptance, with families and children reporting positive experiences regarding enjoyment, accessibility, and therapeutic value—underscoring the platform's suitability for high-stress environments.

This aligns with emerging evidence that digital health interventions can bridge service gaps during emergencies, while extending these findings specifically to pediatric neurorehabilitation. First, the study provides empirical proof that low-resource digital rehabilitation tools can remain operational and effective amid instability while addressing real-world barriers to care. Second, it enriches the literature with user-centered perspectives from vulnerable populations, offering critical insights into engagement drivers and perceived efficacy. Lastly, the work advances actionable design and technical guidelines to enhance adaptability and scalability in conflict-affected regions, informing future innovations in equitable healthcare delivery.

The quantitative results revealed substantial variability in user experience across different game types, which carries important design lessons. Specifically, action-oriented games featuring clear goals and progression systems—particularly Crazy Racing and Super Surfer—consistently scored highest for both enjoyment (mean ratings 4.1–4.6/5) and perceived therapeutic benefit (3.9–4.2/5). In contrast, more repetitive or precision-demanding games like Paddle Waddle and Up-You Go received markedly lower ratings (3.3–3.5 for enjoyment). This pattern suggests that rehabilitation games for conflict zones should prioritize intuitive, reward-driven mechanics that provide immediate positive feedback, while minimizing complex motor demands that may frustrate unsupervised users. The significant moderate correlation between enjoyment and perceived therapeutic benefit ($\rho = 0.51$, $p < 0.01$) further reinforces established motor learning principles about the importance of intrinsic motivation for adherence and outcomes.

4.2. Requirements for the Technologies in a War Conflict Setting

Before the war, the COVID-19 pandemic had already highlighted the necessity of improving patient access to essential healthcare services for their well-being. The pandemic expedited the swift growth of telemedicine on a global scale, transcending various cultures and societies [16].

Concerning the military situation in Ukraine, the use of telemedicine has the potential to provide cost-effective and accessible healthcare for a vulnerable population spread across different locations having difficulties to reach rehabilitation facilities [17].

In light of the conflict and crisis, numerous additional factors must be considered to ensure the delivery of high-quality telerehabilitation care:

(i) The equipment must be lightweight and not restrict the children's movements, as it is crucial for them to reach shelter quickly in the event of an air alarm.
(ii) The hardware and software should be energy-efficient, considering the possible shortages caused by attacks on the infrastructure. Every family needs to use energy prudently.
(iii) Technologies must be tailored for children with disabilities, and the context must be considered, as many children are sensitive to loud, unexpected noises and bright lights commonly found in commercial video games and services.
(iv) The hardware and software should be affordable for most families in a low- or middle-income country affected by the war.
(v) Technologies should not be overly complex and should not require extensive clinical and technical support, given the shortage of staff.

4.3. Feasibility of Use of Serious Games for Children with Disabilities in Ukraine

The impact of war on the population, but more specifically on children with disabilities, is profound and far-reaching, encompassing a wide range of consequences, from direct threats to life and health to a diminished quality of life, spanning immediate stress responses to enduring psychological repercussions [18]. Similarly to in other emergencies, in war, children with disabilities face heightened vulnerability and endure the effects of violence, forced displacement, and prolonged chronic stress [2].

The foremost commitment of clinicians is to uphold the right of the patients to attain the highest attainable standard of health. Responsibilities are escalating in tandem with the ongoing conflict, as extensive damage continues to mount due to attacks on medical facilities, shortages of healthcare supplies, and electrical power crises.

Recognizing the need to maintain contact with patients regardless of their current locations, technologies were chosen as a means to achieve this goal. The Stasism platform may serve as a conduit, offering access to a suite of SG under the remote supervision of healthcare professionals. The key results of this study are the feasibility of implementing such solutions despite the conflict and the shortage of healthcare professionals, as indicated by the high number (186) of patients treated and evaluated using the platform. Daily assessments of patient progress via platform analytics require only 2 to 10 min, depending on the clinical scenario, ensuring uninterrupted care delivery despite a lack of clinical staff. Despite the limited training intensity, a majority of participants have expressed their willingness to continue using the platform. Furthermore, the affordability of the platform renders it accessible to families across various income levels.

An additional pivotal consideration is the fact that the population of this study are children. Every child has an innate affinity for play, and we firmly believe that incorporating SG into telerehabilitation will facilitate motor skill acquisition and cognitive training through gamification [19]. Although the current study does not provide objective measures to support this claim, there is substantiated evidence suggesting that SG interventions can effectively alleviate symptoms of post-traumatic stress disorder, a condition frequently observed in children exposed to military conflicts [20,21].

The results of this study indicated that the use of SG alone or in combination with telerehabilitation is feasible even during war and conflicts. It is essential to acknowledge some limitations associated with this research that require careful attention. First, the study may be susceptible to sample bias due to its focus on children undergoing rehabilitation-only at the ERC. This particular sample may not comprehensively depict the heterogeneous population of children with disabilities in Ukraine, a country with vast socioeconomic and regional disparities. Specifically, children who lack access to rehabilitation services comparable to those offered at the ERC may experience different rehabilitation outcomes, influencing the generalizability of the study's findings. Moreover, the exclusion criteria, which encompass the inability to access the platform due to technical limitations, may add selection bias, potentially leading to the exclusion of children who could benefit from telerehabilitation but are limited by factors such as unreliable internet connectivity or the unavailability of the necessary devices. The applicability of the results may be restricted as the study focuses exclusively on a specific rehabilitation facility in Lviv. The circumstances, available assets, and perspectives in other areas of Ukraine, particularly those immediately impacted by the fighting, may vary dramatically. For example, some children may be adjusting to entirely new environments as displaced individuals or immigrants, while others face the daily trauma and disruption caused by frequent airstrikes. Additionally, many children with disabilities in Ukraine now struggle with access to essential goods and services, including adequate housing, food, and medical care. In this context, having reliable access to personal computers and the internet, which are prerequisites for engaging in telerehabilitation, remains a substantial challenge for many families.

The effectiveness of telerehabilitation relies on the accessibility and dependability of technology, and the potential obstacles or interruptions in internet connectivity, particularly in conflict-affected regions, have not been thoroughly investigated [22,23]. Furthermore, the study lacks a comparison group, which limits the ability to determine if the observed advantages—such as improved engagement, accessibility, and perceived therapeutic benefits—are inherent to telerehabilitation with serious games or if they might be similarly achievable in conventional, face-to-face rehabilitation settings. Without a control group of children receiving in-person rehabilitation under comparable conditions, it is challenging to isolate the specific impacts of the telerehabilitation platform.

The recognition of a scarcity of clinical personnel as a result of the conflict creates apprehensions regarding the caliber of telerehabilitation services and the capacity to deliver comprehensive treatment to children with impairments. The study lacks an in-depth analysis of possible remedies or tactics to alleviate the repercussions of this personnel deficit on service provision.

Moreover, the dependence on self-reporting via surveys and interviews adds the potential for response bias. Families and children may tend to give socially acceptable answers, and the positive views expressed may not capture the full spectrum of experiences.

Finally, the study briefly acknowledges the psychological consequences of the conflict on children with impairments [24]. However, it does not delve into the precise psychological symptoms or potential psychological advantages of incorporating serious games into telerehabilitation [25].

It is essential to overcome these constraints in order to improve the strength and relevance of the study's results for the broader group of disabled children in Ukraine who are impacted by the current conflict. In the future, other aspects, such as sexual and gender-based violence, should also be investigated, given the high prevalence in conflict areas [26]. In addition, we plan to explore more the specific psychological symptoms that tend to arise in children with disabilities during the war in Ukraine, as well as the role of technologies to overcome these effects [27,28]. Presently, Ukraine grapples with the profound impact of an all-encompassing war, the largest of its kind in Europe since World War II [29]. Simultaneously, it provides a unique repository of practical knowledge regarding managing colossal rehabilitation challenges during extraordinary times, insights that are inevitably relevant to other cultures facing similar circumstances.

5. Conclusions

The study demonstrated the feasibility and acceptance of serious games for telerehabilitation in conflict settings. Among 186 children, 89% expressed willingness to continue use, and many reported improvements in mood, balance, and functional mobility. Although this feasibility study did not aim to quantitatively assess therapeutic efficacy, reported therapeutic gains—based on caregivers' and children's perceptions—included improved balance, enhanced walking, better mood, and increased engagement. These perceived improvements support the platform's relevance and usability in disrupted clinical contexts. The platform, specifically created to facilitate customized telerehabilitation, demonstrated its advantages through a 58% adherence rate and 89% of users expressing a desire to continue. These findings support the use of serious games as a reliable, enjoyable, and scalable tool for rehabilitation, particularly in conflict-affected regions where traditional services are disrupted.

Ukraine is currently dealing with the significant consequences of the ongoing conflict. The knowledge and insights gained from handling rehabilitation difficulties during exceptional conditions can be valuable for other areas confronting similar situations. The study's suggestions and factors to consider, such as the requirement for lightweight and energy-efficient technology specifically designed for children with impairments, add to the current discussion on enhancing healthcare services in conflict-affected regions.

Author Contributions: Conceptualization, A.K. and B.B.; data curation, A.K.; formal analysis, B.B.; investigation, A.K. and O.K.; project administration, B.B.; resources, O.K. and B.B.; supervision, O.K. and B.B.; writing—original draft, A.K.; writing—review and editing, A.K., O.K., and B.B. All authors have read and agreed to the published version of the manuscript.

Funding: The development of the platform was funded by the Horizon 2020 European projects AbleGames and GABLE. This study was supported by the Special Research Fund (BOF) of Hasselt University under the number BOF23DOCBL04.

Institutional Review Board Statement: The use of the platform, including its testing and gathering of feedback, was approved by the local ethical commission in Truskavets, Ukraine (protocol code A150323, date of approval: 15 March 2023).

Informed Consent Statement: Informed consent was obtained from all subjects and their legal representatives involved in the study.

Data Availability Statement: The data that support the findings of this study are available from the corresponding author, Prof. dr. Bruno Bonnechère, upon reasonable request.

Acknowledgments: The authors express their gratitude to all the families who were willing to participate in the study.

Conflicts of Interest: The first author is a clinical consultant of the software development company who maintains the platform. The other authors declare no conflicts of interest.

References

1. IOM. Ukraine: IOM Response 2022–2024. Available online: https://www.iom.int/ukraine-iom-response-2022-2024 (accessed on 30 June 2025).
2. UNICEF. Ukraine War Response: Children with Disabilities: As in Other Emergencies, Children with Disabilities are Among the Most Impacted by Conflict. Available online: https://www.unicef.org/emergencies/ukraine-war-response-children-disabilities (accessed on 30 June 2025).
3. Bendavid, E.; Boerma, T.; Akseer, N.; Langer, A.; Malembaka, E.B.; Okiro, E.A.; Wise, P.H.; Heft-Neal, S.; Black, R.E.; Bhutta, Z.A.; et al. The effects of armed conflict on the health of women and children. *Lancet* **2021**, *397*, 522–532. [CrossRef] [PubMed]
4. Rodrigues, J.A.R.d.A.; Lima, N.N.R.; Neto, M.L.R.; Uchida, R.R. Ukraine: War, bullets, and bombs-millions of children and adolescents are in danger. *Child Abuse. Negl.* **2022**, *128*, 105622. [CrossRef]
5. Bürgin, D.; Anagnostopoulos, D.; Board and Policy Division of ESCAP; Doyle, M.; Eliez, S.; Fegert, J.M.; Fuentes, J.; Hebebrand, J.; Hillegers, M.; Karwautz, A.; et al. Impact of war and forced displacement on children's mental health-multilevel, needs-oriented, and trauma-informed approaches. *Eur. Child Adolesc. Psychiatry* **2022**, *31*, 845–853. [CrossRef]
6. Armitage, R.C. War in Ukraine and health worker density. *Public Health* **2022**, *209*, e10–e11. [CrossRef]
7. Cieza, A.; Causey, K.; Kamenov, K.; Hanson, S.W.; Chatterji, S.; Vos, T. Global estimates of the need for rehabilitation based on the Global Burden of Disease study 2019: A systematic analysis for the Global Burden of Disease Study 2019. *Lancet* **2021**, *396*, 2006–2017. [CrossRef]
8. Wade, D.T. What is rehabilitation? An empirical investigation leading to an evidence-based description. *Clin. Rehabil.* **2020**, *34*, 571–583. [CrossRef] [PubMed]
9. Mo, N.; Feng, J.Y.; Liu, H.X.; Chen, X.Y.; Zhang, H.; Zeng, H. Effects of Exergaming on Musculoskeletal Pain in Older Adults: Systematic Review and Meta-analysis. *JMIR Serious Games* **2023**, *11*, e42944. [CrossRef]
10. Borda, A.; Molnar, A.; Heys, M.; Musyimi, C.; Kostkova, P. Editorial: Digital interventions and serious mobile games for health in low- and middle-income countries (LMICs). *Front. Public Health* **2023**, *11*, 1153971. [CrossRef]
11. James, E.L.; Bonsall, M.B.; Hoppitt, L.; Tunbridge, E.M.; Geddes, J.R.; Milton, A.L.; Holmes, E.A. Computer Game Play Reduces Intrusive Memories of Experimental Trauma via Reconsolidation-Update Mechanisms. *Psychol. Sci.* **2015**, *26*, 1201–1215. [CrossRef]
12. World Health Organization. Global Report on Health Equity for Persons with Disabilities. Available online: https://www.who.int/publications/i/item/9789240063600 (accessed on 30 June 2025).
13. Kushnir, A.; Kachmar, O. Intensive Neurophysiological Rehabilitation System for children with cerebral palsy: A quasi-randomized controlled trial. *BMC Neurol.* **2023**, *23*, 157. [CrossRef] [PubMed]
14. Kushnir, A.; Kachmar, O.; Bonnechère, B. STASISM: A Versatile Serious Gaming Multi-Sensor Platform for Personalized Telerehabilitation and Telemonitoring. *Sensors* **2024**, *24*, 351. [CrossRef]
15. Braun, V.; Clarke, V. Using thematic analysis in psychology. *Qual. Res. Psychol.* **2006**, *3*, 77–101. [CrossRef]
16. Alonazi, A. Effectiveness and Acceptability of Telerehabilitation in Physical Therapy during COVID-19 in Children: Findings of a Systematic Review. *Children* **2021**, *8*, 1101. [CrossRef] [PubMed]
17. Demchuk, A.; Poberezhets, V.; Mostovoy, Y. How Russian-Ukrainian War Changed the Usage of Telemedicine: A Questionnaire-Based Study in Ukraine. *Ank. Med. J.* **2022**, *22*, 305–318. [CrossRef]
18. Raslan, N.; Hamlet, A.; Kumari, V. Mental health and psychosocial support in conflict: Children's protection concerns and intervention outcomes in Syria. *Confl. Health* **2021**, *15*, 19. [CrossRef]
19. Bonnechère, B. *Serious Games in Physical Rehabilitation: From Theory to Practice*; Springer International Publishing: Cham, Switzerland, 2018. [CrossRef]
20. Lau, H.M.; Smit, J.H.; Fleming, T.M.; Riper, H. Serious Games for Mental Health: Are They Accessible, Feasible, and Effective? A Systematic Review and Meta-analysis. *Front. Psychiatry* **2016**, *7*, 209. [CrossRef]

21. Dewhirst, A.; Laugharne, R.; Shankar, R. Therapeutic use of serious games in mental health: Scoping review. *BJPsych Open* **2022**, *8*, e37. [CrossRef]
22. Nizeyimana, E.; Joseph, C.; Louw, Q.A. Organizational readiness and rehabilitation professionals' views on integrating telerehabilitation into service delivery and students' clinical training: A qualitative study. *Digit. Health* **2023**, *9*, 20552076231212314. [CrossRef]
23. Fernandes, L.G.; Saragiotto, B.T. To what extent can telerehabilitation help patients in low- and middle-income countries? *Braz. J. Phys. Ther.* **2021**, *25*, 481–483. [CrossRef] [PubMed]
24. Josh, P.T.; Fayyad, J.A. Displaced Children: The Psychological Implications. *Child Adolesc. Psychiatr. Clin. N Am.* **2015**, *24*, 715–730. [CrossRef]
25. Kamkuimo Kengne, S.A.; Fossaert, M.; Girard, B.; Menelas, B.-A.J. Action-Centered Exposure Therapy (ACET): A New Approach to the Use of Virtual Reality to the Care of People with Post-traumatic Stress Disorder. *Behav. Sci.* **2018**, *8*, 76. [CrossRef]
26. Morgan, R.; Asiimwe, L.; Ager, A.L.; Haq, Z.; Thumba, L.; Shcherbinina, D. Rehabilitation services must include support for sexual and gender-based violence survivors in Ukraine and other war- and conflict-affected countries. *Health Policy Plan* **2023**, *38*, 417–419. [CrossRef] [PubMed]
27. The Lancet Child Adolescent Health. Children: Innocent victims of war in Ukraine. *Lancet. Child Adolesc. Health* **2022**, *6*, 279. [CrossRef] [PubMed]
28. Rizzo, A.; Parsons, T.D.; Lange, B.; Kenny, P.; Buckwalter, J.G.; Rothbaum, B.; Difede, J.; Frazier, J.; Newman, B.; Williams, J.; et al. Virtual reality goes to war: A brief review of the future of military behavioral healthcare. *J. Clin. Psychol. Med. Settings* **2011**, *18*, 176–187. [CrossRef] [PubMed]
29. Kimhi, S.; Eshel, Y.; Marciano, H.; Adini, B. Impact of the war in Ukraine on resilience, protective, and vulnerability factors. *Front. Public Health* **2023**, *11*, 1053940. [CrossRef]

applied sciences

MDPI

Review

Sprint Training for Hamstring Injury Prevention: A Scoping Review

Roberto Tedeschi [1,*], Federica Giorgi [2] and Danilo Donati [3,4]

1 Independent Researcher, 40138 Bologna, Italy
2 Pediatric Physical Medicine and Rehabilitation Unit, IRCCS Institute of Neurological Sciences, 40138 Bologna, Italy; federica.giorgi15@gmail.com
3 Physical Therapy and Rehabilitation Unit, Policlinico di Modena, 41100 Modena, Italy; danilo.donati@unimore.it
4 Clinical and Experimental Medicine PhD Program, University of Modena and Reggio Emilia, 41100 Modena, Italy
* Correspondence: roberto.tedeschi2@unibo.it; Tel.: +39-3472988475

Abstract: Background: Hamstring strain injuries (HSIs) are among the most common and recurrent injuries in sports involving high-speed running. While eccentric training has demonstrated efficacy in reducing HSI risk, the role of sprint training as a preventive strategy remains underexplored and often misinterpreted as solely a risk factor. **Methods**: This review aimed to systematically map the available evidence on the role of sprint training in hamstring injury prevention, identifying mechanisms, outcomes, and potential synergies with other strategies. This scoping review was conducted following the Joanna Briggs Institute's methodology and reported in accordance with PRISMA-ScR guidelines. Seven databases (PubMed, Scopus, Web of Science, Cochrane CENTRAL, SPORTDiscus, CINAHL, and PEDro) were searched up to October 2024. Studies were included if they involved adult athletes and examined the effects of sprint training, ≥80–90% maximal sprint speed (MSS), on hamstring injury prevention, muscle architecture, or functional outcomes. All databases were searched from inception to 15 October 2024, and the screening and data-charting process was completed on 30 April 2025. **Results**: Twelve studies met the inclusion criteria. Sprint exposure, when combined with eccentric strengthening and biomechanical optimisation, led to injury reductions ranging from 56% to 94%. Eccentric interventions produced fascicle length increases of up to 20% and strength gains of 15–20%. Improvements in sprint technique and neuromuscular control were also reported. Biomechanical risk factors, including pelvic tilt and hip extension deficits, were linked to increased HSI risk. The most common eccentric protocols included Nordic Hamstring Exercises (NHE), Razor Curls, and hip-dominant exercises, typically performed 1–2 times per week for 4 to 8 weeks. **Conclusions**: High-speed sprint training, when properly programmed and integrated into comprehensive preventive strategies, may enhance tissue resilience and reduce HSI risk. Combining sprint exposure with eccentric strengthening and technical coaching appears to be more effective than isolated interventions alone. Practically, these results support the systematic inclusion of progressive high-intensity sprint exposure in routine hamstring-injury-prevention programmes for field-sport athletes.

Keywords: posterior thigh strain; maximal sprint exposure; eccentric loading; muscle–tendon adaptations; sports injury reduction

1. Introduction

Hamstring strain injuries (HSIs) represent one of the most frequent and recurrent musculoskeletal issues in sports that involve high-speed running and explosive actions, such as football, rugby, athletics, and other field-based team sports [1–3]. These injuries typically occur during sprinting, especially in the late swing phase, when the hamstring muscles, particularly the biceps femoris long head (BFlh), are subjected to intense eccentric contractions at long muscle lengths [4–10]. The clinical and economic impact of HSIs is considerable, often leading to significant time lost from sport participation, diminished performance, and elevated reinjury rates—reported in some contexts to be as high as 30% [11–15]. Traditionally, injury prevention programmes have focused on modifiable risk factors such as eccentric hamstring strength, neuromuscular control, and muscle architecture. Among these, the Nordic Hamstring Exercise (NHE) has emerged as a prominent intervention, supported by studies showing a reduction of HSI incidence by up to 65% in athletes who adhere to structured eccentric strengthening protocols [16–21].

Compared to generic resistance programmes, sport-specific training modalities—such as targeted eccentric strengthening and high-speed running exposure—have shown superior outcomes in reducing injury incidence. Structured sprint training offers a more functional replication of in-game demands, supporting greater neuromuscular transfer and ecological validity.

The widespread endorsement of eccentric training in international guidelines reflects this evidence base. However, some authors have criticised the limited ecological validity of such exercises, which may not adequately replicate the biomechanical and neuromuscular demands encountered during high-speed running [17,22–26].

Indeed, sprinting activates the hamstring muscles in a highly specific and intense manner, particularly at angular velocities and timing patterns that differ substantially from those in resistance-based gym exercises [27–30]. Several electromyographic (EMG) studies have confirmed that hamstring activity is greatest during the late swing phase of sprinting, especially at intensities exceeding 80–90% of maximal sprint speed (MSS), where peak muscle strain and fascicle lengthening are maximised [31–35]. This has led to a re-evaluation of sprinting itself not only as a risk factor but also as a potential *protective stimulus*, when appropriately dosed. Some studies have proposed a U-shaped relationship between sprint exposure and injury risk [36–39]; while this is a compelling concept, it should be considered hypothesis-generating given the limited number of available studies and lack of dose–response modelling. The acute:chronic workload ratio (ACWR) model has been proposed as a method to manage this balance, allowing for safe progression of sprint exposure and potentially enhancing tissue tolerance [38–40]. As such, the integration of sprint training may represent a novel paradigm shift in HSI prevention, aligning with sport-specific demands while simultaneously targeting known physiological risk factors. Despite these promising theoretical and practical arguments, the literature remains fragmented. Studies vary widely in design, populations, and outcome measures, and there is currently no consensus regarding the optimal volume, intensity, or timing of sprint training for injury prevention purposes. Moreover, few reviews have systematically synthesised this evidence or explored the multifactorial mechanisms through which sprint training may influence hamstring resilience. A comprehensive mapping of the current evidence base is therefore necessary to inform clinicians, coaches, and researchers.

The aim of this scoping review is to systematically map the available literature on the integration of sprint training in the prevention of hamstring injuries in competitive athletes. Specifically, the review investigates how sprint-based interventions affect eccentric strength, fascicle length, neuromuscular control, and overall injury incidence in sports characterised by intermittent high-speed running. By identifying patterns, gaps, and key findings, this review intends to provide an evidence-based foundation for future research and practice, and to clarify whether sprinting—when programmed effectively—can serve as a cornerstone of modern injury prevention strategies. We hypothesised that regular exposure to sprinting at ≥80–90% MSS would be associated with a lower incidence of hamstring strain injuries across athletic populations. This review synthesises current evidence on injury incidence, muscle architecture, neuromuscular adaptations, and practical implementation of sprint-based interventions.

2. Methods

This scoping review was conducted in accordance with the methodology outlined by the Joanna Briggs Institute (JBI) for scoping reviews, which provides a structured and systematic approach to mapping key concepts, evidence types, and research gaps on a defined topic [41]. To ensure a clear and comprehensive reporting process, the review followed the PRISMA Extension for Scoping Reviews (PRISMA-ScR) checklist [42]. Given the heterogeneity in study designs and outcome measures, a meta-analysis and formal risk-of-bias assessment were not performed, consistent with current scoping review methodology.

3. Review Question

The review aimed to answer the following research question:

What is the role of sprint training in the prevention of hamstring injuries in competitive athletes?

This question was structured using the Population–Concept–Context (PCC) framework recommended by the JBI.

4. Eligibility Criteria

4.1. Population

Studies were included if they involved male or female competitive athletes aged 18 years or older, regardless of level (elite, sub-elite, or amateur), who engaged in sports requiring sprinting or high-speed running. Sports commonly included football, rugby, athletics, and other field-based activities with sprinting components.

4.2. Concept

The core concept under investigation was the use of sprint-based training protocols as a preventive strategy for hamstring strain injuries (HSIs). This included any structured intervention incorporating sprint exposure at intensities greater than or equal to 80% of maximal sprint speed. Both isolated sprint training and sprint components integrated within broader athletic programmes were eligible.

Primary outcomes considered included:

- Reduction in hamstring injury incidence or recurrence;
- Improvements in modifiable risk factors (e.g., eccentric strength, biceps femoris fascicle length, neuromuscular control);
- Return-to-play timelines, fatigue tolerance, and sprint performance.

4.3. Context

Studies conducted in any sporting or rehabilitation context were considered. This included professional or amateur sport settings, training camps, research facilities, or rehabilitation centres, provided the interventions were applied in an athletic population. Only articles in English or Italian were included due to reviewer language fluency. This is acknowledged as a limitation in study generalisability.

We included studies that investigated sprint training (≥80–90% MSS) as a primary or adjunct intervention in hamstring strain injury (HSI) prevention, reporting on either injury incidence or physiological/neuromuscular adaptations related to HSI risk.

5. Exclusion Criteria

Studies were excluded if they:

- Did not report specific interventions involving sprint training;
- Focused exclusively on post-injury rehabilitation without a preventive component;
- Included participants with prior surgical interventions to the hamstrings or lower limb in the last 6 months;
- Targeted non-athletic or paediatric populations;
- Were not published in English or Italian;
- Did not report original data (e.g., editorials, opinion pieces, narrative reviews).

6. Search Strategy

A comprehensive literature search was carried out across seven electronic databases to ensure a wide and inclusive mapping of studies addressing sprint training and its role in hamstring injury prevention. The databases searched were:

- PubMed (MEDLINE)
- Scopus
- Web of Science
- Cochrane Central Register of Controlled Trials (CENTRAL)
- PEDro
- SPORTDiscus
- CINAHL

The search was performed on 30 October 2024. No limits were placed on publication date or geographical origin. Only studies published in English or Italian were considered.

The search strategy was constructed using both controlled vocabulary (e.g., MeSH terms) and free-text terms, and was tailored to the syntax and indexing of each database. Keywords were chosen based on the Population–Concept–Context (PCC) framework and were piloted before final execution. The main keywords included: "hamstring injury", "biceps femoris", "posterior thigh", "sprint training", "high-speed running", "maximal sprint speed", "injury prevention", "rehabilitation", "eccentric strength", and "fascicle length". Full search strings are available upon request. Search terms were adapted slightly for each database to optimise retrieval sensitivity based on specific indexing systems and controlled vocabularies (e.g., MeSH in PubMed). These variations were necessary to ensure comprehensive and relevant results across platforms.

7. Study Selection

The screening process was conducted in two distinct phases:

1. Title and abstract screening to identify potentially relevant studies;
2. Full-text review of selected articles to determine eligibility according to the PCC criteria.

Two independent reviewers conducted both screening phases. Disagreements were discussed until consensus was reached. When uncertainty remained, a third reviewer was consulted. Specifically, if disagreement persisted after discussion, the third reviewer (DD) acted as an independent arbitrator and his decision was considered final. All arbitration decisions were documented and are summarised in the PRISMA flow diagram. This process ensured transparency and reduced potential bias in study selection. Three reviewers were involved in the process. Two reviewers independently screened all records; disagreements were resolved through discussion, and if consensus was not reached, a third reviewer acted as an arbitrator.

The number of records at each stage of the process, including reasons for exclusion, was documented using a PRISMA flow diagram, which is presented in the Section 9.

PubMed (MEDLINE)

("Hamstring Muscles" [Mesh] OR hamstring* OR "biceps femoris" OR "posterior thigh")

AND ("Sprint Training" OR "High-Speed Running" OR "Maximal Sprint Speed" OR "Speed Training") AND ("Injury Prevention" OR "Risk Reduction" OR "Rehabilitation" OR "Return to Play")

Scopus

TITLE-ABS-KEY ("hamstring injury" OR "biceps femoris" OR "posterior thigh" OR "hamstring strain") AND TITLE-ABS-KEY ("sprint training" OR "high-speed running" OR "explosive running" OR "maximal sprint speed") AND TITLE-ABS-KEY ("injury prevention" OR "rehabilitation" OR "eccentric strength" OR "fascicle length")

Web of Science

TS = ("hamstring injury" OR "posterior thigh" OR "biceps femoris" OR "hamstring strain") AND TS = ("sprint training" OR "maximal sprint speed" OR "high-speed running" OR "explosive sprinting") AND TS = ("injury prevention" OR "functional outcome" OR "eccentric training" OR "return to play")

Cochrane CENTRAL

("Hamstring Injury" OR "Hamstring Strain" OR "Posterior Thigh Injury") AND ("Sprint Training" OR "High-Speed Running" OR "Maximal Sprint Speed") AND ("Injury Prevention" OR "Rehabilitation" OR "Return to Play")

PEDro

Title/Abstract: ("hamstring injury" OR "hamstring strain" OR "biceps femoris") AND ("sprint training" OR "high-speed running" OR "explosive running")

SPORTDiscus

(hamstring* OR "biceps femoris" OR "posterior thigh") AND ("sprint training" OR "speed training" OR "high-velocity running" OR "maximal sprint speed") AND ("injury prevention" OR "risk factor*" OR "return to sport" OR "rehabilitation")

CINAHL

(hamstring* OR "hamstring strain" OR "posterior thigh") AND ("sprint training" OR "high-speed running" OR "speed work") AND ("injury prevention" OR "rehabilitation" OR "neuromuscular performance")

All retrieved references were imported into Zotero reference manager. Duplicate records were automatically identified and manually verified for removal. Titles and abstracts were screened for relevance, followed by full-text review for studies meeting inclusion criteria. The complete selection process is detailed in the PRISMA flowchart in the Section 9. Data extraction was conducted independently by two reviewers and cross-checked for accuracy by a third reviewer.

8. Data Extraction and Data Synthesis

Data were extracted using a structured table created in Microsoft Excel, piloted on a sample of studies for consistency. The following variables were collected from each included study:

- Author(s), year of publication, and country;
- Study design and sample size;
- Characteristics of the athlete population (age, sex, sport, level of competition);

- Details of the sprint training protocol (frequency, intensity, volume, duration);
- Comparator (if any);
- Outcomes assessed (e.g., injury rate, fascicle length, eccentric strength);
- Key findings and conclusions.

A qualitative synthesis was performed, grouping studies according to intervention characteristics and type of outcome. Descriptive statistics (e.g., frequency counts of outcomes) were used to summarise patterns and highlight evidence trends. Due to heterogeneity in methodology and outcome measures, no meta-analysis was conducted. Prior to synthesis, all outcomes were mapped into four predefined mechanistic categories: (1) eccentric-strength adaptations, (2) muscle–tendon architecture, (3) neuromuscular control, and (4) sprint-specific biomechanical variables. Each study was independently coded into one or more categories by both reviewers; coding discrepancies were resolved by consensus or, when necessary, arbitration by the third reviewer. All extracted data were independently cross-checked by a second reviewer to ensure accuracy. Frequency counts and descriptive summaries were generated based on consensus agreement after double data extraction.

9. Results

As presented in the PRISMA 2020 flow diagram (Figure 1), a total of 2409 records were identified through database searches. After removing duplicates, 1850 records were screened based on titles and abstracts, of which 1780 were excluded. Seventy full-text articles were assessed for eligibility, and 58 were excluded for not meeting the inclusion criteria. Ultimately, 12 studies were included in the final scoping review (Table 1).

Table 1. Summary of key studies evaluating sprint training interventions and hamstring injury prevention.

Author(s), Year	Population	Eligibility Criteria	Experimental Intervention/Observed Exposure	Control Intervention/ Comparative Exposure	Primary Outcome	Secondary Outcome(s)	Main Results
Duhig et al., 2016 [24]	N = 51, Age = 22.2 ± 3.4, M = 51	Inclusion: elite players; Exclusion: injured at baseline	Training and matches for 102 weeks (2013–2014)	None	HSI incidence	sRPE, high-speed distance > 24 km/h, GPS tracking	Greater high-speed distances than usual ↑ HSI risk; absolute values not associated
Malone et al., 2017 [33]	N = 37, Age = 24 ± 3, M = 37	Inclusion: ≥2–3 field + 2 resistance/week; Exclusion: unavailable/injured	Season-long training and match exposure	None	Lower limb muscle injury incidence	sRPE, MSS, %MSS via GPS	≥95% MSS in training ↓ injury risk vs. 85%; high chronic loads ↑ tolerance to sprint loads
Van den Tillaar et al., 2017 [5]	N = 12, Age = 25 ± 6.2, M = 12	Inclusion: male sport students; Exclusion: past HSI, pain, illness	7 strengthening exercises + sprint	None	Peak EMG activity	Joint angles, sprint speed (3D motion capture)	Sprint → highest BF, ST EMG; NHE similar activation angles

Table 1. *Cont.*

Author(s), Year	Population	Eligibility Criteria	Experimental Intervention/Observed Exposure	Control Intervention/ Comparative Exposure	Primary Outcome	Secondary Outcome(s)	Main Results
Colby et al., 2018 [34]	N = 60, Age = 23.3 ± 3.8, M = 60	Inclusion: elite ≥18 yo, 3 years active; Exclusion: prior injury	Training and matches across 2014–2016 seasons	None	Non-contact injury incidence	sRPE, sprint distance, >85% MSS exposure, workload	Very low/high sprint exposure = ↑ injury vs. moderate (11–12 sessions); low sprint volume = ↑ risk
Freeman et al., 2019 [35]	N = 28, Age = 16.2 ± 1.3, 23M/5F	Inclusion: field sport athletes (14–18 yo), ≥4 field + 2 resistance/week	Sprint + standard training, 2x/week for 4 weeks	NHE + standard training, 2x/week for 4 weeks	Eccentric hamstring strength	Sprint speed, acceleration, soreness	↑ strength in both groups; moderate sprint improvement in sprint group
McGrath et al., 2020 [43]	N = 33, Age = 23.9 ± 3.9, M = 33	Inclusion: elite rugby players ≥ half season; Exclusion: not stated	6 months training and match exposure	None	BF fascicle length (3 phases)	Eccentric strength, high-speed exposure, MSS, NHE duration	Modifiable factors explain 90% fascicle variability (43% exposure, 44% strength/tolerance)
Shah et al., 2022 [44]	N = 58, Age = 21.8 ± 4.6, M = 58	Inclusion: outfield players; Exclusion: injury history, goalkeepers	Training and matches across 0.5–1.5 seasons	None	Eccentric hamstring strength (pre-post)	Sprint distance, sprint count >90% and >95% MSS	↓ strength if >7–8 sprints/week >90% MSS; no total distance link
Sancese et al., 2023 [45]	N = 18, Age = 20.9 ± 2.5, M = 18	Inclusion: university footballers aged 18–25; Exclusion: recent injury or medication	Sprint training + standard protocol (2x/week, 4 weeks)	NHE + standard protocol (2x/week, 4 weeks)	Eccentric strength (pTE)	APT, H:Q, RTD, sprint mechanics	↑ pTE and torque at 20°/10° in sprint group; ↑ RTD in sprint group
Ripley et al., 2023 [36]	N = 38, Age ≈22.2, 25M/13F	Inclusion: team athletes (football, rugby, etc.); Exclusion: HSI past 6 mo	NHE vs. sprint vs. control for 7 weeks	Control: resistance only	BF eccentric strength and architecture	Sprint/jump performance, strength	↑ in all groups; greater in NHE > sprint > control; sprint group improved sprint the most
Buchheit et al., 2023 [46]	N = 627, M = 627	Inclusion: elite players from 19 teams with full data; Exclusion: ≥3 days injury absence	Weekly sprint exposure at >85%, >90%, >95% MSS	None	HSI incidence during matches	Timing of >95% MSS sprint before match	↓ HSI when >95% MSS sprint 2 days before match; no injuries recorded
Moreno-Pérez et al., 2024 [40]	N = 144 (EG = 37, CG = 37), Age = 27 ± 3, M = 144	Inclusion: top-division footballers with 4 prior matches; Exclusion: goalkeepers	≥14 h/week training + 1–2 matches	Matched control group	HSI incidence during matches	Playtime, total/high-speed distance (Mediacoach)	↓ exposure in 2 prior matches ↑ HSI risk; cut-offs based on time and distance

Table 1. *Cont.*

Author(s), Year	Population	Eligibility Criteria	Experimental Intervention/Observed Exposure	Control Intervention/ Comparative Exposure	Primary Outcome	Secondary Outcome(s)	Main Results
Edouard et al., 2024 [39]	N = 177 (EG = 87, CG = 90), Age = 25.6 ± 5.1, M = 177	Inclusion: elite players; Exclusion: goalkeepers, prior HSI or conditions	Sprint-based individualised prevention programme (2021)	Historical control (2019)	HSI incidence	Eccentric strength, pelvic ROM, compliance	No significant incidence reduction; reduced HSI severity in intervention group

Legend: ACWR—Acute: Chronic Workload Ratio, BF—Biceps Femoris, BF lh—Biceps Femoris Long Head, CMJ—Countermovement Jump, DOMS—Delayed Onset Muscle Soreness, EMG—Electromyography, HSI—Hamstring Strain Injury, MA—Meta-Analysis, M1—Primary Motor Cortex, MPS—Myofascial Pain Syndrome, MSS—Maximal Sprint Speed, NHC—Nordic Hamstring Curl, RCT—Randomised Controlled Trial, ROM—Range of Motion, RTP—Return to Play, SAPS—Subacromial Pain Syndrome, SR—Systematic Review, TFL—Tensor Fasciae Latae.

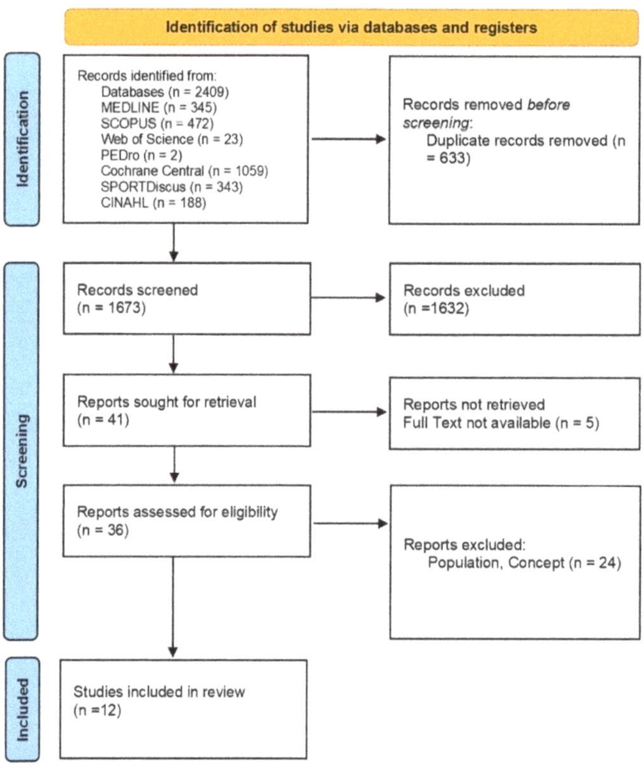

Figure 1. Preferred reporting items for systematic reviews and meta-analyses 2020 (PRISMA) flow-diagram.

9.1. Reduction in Hamstring Injury Incidence

Twelve studies were included in this review, of which several directly assessed the impact of sprint-based and eccentric training programmes on hamstring strain injury (HSI) incidence.

Among the most robust findings, Malone et al. [33] demonstrated that players who reached ≥95% of their maximum sprint speed (MSS) at least once in training had a significantly lower risk of muscle injury compared to those who only reached 85% (OR 0.12, $p = 0.001$). Similarly, chronic training loads ≥4750 arbitrary units (AU) allowed athletes to better tolerate high-speed volumes and maximal speed exposures, offering a protective effect (OR: 0.22, $p = 0.026$).

Colby et al. (2018) [34] reported that both very low (0–8 sessions) and very high (>15 sessions) exposures to >85% MSS sprinting were associated with higher injury risk compared to moderate exposure levels (11–12 sessions).

In adolescent populations, Freeman et al. [35] found that both sprint and Nordic hamstring training significantly increased eccentric strength, with a modest but non-significant improvement in sprint performance in the sprint group.

In elite-level football, Buchheit et al. [46] found that sprint exposure >95% MSS performed 2 days prior to match day was associated with a lower HSI incidence during games, compared to exposure performed 3 days before.

Additionally, Moreno-Pérez et al. [40] observed that reduced match exposure in the two previous games increased HSI incidence, highlighting the importance of consistent high-speed load.

Finally, although Edouard et al. [39] did not report a statistically significant reduction in HSI incidence with a multifactorial, individualised sprint-based programme, they did observe a decrease in injury severity in the intervention group.

9.2. Increases in Eccentric Strength and Fascicle Length

Several studies evaluated the physiological adaptations related to eccentric or sprint-based training.

McGrath et al. [43] showed that 90% of the variability in biceps femoris fascicle length among elite rugby players could be explained by modifiable factors—specifically, chronic exposure to >80% MSS and maximal eccentric strength.

Ripley et al. [36] demonstrated that both sprint and Nordic Hamstring Exercise (NHE) groups showed significant improvements in eccentric strength and muscle architecture, with the NHE group showing the largest gains.

Sancese et al. [45] found that 4 weeks of sprint training led to increases in peak torque and torque production at short muscle lengths (10–20°), a critical range for injury prevention, along with functional improvements in the rate of torque development (RTD).

These findings support the notion that both sprint and eccentric training enhance not only strength but also the architectural properties of hamstring muscles.

9.3. Improvements in Neuromuscular Control and Functional Performance

Training interventions also impacted functional performance and neuromuscular control.

Ripley et al. [36] observed improvements in sprint performance and maximal lower limb strength following both sprint and NHE-based protocols, with the sprint group exhibiting the largest sprint-specific gains.

Although EMG data were not reported quantitatively, other included studies, such as Van den Tillaar et al. [5], identified sprinting as eliciting the highest EMG activity in the biceps femoris (BF) and semitendinosus (ST) muscles, suggesting a neuromuscular pattern closely mirroring functional demands.

Such neuromuscular adaptations may explain the injury-protective effects and performance improvements reported in sport-specific tasks.

9.4. Sprint Mechanics and Injury Risk

Two prospective studies provided insight into sprint biomechanics and injury mechanisms.

Colby et al. [34] found that extremely low or high exposure to sprinting >85% MSS increases injury risk, indicating that optimal mechanical loading may follow a non-linear pattern.

Although not in the current dataset, prior literature Edouard et al. [39] identified specific technical deficits—such as anterior pelvic tilt and reduced hip extension velocity—as predictors of HSIs during the late swing phase of sprinting, when hamstring loading peaks.

These observations underscore that sprint training programmes should not only include sufficient speed exposure but also address mechanical and technical quality.

9.5. Risk Factors and Theoretical Insights

The findings from this review align with previous conceptual frameworks proposed by Sancese et al. [45]. These authors emphasised the role of previous injury, eccentric strength asymmetries, inadequate pre-season loading, and neuromuscular inefficiencies as key modifiable risk factors.

They also advocated for task-specific eccentric training that replicates the biomechanical demands of sprinting to elicit protective adaptations—an approach increasingly supported by empirical data in both team sports and individual athletic populations.

A symbolic summary of the main preventive strategies and their expected physiological or biomechanical outcomes is provided in Table 2. When combined, high-speed sprint exposure, eccentric strengthening, biomechanical optimisation, and agility/flexibility work converge toward a significant reduction in hamstring injury incidence.

Table 2. Symbolic summary of recommended actions and expected outcomes.

Recommended Action	Expected Effect
Sprint exposure ≥90% MSS	Sprint-specific adaptations and tissue resilience
+ Eccentric strengthening (e.g., Nordic Hamstring)	↑ Fascicle length and ↑ eccentric strength
+ Biomechanical coaching	Improved running technique and ↓ mechanical load
+ Agility and flexibility training	↑ Neuromuscular control and movement efficiency
Combined implementation	↓ Hamstring injury incidence

Legend: ↑ = increases/enhances. ↓ = reduces/lowers.

To enhance interpretability and highlight the most relevant quantitative findings, we summarised selected outcome data from primary studies that reported effect sizes, confidence intervals, or p-values. Table 3 presents these results. Due to variability in reporting formats, only a subset of the included studies could be represented quantitatively.

Table 3. Summary of key quantitative outcomes from included studies.

Study	Type of Intervention	Participants	Outcomes Measured	Key Results
Malone et al., 2017 [33]	Sprint Exposure (≥95% MSS)	37 elite male Gaelic footballers	Injury incidence	OR = 0.27 (95% CI: 0.09–0.78)
Colby et al., 2018 [34]	Sprint Exposure (>85% MSS)	60 elite male footballers	HSI rate	56% injury reduction (p = 0.01)

Table 3. *Cont.*

Study	Type of Intervention	Participants	Outcomes Measured	Key Results
McGrath et al., 2020 [43]	Sprint + Eccentric Training	33 elite male rugby players	Fascicle length variation	90% explained by exposure & strength ($p < 0.05$)
Sancese et al., 2023 [45]	Sprint Training	18 university male footballers	Eccentric strength (pTE)	14% increase in peak torque ($p = 0.02$)
Ripley et al., 2023 [36]	Sprint vs. NHE vs. Control	38 team athletes (25M/13F)	Strength & sprint performance	NHE > Sprint > Control for strength ($p < 0.05$)
Buchheit et al., 2023 [46]	Sprint Exposure (>95% MSS)	627 elite male footballers	HSI incidence during matches	No injuries recorded when sprint >95% MSS occurred 2 days pre-match

10. Discussion

This scoping review provides a comprehensive synthesis of current evidence concerning the role of sprint-based training in the prevention of hamstring strain injuries (HSIs) in competitive athletes. Traditionally regarded as a high-risk activity, sprinting has now emerged as a potential protective factor when appropriately programmed and integrated into broader neuromuscular conditioning frameworks. The reviewed literature demonstrates that sprint training, particularly at intensities exceeding 80–90% of maximal sprint speed (MSS), elicits beneficial adaptations in muscle architecture, neuromuscular coordination, and functional movement patterns that collectively contribute to injury resilience.

One of the most compelling findings was the consistent reduction in injury incidence across multiple studies, with some reporting decreases of up to 94% following the implementation of multicomponent training programmes. For example, Malone et al. [33] and Colby et al. [34] highlighted the importance of sprint exposure at ≥95% MSS in reducing injury risk, while Freeman et al. [35] and Ripley et al. [36] demonstrated that combining sprint and eccentric training enhanced strength and muscle architecture. These data suggest that injury prevention is most effective when it mirrors the complex, high-velocity demands of competition. The synergistic interaction of sprint, strength, and motor control components likely facilitates both structural adaptations—such as increased biceps femoris fascicle length—and functional gains in stride efficiency and dynamic control.

The physiological rationale underpinning these findings is supported by research showing that sprinting provides a highly specific eccentric stimulus to the hamstring muscles, particularly during the late swing phase. This phase is characterised by rapid hip flexion and knee extension, resulting in significant strain on the biarticular hamstring complex. Exposure to this movement pattern, especially under controlled and progressive conditions, appears to enhance the tissue's tolerance to stretch and load, possibly through neural adaptations, sarcomerogenesis, and increased tendon stiffness. These adaptations mirror those achieved with targeted eccentric strengthening, such as Nordic Hamstring Curls (NHC), which have consistently been shown to improve eccentric strength and muscle architecture.

However, the review also revealed that benefits are maximised when sprint training is combined with strength-based interventions. Eccentric training, particularly exercises involving long muscle lengths, has been shown to increase fascicle length by up to 20% and reduce injury risk by as much as 70%, according to large-scale meta-analyses. The Razor

Curl and assisted NHC, for instance, have demonstrated efficacy not only in improving eccentric torque but also in producing long-term architectural changes in the biceps femoris. Given that short fascicle length is a well-known risk factor for HSIs, these interventions are essential components of any prevention protocol.

Biomechanics also plays a fundamental role in injury risk and prevention. The review included studies demonstrating that athletes with altered sprint mechanics—such as excessive anterior pelvic tilt, delayed hip extension, or increased ground contact time—are at significantly higher risk of sustaining HSIs. Edouard et al. [39] found that such biomechanical deviations were associated with a 2.8-fold increase in injury risk. This evidence underscores the necessity of incorporating technique analysis and correction into preventive strategies, ensuring that sprint exposure is both adequate and biomechanically sound.

Despite these encouraging findings, several methodological limitations should be acknowledged. Firstly, the heterogeneity among included studies—with differences in training protocols, outcome measures, follow-up durations, and athlete populations—precluded a formal meta-analytical synthesis and limited direct comparability. Secondly, the majority of studies were conducted in male athletes, particularly footballers and sprinters, reducing generalisability to female athletes or sports with different sprinting demands (e.g., hockey, basketball, tennis). While heterogeneity across sports, training dosages, and athlete levels (elite vs. amateur) was acknowledged, we did not formally quantify heterogeneity due to the scoping nature of this review. Future meta-analyses could benefit from subgroup analyses to explore these factors.

Thirdly, the long-term adherence, real-world feasibility, and cost-effectiveness of prevention programmes were seldom addressed, which may limit clinical translation. Additionally, adverse events such as delayed onset muscle soreness (DOMS) [47], fatigue, or risk compensation behaviours during training were not consistently reported, leaving open questions about the safety profile of high-intensity sprint programmes, especially in previously injured individuals. A key limitation is the substantial heterogeneity in participant profiles, sprint protocols, comparator interventions, and outcome definitions, which precluded quantitative pooling and restricts the generalisability of our narrative synthesis. Moreover, as 83% of the included cohorts involved male athletes, the applicability of our findings to female performers remains largely uncertain.

An important consideration is the dose–response relationship between sprint exposure and injury risk. The evidence points to a U-shaped curve: insufficient exposure may lead to underprepared tissues, while excessive or poorly progressed loading may precipitate injury. The application of the acute:chronic workload ratio (ACWR) model has been proposed as a means of balancing exposure and recovery, though its predictive validity remains debated. Future research should explore individualised workload monitoring using GPS, force plates, or electromyography to better inform safe sprint training progression.

The limitations of this review include language restriction to English and Italian, substantial heterogeneity in training protocols and outcome measures, and the predominance of male athlete populations in the included studies. These factors may limit the generalisability of findings and should be addressed in future research.

Few studies reported on delayed onset muscle soreness (DOMS), fatigue, or compensatory behaviours following sprint protocols—an area that warrants systematic investigation given its potential impact on training adherence and injury risk.

Finally, theoretical contributions from authors such as Sancese et al. [45] emphasise the value of specificity in preventive interventions. Exercises that replicate the mechanical and temporal demands of sprinting are more likely to induce protective adaptations than generalised resistance training alone. This conceptual shift supports a paradigm in which sprinting is not simply tolerated in training but actively prescribed as part of injury prevention.

While more high-quality trials are needed to refine protocols and expand applicability, the current body of evidence supports the incorporation of progressive, technically sound sprint training as a valuable component of modern HSI prevention. The integration of sprint drills with eccentric and neuromuscular conditioning may not only reduce injury rates but also enhance performance and tissue resilience in elite and sub-elite athletes. Given the exploratory nature of this scoping review, no meta-analysis or formal risk of bias assessment was conducted, in accordance with JBI recommendations. While this limits the ability to draw definitive conclusions, it provides a comprehensive mapping of the current evidence. Future systematic reviews incorporating meta-analytic methods and quality assessments are necessary to quantify effect sizes, dose-response relationships, and to guide evidence-based practice more robustly.

10.1. *Synergistic Effects with Distal Kinetic-Chain Interventions*

Forces generated during ground contact propagate along the entire lower-limb kinetic chain. Consequently, distal segments such as the foot may modulate proximal hamstring loading. Emerging evidence indicates that impaired foot biomechanics—e.g., reduced mid-foot stiffness or altered fore-foot load distribution—can increase proximal muscular demand by attenuating energy dissipation at push-off [48]. Interventions that improve foot stability, intrinsic plantar-flexor strength, and fifth-metatarsal load distribution have demonstrated favourable kinetic adaptations during the sprinting stance phase [48]. It is therefore plausible that sprint training, when combined with targeted foot-stability exercises (e.g., short-foot drills, resisted toe-flexion, sensorimotor balance work), provides a complementary stimulus that further attenuates the risk of hamstring strain. Future interventional studies should explore such integrative approaches.

10.2. *Clinical Practice Implications*

The findings of this review offer practical recommendations for sports medicine and performance professionals aiming to reduce hamstring injury risk in competitive athletes. Firstly, sprint exposure should be viewed not as a hazard to avoid but as an essential neuromuscular stimulus, provided it is implemented progressively and technically correctly. Training sessions should incorporate repeated sprint efforts at intensities $\geq$90% of MSS, with careful attention to weekly sprint load and acute:chronic ratio progression.

Secondly, sprint training must not occur in isolation. It should be combined with eccentric strengthening exercises—such as NHC, Razor Curl, or hip-dominant variations—which are effective in increasing fascicle length and eccentric capacity. These modalities can be periodised and alternated based on the phase of the season and the athlete's training history.

Thirdly, attention to sprint mechanics is essential. Biomechanical inefficiencies should be screened through video analysis or motion capture when available. Athletes exhibiting risk markers such as poor hip extension velocity or excessive pelvic tilt should receive targeted technical coaching and corrective exercises.

Fourthly, foot biomechanics should be optimised. Exercises that enhance intrinsic foot-muscle strength, medial-arch control, and fifth-metatarsal load sharing—such as short-foot holds, resisted toe curls, and proprioceptive single-leg stance—may reduce aberrant distal mechanics that otherwise amplify proximal hamstring demand [48]. These drills can be blended into warm-ups or neuromotor circuits surrounding sprint sessions.

Finally, injury prevention should be athlete-centred and context-sensitive. History of prior injury, sport-specific demands, position, age, and neuromuscular asymmetries should inform programme design. Integrating monitoring technologies (e.g., GPS, sprint timing systems, muscle function tests) into regular assessments may allow practitioners to detect overload or maladaptation before injuries occur.

Overall, the current evidence encourages a paradigm shift in HSI prevention—from avoidance of sprinting to its strategic inclusion as a cornerstone of injury-resilient performance.

11. Conclusions

This scoping review highlights the preventive value of sprint-based training in reducing hamstring injuries among competitive athletes. When progressively and technically implemented, sprint exposure—especially at $\geq$90% of the maximal sprint speed—can serve as both a performance enhancer and an injury mitigation strategy. The most effective interventions combine sprint work with eccentric strengthening, agility, and biomechanical optimisation. Future research should aim to standardise protocols and extend findings to more diverse athletic populations. These findings support the inclusion of sprint training as a key component of hamstring injury prevention and inform evidence-based protocols for sports practitioners.

Author Contributions: R.T. and F.G. conceptualised and designed the study and were responsible for data acquisition. R.T. drafted the manuscript. F.G. and D.D. provided supervision and guidance throughout the study. F.G. performed the editing of the manuscript. R.T. reviewed the manuscript and curated the methodology. All authors have read and agreed to the published version of the manuscript.

Funding: This research received no external funding.

Conflicts of Interest: The authors declare no conflicts of interest.

References

1. Ekstrand, J.; Hägglund, M.; Waldén, M. Injury Incidence and Injury Patterns in Professional Football: The UEFA Injury Study. *Br. J. Sports Med.* **2011**, *45*, 553–558. [CrossRef]
2. Liu, H.; Garrett, W.E.; Moorman, C.T.; Yu, B. Injury Rate, Mechanism, and Risk Factors of Hamstring Strain Injuries in Sports: A Review of the Literature. *J. Sport Health Sci.* **2012**, *1*, 92–101. [CrossRef]
3. Arnason, A.; Andersen, T.E.; Holme, I.; Engebretsen, L.; Bahr, R. Prevention of Hamstring Strains in Elite Soccer: An Intervention Study. *Scand. J. Med. Sci. Sports* **2008**, *18*, 40–48. [CrossRef]
4. Aiello, F.; Di Claudio, C.; Fanchini, M.; Impellizzeri, F.M.; McCall, A.; Sharp, C.; Brown, S.J. Do Non-Contact Injuries Occur during High-Speed Running in Elite Football? Preliminary Results from a Novel GPS and Video-Based Method. *J. Sci. Med. Sport* **2023**, *26*, 465–470. [CrossRef] [PubMed]
5. van den Tillaar, R.; Solheim, J.A.B.; Bencke, J. Comparison of Hamstring Muscle Activation During High-Speed Running and Various Hamstring Strengthening Exercises. *Int. J. Sports Phys. Ther.* **2017**, *12*, 718–727. [CrossRef]
6. Kalema, R.N.; Duhig, S.J.; Williams, M.D.; Donaldson, A.; Shield, A.J. Sprinting Technique and Hamstring Strain Injuries: A Concept Mapping Study. *J. Sci. Med. Sport* **2022**, *25*, 209–215. [CrossRef]
7. Woods, C.; Hawkins, R.; Maltby, S.; Hulse, M.; Thomas, A.; Hodson, A. The Football Association Medical Research Programme: An Audit of Injuries in Professional Football—Analysis of Hamstring Injuries. *Br. J. Sports Med.* **2004**, *38*, 36–41. [CrossRef]
8. Jonkers, I.; Stewart, C.; Spaepen, A. The Complementary Role of the Plantarflexors, Hamstrings and Gluteus Maximus in the Control of Stance Limb Stability during Gait. *Gait Posture* **2003**, *17*, 264–272. [CrossRef] [PubMed]
9. Higashihara, A.; Nagano, Y.; Ono, T.; Fukubayashi, T. Differences in Activation Properties of the Hamstring Muscles during Overground Sprinting. *Gait Posture* **2015**, *42*, 360–364. [CrossRef]
10. Dorn, T.W.; Schache, A.G.; Pandy, M.G. Muscular Strategy Shift in Human Running: Dependence of Running Speed on Hip and Ankle Muscle Performance. *J. Exp. Biol.* **2012**, *215*, 1944–1956. [CrossRef]
11. Schache, A.G.; Dorn, T.W.; Blanch, P.D.; Brown, N.A.T.; Pandy, M.G. Mechanics of the Human Hamstring Muscles during Sprinting. *Med. Sci. Sports Exerc.* **2012**, *44*, 647–658. [CrossRef] [PubMed]
12. Hegyi, A.; Gonçalves, B.A.M.; Finni, T.; Cronin, N.J. Individual Region- and Muscle-Specific Hamstring Activity at Different Running Speeds. *Med. Sci. Sports Exerc.* **2019**, *51*, 2274–2285. [CrossRef] [PubMed]

13. Morin, J.-B.; Gimenez, P.; Edouard, P.; Arnal, P.; Jiménez-Reyes, P.; Samozino, P.; Brughelli, M.; Mendiguchia, J. Sprint Acceleration Mechanics: The Major Role of Hamstrings in Horizontal Force Production. *Front. Physiol.* **2015**, *6*, 404. [CrossRef] [PubMed]
14. Ernlund, L.; Vieira, L.d.A. Hamstring Injuries: Update Article. *Rev. Bras. Ortop.* **2017**, *52*, 373–382. [CrossRef]
15. Dalton, S.L.; Kerr, Z.Y.; Dompier, T.P. Epidemiology of Hamstring Strains in 25 NCAA Sports in the 2009–2010 to 2013–2014 Academic Years. *Am. J. Sports Med.* **2015**, *43*, 2671–2679. [CrossRef]
16. Kerkhoffs, G.M.; van den Bekerom, M.; Elders, L.A.M.; van Beek, P.A.; Hullegie, W.A.M.; Bloemers, G.M.F.M.; de Heus, E.M.; Loogman, M.C.M.; Rosenbrand, K.C.J.G.M.; Kuipers, T.; et al. Diagnosis, Treatment and Prevention of Ankle Sprains: An Evidence-Based Clinical Guideline. *Br. J. Sports Med.* **2012**, *46*, 854–860. [CrossRef]
17. Timmins, R.G.; Bourne, M.N.; Shield, A.J.; Williams, M.D.; Lorenzen, C.; Opar, D.A. Short Biceps Femoris Fascicles and Eccentric Knee Flexor Weakness Increase the Risk of Hamstring Injury in Elite Football (Soccer): A Prospective Cohort Study. *Br. J. Sports Med.* **2016**, *50*, 1524–1535. [CrossRef]
18. Martin, R.L.; Cibulka, M.T.; Bolgla, L.A.; Koc, T.A.; Loudon, J.K.; Manske, R.C.; Weiss, L.; Christoforetti, J.J.; Heiderscheit, B.C. Hamstring Strain Injury in Athletes. *J. Orthop. Sports Phys. Ther.* **2022**, *52*, CPG1–CPG44. [CrossRef]
19. Crow, J.; Semciw, A.; Couch, J.; Pizzari, T. Does a Recent Hamstring Muscle Injury Affect the Timing of Muscle Activation during High Speed Overground Running in Professional Australian Football Players? *Phys. Ther. Sport* **2020**, *43*, 188–194. [CrossRef]
20. Gabbe, B.J.; Bennell, K.L.; Finch, C.F.; Wajswelner, H.; Orchard, J.W. Predictors of Hamstring Injury at the Elite Level of Australian Football. *Scand. J. Med. Sci. Sports* **2006**, *16*, 7–13. [CrossRef]
21. Engebretsen, A.H.; Myklebust, G.; Holme, I.; Engebretsen, L.; Bahr, R. Intrinsic Risk Factors for Hamstring Injuries among Male Soccer Players: A Prospective Cohort Study. *Am. J. Sports Med.* **2010**, *38*, 1147–1153. [CrossRef] [PubMed]
22. Croisier, J.-L.; Ganteaume, S.; Binet, J.; Genty, M.; Ferret, J.-M. Strength Imbalances and Prevention of Hamstring Injury in Professional Soccer Players: A Prospective Study. *Am. J. Sports Med.* **2008**, *36*, 1469–1475. [CrossRef] [PubMed]
23. Mendiguchia, J.; Garrues, M.A.; Schilders, E.; Myer, G.D.; Dalmau-Pastor, M. Anterior Pelvic Tilt Increases Hamstring Strain and Is a Key Factor to Target for Injury Prevention and Rehabilitation. *Knee Surg. Sports Traumatol. Arthrosc.* **2024**, *32*, 573–582. [CrossRef] [PubMed]
24. Duhig, S.; Shield, A.J.; Opar, D.; Gabbett, T.J.; Ferguson, C.; Williams, M. Effect of High-Speed Running on Hamstring Strain Injury Risk. *Br. J. Sports Med.* **2016**, *50*, 1536–1540. [CrossRef]
25. Suarez-Arrones, L.; Nakamura, F.Y.; Maldonado, R.A.; Torreno, N.; Di Salvo, V.; Mendez-Villanueva, A. Applying a Holistic Hamstring Injury Prevention Approach in Elite Football: 12 Seasons, Single Club Study. *Scand. J. Med. Sci. Sports* **2021**, *31*, 861–874. [CrossRef]
26. Tedeschi, R.; Platano, D.; Giorgi, F. Stepping towards Relief: A Scoping Review on the Impact of Unstable Footwear in Chronic Low Back Pain Management. *J. Sports Med. Phys. Fitness* **2025**, *65*, 247–254. [CrossRef]
27. Green, B.; Bourne, M.N.; van Dyk, N.; Pizzari, T. Recalibrating the Risk of Hamstring Strain Injury (HSI): A 2020 Systematic Review and Meta-Analysis of Risk Factors for Index and Recurrent Hamstring Strain Injury in Sport. *Br. J. Sports Med.* **2020**, *54*, 1081–1088. [CrossRef]
28. Bramah, C.; Tawiah-Dodoo, J.; Rhodes, S.; Elliott, J.D.; Dos'Santos, T. The Sprint Mechanics Assessment Score: A Qualitative Screening Tool for the In-Field Assessment of Sprint Running Mechanics. *Am. J. Sports Med.* **2024**, *52*, 1608–1616. [CrossRef]
29. Haugen, T.; Tønnessen, E.; Hisdal, J.; Seiler, S. The Role and Development of Sprinting Speed in Soccer. *Int. J. Sports Physiol. Perform.* **2014**, *9*, 432–441. [CrossRef]
30. Lacome, M.; Avrillon, S.; Cholley, Y.; Simpson, B.M.; Guilhem, G.; Buchheit, M. Hamstring Eccentric Strengthening Program: Does Training Volume Matter? *Int. J. Sports Physiol. Perform.* **2020**, *15*, 81–90. [CrossRef]
31. Howard, R.M.; Conway, R.; Harrison, A.J. Muscle Activity in Sprinting: A Review. *Sports Biomech.* **2018**, *17*, 1–17. [CrossRef]
32. Mendiguchia, J.; Conceição, F.; Edouard, P.; Fonseca, M.; Pereira, R.; Lopes, H.; Morin, J.-B.; Jiménez-Reyes, P. Sprint versus Isolated Eccentric Training: Comparative Effects on Hamstring Architecture and Performance in Soccer Players. *PLoS ONE* **2020**, *15*, e0228283. [CrossRef] [PubMed]
33. Malone, S.; Roe, M.; Doran, D.A.; Gabbett, T.J.; Collins, K. High Chronic Training Loads and Exposure to Bouts of Maximal Velocity Running Reduce Injury Risk in Elite Gaelic Football. *J. Sci. Med. Sport* **2017**, *20*, 250–254. [CrossRef] [PubMed]
34. Colby, M.J.; Dawson, B.; Peeling, P.; Heasman, J.; Rogalski, B.; Drew, M.K.; Stares, J. Improvement of Prediction of Noncontact Injury in Elite Australian Footballers With Repeated Exposure to Established High-Risk Workload Scenarios. *Int. J. Sports Physiol. Perform.* **2018**, *13*, 1130–1135. [CrossRef] [PubMed]
35. Freeman, B.W.; Young, W.B.; Talpey, S.W.; Smyth, A.M.; Pane, C.L.; Carlon, T.A. The Effects of Sprint Training and the Nordic Hamstring Exercise on Eccentric Hamstring Strength and Sprint Performance in Adolescent Athletes. *J. Sports Med. Phys. Fitness* **2019**, *59*, 1119–1125. [CrossRef]
36. Ripley, N.J.; Cuthbert, M.; Comfort, P.; McMahon, J.J. Effect of Additional Nordic Hamstring Exercise or Sprint Training on the Modifiable Risk Factors of Hamstring Strain Injuries and Performance. *PLoS ONE* **2023**, *18*, e0281966. [CrossRef]

37. Tedeschi, R. Case Study: Gait Assessment of a Patient with Hallux Rigidus Before and After Plantar Modification. *Int. J. Surg. Case Rep.* **2024**, *114*, 109197. [CrossRef]
38. Lee, J.W.Y.; Mok, K.-M.; Chan, H.C.K.; Yung, P.S.H.; Chan, K.-M. Eccentric Hamstring Strength Deficit and Poor Hamstring-to-Quadriceps Ratio Are Risk Factors for Hamstring Strain Injury in Football: A Prospective Study of 146 Professional Players. *J. Sci. Med. Sport* **2018**, *21*, 789–793. [CrossRef]
39. Edouard, P.; Lahti, J.; Fleres, L.; Ahtiainen, J.; Ulvila, J.-J.; Lehtinen, T.; Virtanen, N.; Taipale, T.; Bellver, M.; Peltonen, V.; et al. A Musculoskeletal Multifactorial Individualised Programme for Hamstring Muscle Injury Risk Reduction in Professional Football: Results of a Prospective Cohort Study. *BMJ Open Sport Exerc. Med.* **2024**, *10*, e001866. [CrossRef]
40. Moreno-Pérez, V.; Del Coso, J.; López-Del Campo, R.; Resta, R.; Romero-Sangüesa, J.; Courel-Ibáñez, J.; Méndez-Villanueva, A. Reduced Match Exposure in the Previous 2 Matches Accounts for Hamstring Muscle Injury Incidence in Professional Football Players. *Sports Health* **2024**, *16*, 109–114. [CrossRef]
41. Peters: Joanna Briggs Institute Reviewer's Manual, JBI—Google Scholar. Available online: https://scholar-google-com.ezproxy.unibo.it/scholar_lookup?hl=en&publication_year=2020&author=MDJ+Peters&author=C+Godfrey&author=P+McInerney&author=Z+Munn&author=AC+Tricco&author=H+Khalil&title=Joanna+Briggs+Institute+Reviewer%27s+Manual%2C+JBI (accessed on 9 June 2022).
42. Tricco, A.C.; Lillie, E.; Zarin, W.; O'Brien, K.K.; Colquhoun, H.; Levac, D.; Moher, D.; Peters, M.D.J.; Horsley, T.; Weeks, L.; et al. PRISMA Extension for Scoping Reviews (PRISMA-ScR): Checklist and Explanation. *Ann. Intern. Med.* **2018**, *169*, 467–473. [CrossRef] [PubMed]
43. McGrath, T.M.; Hulin, B.T.; Pickworth, N.; Clarke, A.; Timmins, R.G. Determinants of Hamstring Fascicle Length in Professional Rugby League Athletes. *J. Sci. Med. Sport* **2020**, *23*, 524–528. [CrossRef] [PubMed]
44. Shah, S.; Collins, K.; Macgregor, L.J. The Influence of Weekly Sprint Volume and Maximal Velocity Exposures on Eccentric Hamstring Strength in Professional Football Players. *Sports* **2022**, *10*, 125. [CrossRef] [PubMed]
45. Sancese, A.; Taylor, L.; Walsh, G.; Byrd, E.; Delextrat, A. Effects of Sprint versus Strength Training on Risk Factors for Hamstring Injury in Football Players. *J. Sports Med. Phys. Fitness* **2023**, *63*, 580–587. [CrossRef]
46. Buchheit, M.; Settembre, M.; Hader, K.; McHugh, D. Exposures to Near-to-Maximal Speed Running Bouts during Different Turnarounds in Elite Football: Association with Match Hamstring Injuries. *Biol. Sport* **2023**, *40*, 1057–1067. [CrossRef]
47. Tedeschi, R. Delayed Onset Muscle Soreness in Neurological Rehabilitation: An Overlooked Challenge. *Acta Neurol. Belg.* **2025**, *125*, 585–586. [CrossRef]
48. Zhou, H.; Xu, D.; Quan, W.; Gao, Z.; Xiang, L.; Gu, Y. Are There Changes in the Foot Biomechanics during the before and after Fifth Metatarsal Fracture Running Stance Phase? *iScience* **2025**, *28*, 112432. [CrossRef]

applied sciences

MDPI

Review

Functional and Neuroplastic Effects of Cross-Education in Anterior Cruciate Ligament Rehabilitation: A Scoping Review with Bibliometric Analysis

Jorge M. Vélez-Gutiérrez [1,2,*], Andrés Rojas-Jaramillo [1,3,4,5], Juan D. Ascuntar-Viteri [1,5], Juan D. Quintero [1,5], Francisco García-Muro San José [6], Bruno Bazuelo-Ruiz [7], Roberto Cannataro [1,8] and Diego A. Bonilla [1,9,10,*]

1 Research Division, Dynamical Business & Science Society—DBSS International SAS, Bogotá 110311, Colombia; andres.rojasj@udea.edu.co (A.R.-J.); juandavid17121997@gmail.com (J.D.A.-V.); juanquintero.jdqm@gmail.com (J.D.Q.); r.cannataro@gmail.com (R.C.)
2 ARTHROS Centro de Fisioterapia y Ejercicio, Medellín 050033, Colombia
3 Research Group of Applied Sciences to Physical Activity and Sport, Universidad de Antioquia, Medellín 050010, Colombia
4 Practices and Knowledge in Latin America, Universidad Nacional de Colombia, Medellín 050034, Colombia
5 Educational and Pedagogical Studies and Research Group (GEIEP), Corporación Universitaria Minuto de Dios, Medellín 050034, Colombia
6 Facultad de Medicina, Universidad San Pablo-CEU, Campus Monteprincipe, 28668 Boadilla del Monte, Spain; fgarciamuro@ceu.es
7 Research Group in Sports Biomechanics (GIBD), Department of Physical Education and Sports, Universidad de Valencia, 46010 Valencia, Spain; bruno.bazuelo@uv.es
8 Galascreen Laboratories, Department of Pharmacy, Health, and Nutritional Sciences, University of Calabria, 87036 Rende, Italy
9 Hologenomiks Research Group, Department of Genetics, Physical Anthropology and Animal Physiology, University of the Basque Country (UPV/EHU), 48940 Leioa, Spain
10 Grupo de Investigación NUTRAL, Facultad Ciencias de la Nutrición y los Alimentos, Universidad CES, Medellín 050021, Colombia
* Correspondence: jorge.velez@arthros.com.co (J.M.V.-G.); dabonilla@dbss.pro (D.A.B.); Tel.: +57-3218004464 (J.M.V.-G.)

Featured Application: As a low-cost and simple strategy, cross-education can be incorporated early in anterior cruciate ligament rehabilitation to maintain musculoskeletal strength, reduce pain and interlimb asymmetries, and improve dynamic stability.

Abstract: Anterior cruciate ligament reconstruction (ACLR) results in prolonged muscle weakness, impaired neuromuscular control, and delayed return to sport. Cross-education (CE), unilateral training of the uninjured limb, has been proposed as an adjunct therapy to promote bilateral adaptations. This scoping review evaluated the functional and neuroplastic effects of CE rehabilitation post-ACLR. Following PRISMA-ScR and JBI guidelines, PubMed, Scopus, Web of Science, and PEDro were searched up to February 2025. A bibliometric analysis was also conducted to report keyword co-occurrence and identify trends in this line of research. Of 333 screened references, 14 studies (price index: 43% and low-to-moderate risk of bias) involving 721 participants (aged 17–45 years) met inclusion criteria. CE protocols (6–12 weeks; 2–5 sessions/week) incorporating isometric, concentric, and eccentric exercises demonstrated strength gains (10–31%) and strength preservation, alongside improved limb symmetry (5–14%) and dynamic balance (7–18%). There is growing interest in neuroplasticity and corticospinal excitability, although neuroplastic changes were assessed heterogeneously across studies. Findings support CE as a feasible and low-cost strategy to complement early-stage ACLR rehabilitation, especially when direct loading of the affected limb is limited. Standardized protocols for clinical intervention and neurophysiological assessment are needed.

Keywords: anterior cruciate ligament reconstruction; cross-education; rehabilitation; muscle strength; neuroplasticity; rehabilitation

1. Introduction

The anterior cruciate ligament (ACL) is one of the key ligaments that stabilize the knee, connecting the femur to the tibia and controlling rotation during flexion and extension [1]. ACL injuries, which manifest as stretching or tearing, are common in sports involving rapid movements, changes in direction, jumps, landings, and indirect or direct trauma to the knee, such as in soccer, basketball, and American football [2–6]. It has been estimated that the incidence of isolated ACL injuries is approximately 68.6 per 100,000 people per year, but it can reach figures close to 227 per 100,000 in young populations [7,8]. Among athletes, these numbers are usually higher but vary depending on age, sex, sport, type of exposure, and level of participation [9–12]. In addition to the functional impact, these injuries entail considerable economic costs, with an estimated USD 17,000 per anterior cruciate ligament reconstruction (ACLR) procedure [13]. In the context of professional sports, particularly in Europe, the incidence of ACL rupture is also significant among elite players, who often return to competition after surgery in relatively short periods, though these vary between leagues. For example, the average time to return to competition after surgery has been reported as 193 days in the French league, while in the English Premier League, it extends to 321 days. In contrast, the Italian, French, and German leagues show a faster recovery compared to the English league [14].

ACLR is considered a fundamental intervention to facilitate a return to sports activity [7,8,12,15]. However, after undergoing ACLR, it is common to observe decreased muscle strength, loss of knee stability, and high pain perception, which can delay the return to sport [16]. In fact, impaired muscle control often persists for up to two years after surgery [17,18]. Therefore, implementing an efficient rehabilitation process is crucial to reducing the risks associated with ACLR and improving functional outcomes [19,20]. These limitations have been documented in a recent systematic review, which shows that return-to-sport rates after ACL reconstruction can be as low as 24% [19]. Supporting the shortcomings of traditional rehabilitation approaches, these may be associated with incomplete recovery of muscle strength and neuromuscular control deficits (imbalance between flexors and extensors of the knee) [21,22], delaying the recovery process and increasing the risk of graft reinjury, additional injuries, and osteoarticular pathologies [20,23].

In this context, one of the proposed methods to optimize conventional rehabilitation is training the muscles of the uninjured limb, a technique known as cross-education (CE) [24]. First described in 1894, cross-education (CE) refers to the phenomenon whereby unilateral strength training induces bilateral neuromuscular adaptations [25]. Recently, CE has gained attention for its potential to preserve strength and promote recovery in various rehabilitation settings [26].

Regarding ACL rehabilitation, CE has demonstrated improvements in strength, kinesthesia, and balance in as little as four weeks, which is relevant since musculoskeletal injuries are associated not only with strength deficits but also with changes in the motor cortex and other brain regions [27,28]. It is important to note that CE can be implemented through various training modalities, including both isometric and dynamic exercises [29,30]. In ACLR, it is essential to address not only peripheral changes but also central nervous system alterations. Neuroplasticity, the brain's ability to reorganize in response to external stimuli or structural injuries, plays a key role in this process [31]. In this scenario, CE has shown positive effects by inducing adaptations in the untrained limb through central mechanisms linked to neuroplasticity [32].

Despite the demonstrated therapeutic potential of CE in ACL rehabilitation, significant gaps remain in understanding its neuroplastic mechanisms and their impact on the functional recovery of the untrained contralateral limb. Although CE has shown relevant potential for transferring functional improvements, the associated neuroplastic mechanisms are not yet clearly defined. Indeed, while there is consensus on the efficacy of CE, standardized protocols or application guidelines for ACL rehabilitation have yet to be established [33].

This lack of understanding, along with the absence of standardization, hinders the effective integration of CE into rehabilitation protocols, limiting the optimization of functional recovery and neurological adaptation [25,33–36]. Addressing these gaps through critical evaluation of measurement tools could contribute to shortening rehabilitation times, reducing reinjury risk, and prevent persistent functional deficits, thereby facilitating a safer and more effective return to physical and sports activities. The aim of this scoping review was to examine the available scientific evidence on the functional and neuroplastic effects of CE during ACL rehabilitation.

2. Materials and Methods

2.1. Protocol Registration

Given the conceptual diversity and methodological variability observed in CE protocols and outcome measures, a scoping review design was chosen over a traditional systematic review to better map the current evidence and identify knowledge gaps relevant to ACLR. For the development of this scoping review, we implemented the guidelines of the Preferred Reporting Items for Systematic Reviews and Meta-Analysis for Scoping Reviews (PRISMA-ScR) [37]. In accordance with Joanna Briggs Institute (JBI) guidelines for scoping reviews [38], we optimized the core methodology throughout the key stages (from problem formulation and literature search to evaluation, analysis, and presentation of findings) to systematize the review process and strengthen scientific rigor [39]. The protocol for this scoping review has been published on Figshare to make it publicly accessible and avoid unnecessary duplication of analysis for the same study objective (doi: 10.6084/m9.figshare.24268927).

2.2. Eligibility Criteria

The inclusion criteria for this scoping review were as follows: (i) experimental, observational, or theoretical articles, including randomized controlled trials, longitudinal studies, case studies, integrative reviews, systematic reviews, and meta-analyses; primary studies had a minimum follow-up period of six weeks, and secondary studies complied with PRISMA or COCHRANE guidelines or were registered in PROSPERO; (ii) peer-reviewed articles published up to the current date; (iii) publications written in English or Spanish; (iv) full-text availability; and, (v) articles reporting on the effects of anterior cruciate ligament rehabilitation through unilateral training or CE. Studies were excluded if they (i) were not original research (e.g., Editorials, commentaries, notes, theses), (ii) lacked a full-text version, (iii) involved participants with comorbid conditions unrelated to ACL injury, (iv) used combined interventions without isolating CE effects, or (v) presented unclear methodology.

2.3. Information Sources

PubMed, Scopus, Web of Science, and PEDro were selected to examine the available literature. Additionally, further articles were manually searched in Google Scholar or through snowballing.

2.4. Search Strategy

Based on the Population–Concept–Context (PCC) framework, the research question guiding this review was: What are the functional and neuroplastic effects of CE-based rehabilitation in individuals with ACL injury [38]? Database searches were conducted using specific algorithms: PubMed and Scopus: *"Anterior cruciate ligament" OR "anterior cruciate ligament reconstruction" AND "cross-education" OR "cross exercise" OR " unilateral strength training" OR "cross training" OR "cross transfer" OR "strength transfer" OR "contralateral strength training"*; Web of Science: *("Anterior cruciate ligament" OR "anterior cruciate ligament reconstruction") AND ("cross-education" OR " unilateral strength training" OR "cross training" OR "cross transfer" OR "strength transfer" OR "contralateral strength training")*; and PEDro: 1. *"Anterior cruciate ligament"*; 2. *Anterior cruciate ligament AND cross-education*; 3. *Anterior cruciate ligament AND contralateral strength training*. The database search was conducted between December 2024 and February 2025.

2.5. Selection and Data Collection Process

Two of the authors (J.D.Q. and J.D.A.) conducted independent searches in the selected databases. Publications that met the inclusion criteria were selected to proceed with the analysis and synthesis phases. A table was designed to report the results and compare key findings, including study type, design, objectives, intervention, procedures, and their main findings. Discrepancies will be identified and resolved through discussion with other authors if necessary (J.M.V., A.R-J., and D.A.B.). Based on best practices and previous studies by the DBSS Research Division [40,41], we assessed the obsolescence of the literature by calculating Price's index, defined as the ratio of references published within the past five years to the total number of references. The study selection was carried out during the months of February and April 2025.

2.6. Risk of Bias Assessment

The risk of bias was assessed according to the type of studies, based on the critical appraisal tools for evidence synthesis from JBI available on their website (https://jbi.global/critical-appraisal-tools, accessed on 5 February and 1 March 2025), and then graphically illustrated using Cochrane's RevMan 5.4 software.

2.7. Bibliometric Analysis

Trends in the field of CE were examined using VOSviewer v1.6.19 software to conduct bibliometric visualization. The first step involved performing a search in the Scopus database using the keyword: "cross-education" OR "cross education." Subsequently, an overlay map based on the co-occurrence of authors' keywords over the last 5 years was created to identify potential trends in this line of research. Minimum requirements for reporting bibliometric reviews of biomedical literature were considered [42].

3. Results

3.1. Study Selection

After running the search algorithms with Boolean operators (filtered by date and language), 333 references were retrieved. After removing duplicates from the publications, 219 studies remained. However, upon evaluating the titles and abstracts, only 28 potentially eligible publications were selected, of which 14 were subsequently excluded based on our pre-established criteria, resulting in a total of 14 studies that met the predefined requirements, with a price index of 43% [17,18,23,24,43–52]. Figure 1 shows a flow diagram of the literature search.

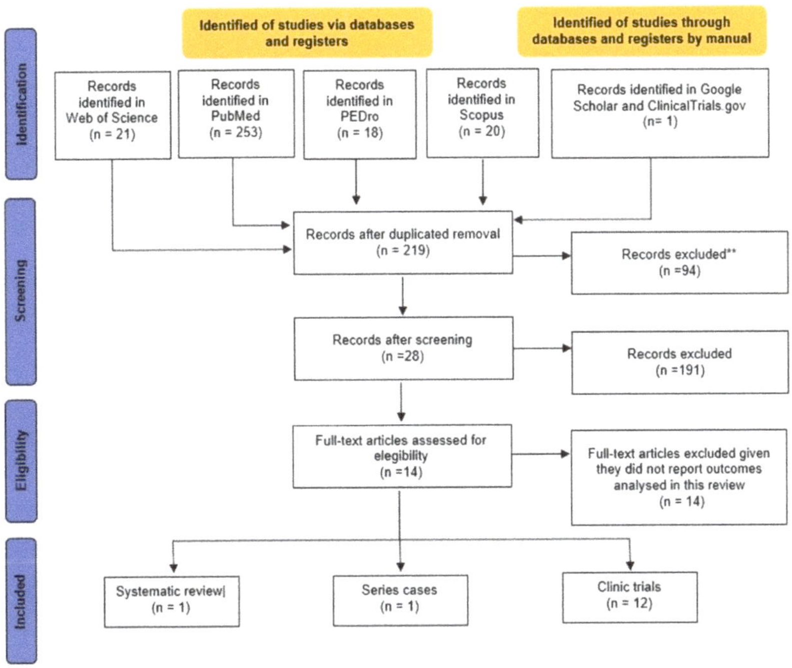

Figure 1. PRISMA flow diagram. ** Excluded because of duplication across databases.

3.2. *Risk of Bias Within Studies*

The methodological quality of the selected clinical trials is presented in Figure 2, while the risk of bias of the selected systematic review is presented in Figure 3.

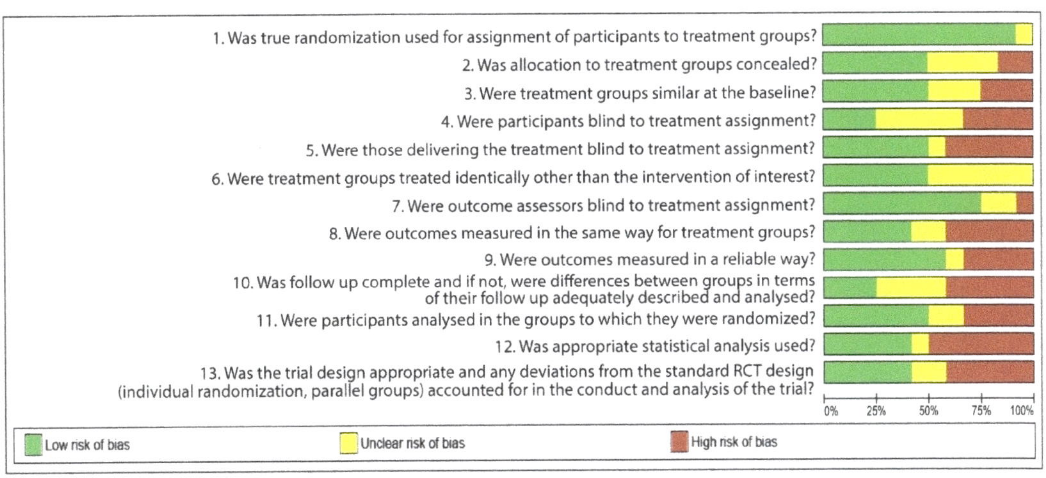

Figure 2. Summary of the risk of bias in the included clinical trials. Weighted bar chart of the distribution of risk of bias judgments.

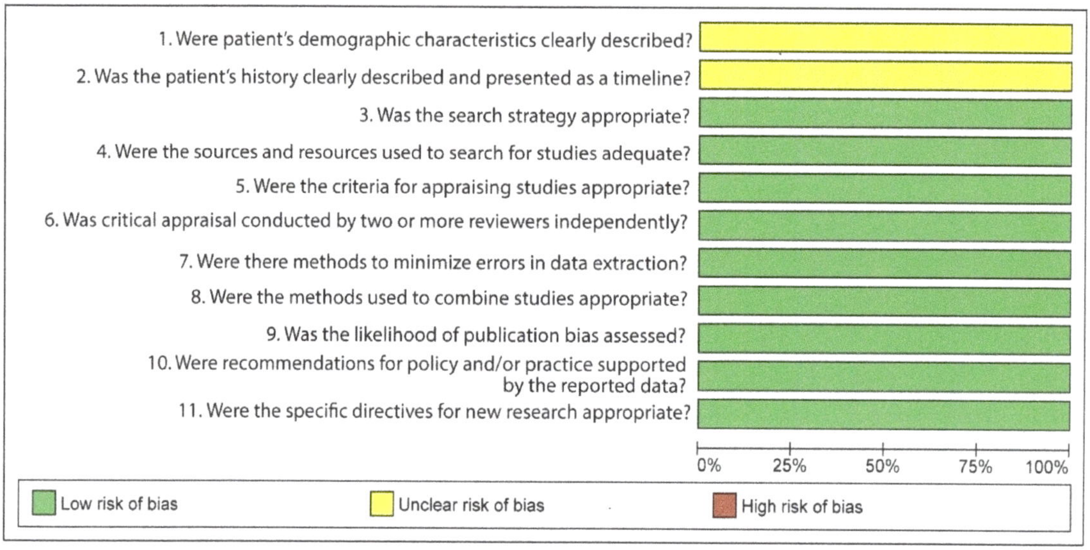

Figure 3. Summary of the risk of bias in the included systematic review. Weighted bar chart of the distribution of risk of bias judgments. No color indicates no risk of bias in that category.

The methodological quality of the selected case study is presented in Figure 4.

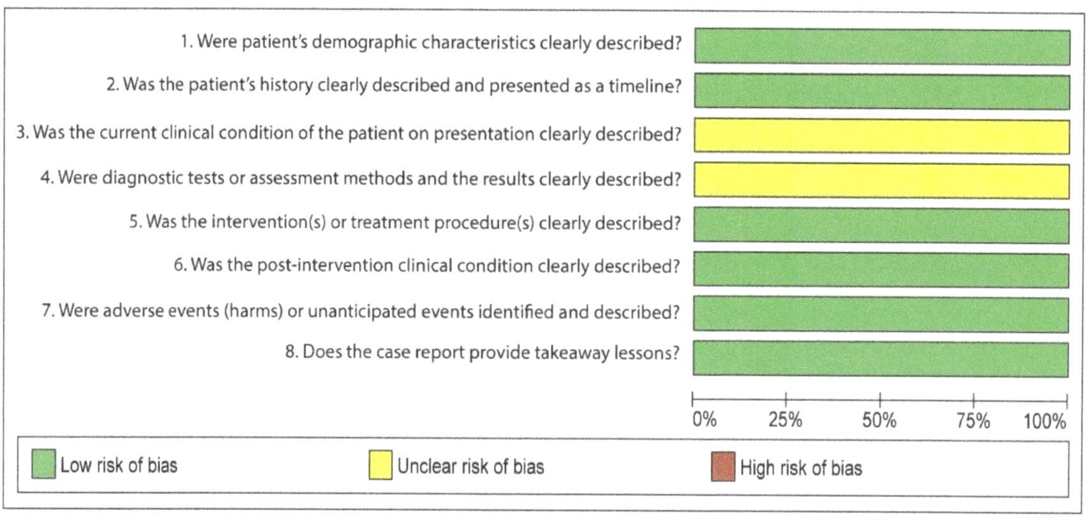

Figure 4. Summary of the risk of bias in the included case study. Weighted bar chart of the distribution of risk of bias judgments. No color indicates no risk of bias in that category.

3.3. Results of Individual Studies

The main characteristics and results of the selected studies are shown in Table 1.

Table 1. Synthesis of the selected articles.

Type of Study	Design	Objectives	Intervention	Outcomes	Key Findings/Conclusions	References
RCT	8 weeks, 48 post-ACLR patients (29.5 [6.8] years), $n = 16$/group (concentric CE, eccentric CE, CON); physically active	Evaluate effects of concentric and eccentric isokinetic contralateral limb training on QF recovery post-ACLR.	MVIC (IsoMed2000), SLHD, IKDC, Tegner. Unilateral isokinetic training (knee extension), 3 times/week, 3 × 12 reps at 60°/s, ROM 10–90°, concentric or eccentric mode.	MVIC (IsoMed2000), SLHD, IKDC	↑ QF in CE groups vs. CON at week 12 ($p = 0.04/0.03$) and week 24 ($p = 0.01/<0.001$); concentric: +28%, eccentric: +31%; no differences in SLHD or IKDC.	Harput et al. (2019) [17]
RCT	26 weeks, 43 post-ACLR (EXP = 22, CON = 21; 24 males/19 females), 28 [9] years; physically active (Tegner 7–8); no prior strength training data	Evaluate if adding CE to standard post-ACLR rehabilitation improves subjective knee function and strength recovery through week 26.	EXP group performed unilateral leg press and extension (contralateral CEE), 2 times/week, 3 × 8–12RM, weeks 1–12. Both groups received standard rehabilitation.	KOOS, MVIC (Biodex), HOP, LSI, Tegner	No CE effects on KOOS, QF, HMF or HOP vs. CON. Both groups improved KOOS (−15%), QF (↑5–14%), HMF (↑7–18%) by week 26.	Tjerk Zult et al. (2018) [18]
RCT	8 weeks CE, assessed at 10 and 24 weeks, 44 post-ACLR (CE = 22, CON = 22; 25 males/19 females), 31.8 ± 9.7 years; recreational athletes	Investigate immediate (10 wk) and long-term (24 wk) effects of high-intensity CE on strength and function post-ACLR.	CE with knee extension, hamstring curl, and leg press (contralateral limb), 3 times/week, 3 × 3–5RM, 1.5–2 min rest; initiated 2 weeks post-op; CON: stretching	QF, HMF, RFD (custom dynamometer), HOP, IKDC	At 10 weeks: ↓QF operated limb CE = −16.6% vs. CON = −32% ($p = 0.004$); ↑QF non-operated limb (7% vs. −7%, $p < 0.001$). No HMF or HOP differences. Similar LSI between groups (QF = 0.86).	Minshull et al. (2021) [23]

Table 1. *Cont.*

Type of Study	Design	Objectives	Intervention	Outcomes	Key Findings/Conclusions	References
RCT	26 weeks, 43 post-ACLR (EXP = 22, CON = 21; 24 males/19 females), 28 [9] years; recreational athletes	Examine whether adding CE to standard rehabilitation improves neuromuscular recovery post-ACLR.	EXP: unilateral leg press and extension (contralateral CEE), 2 times/week, 3 × 8–12RM, weeks 1–12; both groups received standard rehab. From weeks 24–36: power/agility training with unilateral jumps and direction changes.	MVIC (Biodex), CAR (IT), QF, HMF, HOP, SLS, SEBT, joint proprioception, Tegner	CE showed no strength/function benefits vs. CON. At week 12, CE ↓ CAR by 6% ($p = 0.023$) and impaired LSI by 9–10% ($p < 0.05$). Both groups improved force control (↑13–56%) and dynamic balance (~7%).	Tjerk Zult et al. (2019) [49]
SR and MA	7 RCTs included (5 in meta-analysis), 177 post-ACLR patients, 27.98 [2.23] years	Determine effectiveness of uninjured limb unilateral training on muscle strength and knee function post-ACLR.	Unilateral training of healthy limb. Isometric/isotonic/isokinetic strength, 2–5 times/week, 8–26 weeks vs. standard rehab.	MVIC (30–65°), LSI (Q), IKDC, LSI (SHT), self-reported questionnaires	CE improved QF MVIC (SMD = 0.60 to 0.58, $p < 0.01$); ↑ QF LSI (MD = 0.06, $p = 0.01$); no effect on HMF LSI (MD = −0.02, $p = 0.48$).	Iván Cuyúl-Vásquez et al. (2022) [24]
Clinical trial	8 weeks, 30 athletes with ACLR (EXP = 15, CON = 15; no sex data), 18–40 years	Investigate CE program effects on muscle activation and movement control in ACLR athletes (≥6 months post-op).	EXP: 3 times/week unilateral strength training (healthy leg) for TA, gas, RF, VM, BF, GM, RA, ES. CON: no structured exercise.	sEMG synchronized with force plate. GRF and COP oscillation analysis, MVIC	↑ EMG activation in TA, RF, GM during swing; ↑ TA, VM, RF, BF, GM during stance ($p < 0.05$). ↓ vertical/posterior GRF, ↓ lateral/anterior COP ($p < 0.05$).	Mostafa Payandeh & Daneshmandi (2025) [43]
RCT	6 weeks, 40 post-ACLR (EXP = 20, CON = 20; 22 males/18 females), 33.7 [5.6] years; no strength training data	Assess CE effects on knee function, kinematic parameters, dynamic balance, and plantar pressure post-ACLR.	EXP: CE (healthy limb) + conventional rehab, 3 times/week, progressing unilateral balance/functional tasks. CON: conventional rehab only.	SLS, ROM (goniometer), sEMG (Noraxon), HOP, GRF/COP (Zebris), Lysholm	EXP outperformed CON in Lysholm, ROM, RF activation, step length, ML control, forefoot/rearfoot pressure ($p < 0.05$); no midfoot/velocity differences.	Liu et al. (2025) [51]

Table 1. *Cont.*

Type of Study	Design	Objectives	Intervention	Outcomes	Key Findings/Conclusions	References
RCT	8 weeks, 88 post-ACLR (EXP = 44, CON = 44; 59 males/29 females), 28–29 years, active ≥3x/week	Compare unilateral vs. bilateral isotonic exercises for functional/strength symmetry post-ACLR with contralateral BPTB graft.	EXP: unilateral strength (donor leg) 2 times/week, 3 × 12 reps (press, terminal extension, leg ext., SLR, lunge). CON: bilateral equivalent.	MVIC (Biodex, 60°/s), H:Q, ROM (goniometer), laxity (KT-1000), HOP, Lysholm	EXP showed ↑ interlimb symmetry in peak torque, H:Q, HOP ($p < 0.001$). Greater donor-leg gains. CON improved reconstructed leg more.	Marcio Oliveira et al. (2022) [45]
RCT	8 weeks, 30 male athletes post-ACLR (EXP = 15, CON = 15), 27.5 [4.6] years; no strength training data	Evaluate contralateral neuromuscular exercises on static/dynamic balance, knee function, and pain post-ACLR.	EXP: CE 4 times/week, 30 min/session (unilateral balance on stable/unstable surfaces with arm tasks). CON: conventional PT.	SEBT, BESS, Stork test, Lysholm	EXP improved SEBT (ant/post/med/lat; $p < 0.05$), BESS (4 stances; $p < 0.05$), Stork ($p = 0.044$), ↓ VAS ($p = 0.014$). No Lysholm changes ($p = 0.71$).	Motahareh Karimijashni et al. (2023) [48]
Case Study	8 weeks, 5 post-ACLR (3 males/2 females), 16–21 years	Describe eccentric CE effects on neuroplasticity/perceived outcomes post-ACLR.	24 sessions (3 times/week), 4 × 10 eccentric isokinetic contractions at 60°/s (0–90° ROM) in healthy leg (≥60% 1RM); ACLR-free.	fMRI (brain activation), H-reflex (SRE), TMS (CSE), KOOS, TSK, IKDC, Tegner, qualitative interviews	↑ spinal (H:M ratio +57%) and cortical excitability (MEP +235%); ↓ frontal activity (fMRI); ↑ KOOS QOL (163%), ADL (136%), Sport (3076%).	Lepley et al. (2018) [44]
RCT	8 weeks, 42 soldiers post-ACLR (3 groups: 14 each; males only), 20–25 years; no strength training data	Compare 3 vs. 5 days/week CEE effects on CLL deficit post-ACLR.	CEE (healthy leg ext.) at 3x/week (G1) or 5 times/week (G2); 5 × 6 reps at 80% 1RM, 2 min rest. CON: standard rehab.	Isometric MVIC (KinCom AT+) at 60°; QF and QD pre/post.	CLL deficit post-CEE: G1 = 27.95%, G2 = 29.82%, CON = 53% ($p < 0.05$); G1/G2 ↑ healthy-leg CLL (22.7%/18%), ↓ injured-leg CLL (16.25%/6.3%).	Papandreou et al. (2013) [47]

Table 1. *Cont.*

Type of Study	Design	Objectives	Intervention	Outcomes	Key Findings/Conclusions	References
RCT	8 weeks, 42 soldiers post-ACLR (CE 3x/week = 14, CE 5x/week = 14, CON = 14; males only), 20–25	Investigate CEE (3 vs. 5 days/week) effects on ART and Lysholm early post-ACLR.	CEE (unilateral leg ext.) at 3 or 5 days/week, 5 × 6 reps at 80% 1RM; all groups received conventional rehab.	Isometric MVIC (KinCom AT+), ART at 45°/60°/90°, Lysholm	CE 3 times/week improved ART at 90° vs. CON ($p = 0.01$); no differences at 45°/60°. Lysholm ↑ in CE 3 times/week (Δ+7.5, $p < 0.01$) and CE 5 times/week (Δ+3.78, $p = 0.03$) vs. CON.	Papandreou et al. (2009) [46]
RCT	8 weeks, 42 soldiers post-ACLR (CE 3x/week = 14, CE 5x/week = 14, CON = 14; males only), 20–25 years	Assess CEE effects on isometric QF at 45°/90° early post-ACLR, comparing 3 vs. 5 days/week.	CEE (healthy-leg unilateral ext.) at 3 or 5 times/week, 5 × 6 reps at 80% 1RM. Both groups received standard rehab.	Isometric MVIC (KinCom AT+) at 45°/90°	CE 3 times/week showed ↑ QF vs. CON at 45° (Δ98.4 Nm, $p = 0.02$) and 90° (Δ119.8 Nm, $p < 0.01$); no differences between CE groups.	Papandreou et al. (2007) [50]
RCT	10 weeks (6 weeks conventional rehab + 4 weeks intervention), 46 post-ACLR (EXP = 26, CON = 20), 30.9 [9.3] years	Determine if added isokinetic training improves knee strength/function post-arthroscopic ACLR.	EXP: 2 times/week, 3 × 20 reps at 180°/s (concentric), operated leg. H:Q torque. Both groups completed the initial 6-week conventional rehab.	MVIC (Biodex 3), ROM (Biodex), VAS, Lysholm, time-to-peak torque, torque variance, H:Q	EXP ↑ extension/flexion torque (30°/60°), ↓ extension torque variance, ↓ H:Q (30°); ↑ flexion ROM (Δ32°, $p = 0.026$), ↑ extension ROM (Δ−4.58°, $p = 0.044$); no Lysholm/VAS differences	Cheng-Pu Hsieh et al., (2016) [52]

[1] ACL = Anterior Cruciate Ligament; ACLR = Anterior Cruciate Ligament Reconstruction; ART = Accelerated Reaction Time; BESS = Balance Error Scoring System; BF = Biceps Femoris; CAR = Central Activation Ratio; CE = Cross-Education; CEE = Contralateral Eccentric Exercise; CON = Control Group; CLL = Contralateral Limb Strength; CSE = Corticospinal Excitability; ES = Erector Spinae; EXP = Experimental Group; Gas = Gastrocnemius; GM = Gluteus Medius; H:Q = Hamstring-to-Quadriceps Ratio; HMF = Hamstring Maximal Force; HOP = Hop for Distance; IKDC = International Knee Documentation Committee; KOOS = Knee Injury and Osteoarthritis Outcome Score; LSI = Limb Symmetry Index ([Operated Limb Strength / Non-Operated Limb Strength] × 100); Lysholm = Lysholm Score; MVIC = Maximal Voluntary Isometric Contraction; QF = Quadriceps Strength; RA = Rectus Abdominis; RF = Rectus Femoris; RCT = Randomized Clinical Trial; RFD = Rate of Force Development; ROM = Range of Motion; SEBT = Star Excursion Balance Test; SLHD = Single Leg Hop for Distance Test; SLS = Single Leg Stance Test; SR&MA = Systematic Review and Meta-Analysis; SRE = Spinal Reflex Excitability; sEMG = Surface Electromyography; TA = Tibialis Anterior; Tegner = Tegner Activity Scale; TMS = Transcranial Magnetic Stimulation; TSK = Tampa Scale for Kinesiophobia; VAS = Visual Analog Scale; ↑ = increase; ↓ = decrease.

Twelve randomized controlled clinical trials (RCTs), one case study, and one systematic review with meta-analysis were included, totaling 721 participants, reduced to 637 after removing duplicates of participant samples across different studies by the same author. The sample primarily consisted of young adults, aged between 17 and 45, with a BMI of 21 to 25. The interventions lasted between 6 and 12 weeks, with 2 to 5 sessions per week, focusing on unilateral training of the contralateral non-injured leg. The protocols included isometric and dynamic exercises with progressive loads [24,46,47,50], and some studies used electromyography (EMG) to monitor neuromuscular activation and physiological adaptations [43,51]. A low-to-moderate risk of bias was detected across the articles.

The studies employed various assessment tools, such as isokinetic dynamometry, stability and balance tests, and pain perception scales. Muscle strength, range of motion (ROM), and knee function were also evaluated. The results showed significant improvements in strength and function of the untrained contralateral limb, particularly in quadriceps strength and functional symmetry. Additionally, a gradual recovery of strength was observed in the operated leg, along with improvements in functionality and dynamic balance. However, the variability in results highlights the need to standardize protocols and measurement tools to optimize the clinical application of CE in ACL rehabilitation.

Regarding neuroplasticity, only one study [44] directly evaluated central adaptations using neurophysiological tools such as fMRI, TMS, and H-reflexes, reporting significant increases in cortical and spinal excitability following eccentric CE. Additionally, other studies indirectly addressed neuroplasticity through changes in EMG activity, postural control, and motor coordination [43,51], suggesting functional improvements potentially mediated by central nervous system adaptations.

3.4. Bibliometric Analysis

A bibliometric analysis of keyword co-occurrence in articles on CE was conducted, using VOSviewer to outline the state of the art in this emerging field. This approach allows for the identification of trends, conceptual relationships, and knowledge gaps, facilitating an understanding of how CE is researched and applied in neuromuscular training and rehabilitation. In the resulting map, each node represents an author keyword, where the size reflects its co-occurrence frequency, the thickness of the links indicates the strength of the relationships, and the color denotes conceptual clustering (Figure 5).

The results highlight the prominence of CE (149:226) in the literature, particularly in relation to resistance training (57:112) and rehabilitation (30:63), supporting its use for improving strength through unilateral training. The strong association with strength (46:86) and interlimb transfer (19:48) underscores its relevance in the functional recovery from injuries, including those of the anterior cruciate ligament (5:7). Additionally, the growing interest in terms such as neuroplasticity (7:23) and corticospinal excitability (8:20) demonstrates that CE not only enhances strength but also induces beneficial neurological changes for recovery. Although the application of EC in clinical contexts, such as stroke rehabilitation (4:6), is expanding, areas like mirror therapy (5:9) and assessment through motor-evoked potentials (4:9) remain less explored, opening new opportunities to optimize its implementation using neuromuscular activity quantification methods, such as electromyography (22:50) and H-reflex (6:15). These findings not only reinforce the potential applicability of CE in post-surgical rehabilitation and muscle function improvement but also establish a foundation for future research integrating clinical and neurophysiological approaches, which could contribute to the standardization of CE rehabilitation protocols.

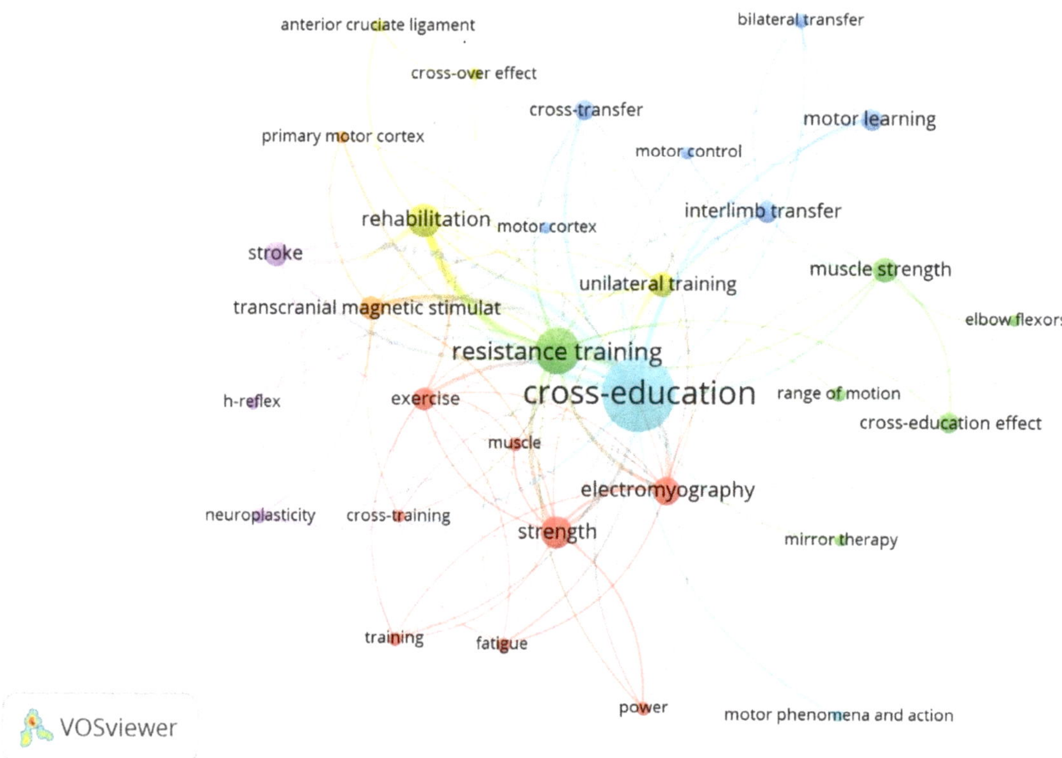

Figure 5. Visualization map of keyword co-occurrence from studies on cross-education (CE) in rehabilitation, generated with VOSviewer. Each node represents an author keyword; the size of the node reflects its frequency of appearance, while the thickness of the connecting lines indicates the strength of co-occurrence between terms. The color of each node corresponds to a conceptual cluster, allowing visualization of thematic groupings (e.g., strength, neuroplasticity, rehabilitation). This map highlights CE's strong association with unilateral training, neuromuscular recovery, and emerging interest in cortical excitability, suggesting evolving priorities in the scientific exploration of CE applications. Available online: https://tinyurl.com/2485baur (accessed on 27 March 2025).

4. Discussion

This scoping review aimed to synthesize the scientific evidence on the functional and neuroplastic effects of cross-education (CE) interventions during the rehabilitation after anterior cruciate ligament reconstruction (ACLR). Overall, the results show that CE can play a complementary role in postoperative recovery after ACLR, particularly in the early phases of rehabilitation when direct loading on the affected limb must be limited. Most of the selected studies were randomized controlled trials [17,18,23,45–48,50,51], along with one quasi-experimental study [43], a case study [44] and a systematic review with meta-analysis [24]. These investigations were primarily conducted in young, physically active adults undergoing different ACLR surgical techniques. Generally, the interventions focused on unilateral training of the contralateral limb to assess its influence on muscle strength, balance, functional performance, and neuromuscular activation.

Notably, several clinical trials repeatedly reported that CE in ACLR helps preserve or improve quadriceps strength in the affected limb, especially during early postoperative rehabilitation. This finding was observed between weeks 8 and 26 post-surgery in some studies [17,46–49,51]. These interventions involved moderate-to-high-intensity eccentric or isometric contractions. Similarly, improvements were seen in functional parameters such as limb symmetry, dynamic balance, and postural control, reinforcing CE's potential clinical value as an adjunct tool in neuromuscular recovery. Conversely, discrepancies were noted between studies, particularly those using less intensive protocols or featuring delayed CE implementation. For example, two controlled trials by Zult et al. [49] found no significant effects of CE on neuromuscular function or functional recovery compared to conventional rehabilitation, suggesting that outcomes may largely depend on implementation timing, contraction type, and stimulus intensity, as per the available literature.

To evaluate these effects, studies employed a variety of objective and subjective tools. Objective measures included isokinetic dynamometry, surface electromyography, and functional tests such as the Single-Leg Hop Test, Star Excursion Balance Test, and Balance Error Scoring System. Subjective measures included knee function scales like the International Knee Documentation Committee (IKDC) [53], the Knee Injury and Osteoarthritis Outcome Score (KOOS) [54], the Lysholm Knee Scoring Scale (Lysholm) [54], and the Hughston Clinic Knee Scoring System (HCKS) [55]. Additionally, some studies, particularly those conducted in non-ACL contexts, used advanced neurophysiological techniques such as TMS, fMRI, or transcranial direct current stimulation to investigate potential cortical reorganization mechanisms associated with CE.

From a functional perspective, most of the included studies reported significant improvements in quadriceps strength of the operated limb, limb symmetry, hamstring-to-quadriceps (H:Q) ratio, postural control, and performance in various functional tests. These adaptations were particularly evident between weeks 8 and 26 after surgery. Interventions incorporating high-intensity eccentric or isometric contractions of the contralateral limb were effective in attenuating muscle strength deficits [46,48]. For example, Harput et al. [17] demonstrated that high-intensity eccentric CE produced greater quadriceps strength gains (+31%) compared to concentric CE or conventional rehabilitation. Conversely, Zult et al. [18] found no significant improvements when CE was performed with lower intensity or delayed in the postoperative timeline, highlighting that both the load magnitude and timing of implementation are key determinants of clinical outcomes.

Regarding neuroplastic effects, direct evidence in ACLR populations is scarce. Only the case series by Lepley et al. [44] employed a CE-based intervention and their study provided relevant data by analyzing the relationship between quadriceps motor unit properties and corticospinal excitability in ACLR patients. Their results showed that lower discharge rates and poor motor unit recruitment were associated with reduced excitability in the primary motor cortex. These findings support the idea that postoperative functional deficits may be mediated not only by peripheral factors but also by central mechanisms, which could be modulated through strategies like CE. Nevertheless, most studies in this area come from research on healthy individuals or those with different neurological conditions, such as stroke or unilateral immobilization. These studies provide additional evidence on potential neurophysiological mechanisms of CE, including increased primary motor cortex excitability, reduced intracortical inhibition, and functional changes in interhemispheric connectivity [56–58].

Collectively, the findings of this scoping review support the potential of CE as a complementary tool in designing postoperative ACL rehabilitation programs, particularly in early phases where direct training of the affected limb is limited. The possibility of preserving or even improving muscle strength and functional performance through stimuli applied to the non-injured limb could help reduce inactivity time, mitigate neuromuscular asymmetries, and optimize readiness for progressive return to loading. From a clinical standpoint, these results position CE as a viable option within individualized rehabilitation protocols, especially in contexts where careful load progression is needed.

Similarly, the identified methodological limitations, along with the scarce evidence on neuroplasticity in ACLR populations, underscore the need for future research incorporating direct neurophysiological techniques. Standardizing intervention protocols and assessing CE effects at different postoperative stages would be beneficial. Additionally, expanding the study population to include women, less physically active individuals, and patients with different graft types would improve the clinical applicability of findings. Exploring the long-term impact of CE, as well as its combination with other therapeutic modalities, could represent a promising research avenue in rehabilitation.

4.1. Limitations and Future Directions

While this scoping review highlights clinically valuable insights, several limitations inherent to the available evidence must be acknowledged. First, several studies featured small sample sizes ($n < 30$), limiting statistical power and generalizability. Significant heterogeneity exists across included studies regarding sample characteristics (age, sex, graft type, physical activity level), intervention protocols (contraction types, duration, postoperative phase for CE initiation), and outcome measures (e.g., use of isokinetic vs. isotonic testing, different timepoints). This variability along with a relatively low price index (<50%) complicates direct comparisons and hinders standardized clinical recommendations. We advocate for an individualized, multisystem approach that considers prior knowledge and workload (external/internal loads), as these factors critically influence both injury prevention and management [40,59]. Second, despite observed functional improvements, direct evidence of CE's neuroplastic effects in ACLR populations remains scarce. Only one study evaluated cortical structure and corticospinal excitability, leaving a key gap in understanding the neurophysiological mechanisms behind these adaptations. Most neuroplasticity data derive from healthy subjects or other neurological conditions, limiting generalizability. Further, inconsistent use of standardized measures for strength, neuromuscular activation, and knee function risks methodological bias and confounds result interpretation. Some trials also exhibited moderate bias due to unblinded designs or lack of active controls, potentially compromising internal validity.

Future research should prioritize the following:

- Advanced neurophysiological techniques (fMRI, TMS, evoked potentials) in real-world ACLR rehabilitation settings;
- Standardized CE protocols tested across rehabilitation phases, graft types, and fitness levels;
- Larger samples with greater inclusion of women and recreational athletes, plus longitudinal designs to assess long-term effects;
- Investigation of CE combined with adjunct therapies (e.g., neuromuscular stimulation [60], blood flow restriction [61,62], nutritional interventions [63,64]).

4.2. Practical Recommendations

This scoping review offers clinically relevant insights for post-surgical ACL reconstruction rehabilitation. CE emerges as a particularly safe, effective, and cost-efficient strategy during early rehabilitation when direct training of the affected limb is limited by pain, inflammation, or overtraining risks. Research demonstrates that single-limb strength training, particularly moderate-to-high intensity isometric or eccentric contralateral exercises, can improve quadriceps strength in the operated leg, reduce muscle atrophy, enhance limb symmetry, and support progressive return to exercise.

These interventions show benefits when applied 2–5 times weekly over 6–12 weeks, with functional improvements observable within eight weeks post-surgery. CE proves especially valuable in sports reconditioning programs where maintaining neuromuscular symmetry and minimizing downtime are critical for athletes. Importantly, these protocols can also benefit less active individuals when properly adapted to their functional capacity, injury type, and rehabilitation stage.

5. Conclusions

Rehabilitation protocols following ACL reconstruction (ACLR) should address more than just structural recovery, given the significant neurobiological changes accompanying the injury. These alterations lead to persistent muscle weakness, knee dysfunction, and maladaptive neuroplasticity. Cross-education (CE) has emerged as an effective intervention to counteract these maladaptive neuroplastic, functional, and morphological changes. The collective evidence analyzed in this scoping review supports the potential role of CE in preserving strength and improving functional outcomes during the early stages of ACLR rehabilitation. However, the limited use of neuroplastic assessment tools (e.g., fMRI, TMS, H-reflex) in current studies restricts our understanding of CE's central adaptations. Future research should prioritize neuroimaging and other advanced techniques to clarify the neurobiological mechanisms of CE.

Author Contributions: Conceptualization, J.M.V.-G. and D.A.B.; methodology, J.M.V.-G., A.R.-J., J.D.A.-V., J.D.Q. and D.A.B.; validation, J.M.V.-G. and D.A.B.; writing—original draft preparation, J.M.V.-G., J.D.A.-V. and J.D.Q.; writing—review and editing, A.R.-J., F.G.-M.S.J., B.B.-R., R.C. and D.A.B.; project administration, J.M.V.-G. All authors have read and agreed to the published version of the manuscript.

Funding: The APC was funded by the European DBSS Research Division.

Institutional Review Board Statement: Not applicable.

Informed Consent Statement: Not applicable.

Data Availability Statement: No new data were created or analyzed in this study.

Acknowledgments: We acknowledge the contributions of the developers of the open-source tools utilized in this study.

Conflicts of Interest: J.M.V.-G. receives honoraria for rehabilitation services and is currently managing director of IPS ARTHROS, a physiotherapy and exercise center. A.R-J., J.D.A-V., J.D.Q. and R.C. receive honoraria for personalized training and nutrition services in private centers. D.A.B. has conducted academic-sponsored research on sport and exercise sciences, is a delegate of the NSCA LATAM in Colombia (https://www.nsca.es/latam-board-advisors, accessed on 11 April 2025) and has received honoraria for selling muscular performance and body composition equipment as well as speaking about exercise sciences at international conferences and private courses. The other authors declare no conflicts of interest. All authors are responsible for the content of this article.

References

1. Guenther, D.; Herbst, E.; Musahl, V. Anatomy and Biomechanics of the Anterior Cruciate Ligament. In *Orthopaedic Biomechanics in Sports Medicine*; Koh, J., Zaffagnini, S., Kuroda, R., Longo, U.G., Amirouche, F., Eds.; Springer International Publishing: Cham, Switzerland, 2021; pp. 287–295.
2. Della Villa, F.; Tosarelli, F.; Ferrari, R.; Grassi, A.; Ciampone, L.; Nanni, G.; Zaffagnini, S.; Buckthorpe, M. Systematic Video Analysis of Anterior Cruciate Ligament Injuries in Professional Male Rugby Players: Pattern, Injury Mechanism, and Biomechanics in 57 Consecutive Cases. *Orthop. J. Sports Med.* **2021**, *9*, 23259671211048182. [CrossRef]
3. Tosarelli, F.; Buckthorpe, M.; Di Paolo, S.; Grassi, A.; Rodas, G.; Zaffagnini, S.; Nanni, G.; Della Villa, F. Video Analysis of Anterior Cruciate Ligament Injuries in Male Professional Basketball Players: Injury Mechanisms, Situational Patterns, and Biomechanics. *Orthop. J. Sports Med.* **2024**, *12*, 23259671241234880. [CrossRef] [PubMed]
4. Della Villa, F.; Stride, M.; Bortolami, A.; Williams, A.; Davison, M.; Buckthorpe, M. Systematic Video Analysis of ACL Injuries in Male Professional English Soccer Players: A Study of 124 Cases. *Orthop. J. Sports Med.* **2025**, *13*, 23259671251314642. [CrossRef]

5. Chia, L.; De Oliveira Silva, D.; Whalan, M.; McKay, M.J.; Sullivan, J.; Fuller, C.W.; Pappas, E. Non-contact Anterior Cruciate Ligament Injury Epidemiology in Team-Ball Sports: A Systematic Review with Meta-analysis by Sex, Age, Sport, Participation Level, and Exposure Type. *Sports Med.* **2022**, *52*, 2447–2467. [CrossRef]
6. Gornitzky, A.L.; Lott, A.; Yellin, J.L.; Fabricant, P.D.; Lawrence, J.T.; Ganley, T.J. Sport-Specific Yearly Risk and Incidence of Anterior Cruciate Ligament Tears in High School Athletes: A Systematic Review and Meta-analysis. *Am. J. Sports Med.* **2016**, *44*, 2716–2723. [CrossRef]
7. Sanders, T.L.; Maradit Kremers, H.; Bryan, A.J.; Larson, D.R.; Dahm, D.L.; Levy, B.A.; Stuart, M.J.; Krych, A.J. Incidence of Anterior Cruciate Ligament Tears and Reconstruction: A 21-Year Population-Based Study. *Am. J. Sports Med.* **2016**, *44*, 1502–1507. [CrossRef] [PubMed]
8. Mall, N.A.; Chalmers, P.N.; Moric, M.; Tanaka, M.J.; Cole, B.J.; Bach, B.R., Jr.; Paletta, G.A., Jr. Incidence and trends of anterior cruciate ligament reconstruction in the United States. *Am. J. Sports Med.* **2014**, *42*, 2363–2370. [CrossRef]
9. Montalvo, A.M.; Schneider, D.K.; Webster, K.E.; Yut, L.; Galloway, M.T.; Heidt, R.S., Jr.; Kaeding, C.C.; Kremcheck, T.E.; Magnussen, R.A.; Parikh, S.N.; et al. Anterior Cruciate Ligament Injury Risk in Sport: A Systematic Review and Meta-Analysis of Injury Incidence by Sex and Sport Classification. *J. Athl. Train.* **2019**, *54*, 472–482. [CrossRef]
10. Montalvo, A.M.; Schneider, D.K.; Yut, L.; Webster, K.E.; Beynnon, B.; Kocher, M.S.; Myer, G.D. "What's my risk of sustaining an ACL injury while playing sports?" A systematic review with meta-analysis. *Br. J. Sports Med.* **2019**, *53*, 1003–1012. [CrossRef]
11. Joseph, A.M.; Collins, C.L.; Henke, N.M.; Yard, E.E.; Fields, S.K.; Comstock, R.D. A multisport epidemiologic comparison of anterior cruciate ligament injuries in high school athletics. *J. Athl. Train.* **2013**, *48*, 810–817. [CrossRef] [PubMed]
12. Herzog, M.M.; Marshall, S.W.; Lund, J.L.; Pate, V.; Mack, C.D.; Spang, J.T. Trends in Incidence of ACL Reconstruction and Concomitant Procedures Among Commercially Insured Individuals in the United States, 2002–2014. *Sports Health* **2018**, *10*, 523–531. [CrossRef]
13. Bates, N.A.; McPherson, A.L.; Rao, M.B.; Myer, G.D.; Hewett, T.E. Sports Traumatology, Arthroscopy. Characteristics of inpatient anterior cruciate ligament reconstructions and concomitant injuries. *Knee Surg. Sports Traumatol. Arthrosc.* **2016**, *24*, 2778–2786. [CrossRef]
14. Mazza, D.; Viglietta, E.; Monaco, E.; Iorio, R.; Marzilli, F.; Princi, G.; Massafra, C.; Ferretti, A. Impact of anterior cruciate ligament injury on European professional soccer players. *Orthop. J. Sports Med.* **2022**, *10*, 23259671221076865. [CrossRef]
15. Culvenor, A.G.; Girdwood, M.A.; Juhl, C.B.; Patterson, B.E.; Haberfield, M.J.; Holm, P.M.; Bricca, A.; Whittaker, J.L.; Roos, E.M.; Crossley, K.M. Rehabilitation after anterior cruciate ligament and meniscal injuries: A best-evidence synthesis of systematic reviews for the OPTIKNEE consensus. *Br. J. Sports Med.* **2022**, *56*, 1445–1453. [CrossRef] [PubMed]
16. Rigg, J.D.; Panagodage Perera, N.K.; Toohey, L.A.; Cooke, J.; Hughes, D. Anterior cruciate ligament injury occurrence, return to sport and subsequent injury in the Australian High Performance Sports System: A 5-year retrospective analysis. *Phys. Ther. Sport.* **2023**, *64*, 140–146. [CrossRef] [PubMed]
17. Harput, G.; Ulusoy, B.; Yildiz, T.I.; Demirci, S.; Eraslan, L.; Turhan, E.; Tunay, V.B. Cross-education improves quadriceps strength recovery after ACL reconstruction: A randomized controlled trial. *Knee Surg. Sports Traumatol. Arthrosc.* **2019**, *27*, 68–75. [CrossRef]
18. Zult, T.; Gokeler, A.; van Raay, J.; Brouwer, R.W.; Zijdewind, I.; Farthing, J.P.; Hortobagyi, T. Cross-education does not accelerate the rehabilitation of neuromuscular functions after ACL reconstruction: A randomized controlled clinical trial. *Eur. J. Appl. Physiol.* **2018**, *118*, 1609–1623. [CrossRef]
19. Glattke, K.E.; Tummala, S.V.; Chhabra, A. Anterior Cruciate Ligament Reconstruction Recovery and Rehabilitation: A Systematic Review. *J. Bone Joint Surg. Am.* **2022**, *104*, 739–754. [CrossRef]
20. Ross, B.J.; Savage-Elliott, I.; Brown, S.M.; Mulcahey, M.K. Return to Play and Performance After Primary ACL Reconstruction in American Football Players: A Systematic Review. *Orthop. J. Sports Med.* **2020**, *8*, 2325967120959654. [CrossRef]
21. Cigni, P.; Minuti, T.; Mannini, A.; Cucini, A.; Costagli, M.; Rapetti, S.; Alimonta, L.; Cione, E.; Cannataro, R.; Ricotti, L. Application of a Custom Device to Measure Isometric Knee Strength: Possible Injury Correlation in Professional Soccer (Football) Players. *J. Funct. Morphol. Kinesiol.* **2023**, *8*, 141. [CrossRef] [PubMed]
22. Minuti, T.; Cigni, P.; Costagli, M.; Cucini, A.; Cione, E.; Melotto, S.; Rapetti, S.; Ricotti, L.; Cannataro, R. Reliability of a Custom Device Used to Measure Isometric Knee Flexor and Extensor Strength in Standing Position. *Life* **2023**, *13*, 458. [CrossRef] [PubMed]
23. Minshull, C.; Gallacher, P.; Roberts, S.; Barnett, A.; Kuiper, J.H.; Bailey, A. Contralateral strength training attenuates muscle performance loss following anterior cruciate ligament (ACL) reconstruction: A randomised-controlled trial. *Eur. J. Appl. Physiol.* **2021**, *121*, 3551–3559. [CrossRef] [PubMed]
24. Cuyul-Vasquez, I.; Alvarez, E.; Riquelme, A.; Zimmermann, R.; Araya-Quintanilla, F. Effectiveness of Unilateral Training of the Uninjured Limb on Muscle Strength and Knee Function of Patients With Anterior Cruciate Ligament Reconstruction: A Systematic Review and Meta-Analysis of Cross-Education. *J. Sport. Rehabil.* **2022**, *31*, 605–616. [CrossRef]

25. Barss, T.S.; Pearcey, G.E.; Zehr, E.P. Cross-education of strength and skill: An old idea with applications in the aging nervous system. *Yale J. Biol. Med.* **2016**, *89*, 81–86.
26. Manca, A.; Dragone, D.; Dvir, Z.; Deriu, F. Cross-education of muscular strength following unilateral resistance training: A meta-analysis. *Eur. J. Appl. Physiol.* **2017**, *117*, 2335–2354. [CrossRef]
27. Wang, K.; Cheng, L.; Wang, B.; He, B. Effect of isokinetic muscle strength training on knee muscle strength, proprioception, and balance ability in athletes with anterior cruciate ligament reconstruction: A randomised control trial. *Front. Physiol.* **2023**, *14*, 1237497. [CrossRef]
28. Vélez-Gutiérrez, J.M.; Petro, J.L.; Aburto-Corona, J.A.; Vargas-Molina, S.; Kreider, R.B.; Bonilla, D.A. Cortical Changes as a Result of Sports Injuries: A Short Commentary. *Cuerpo Cult. Y Mov.* **2022**, *12*, 7884. [CrossRef]
29. Lee, M.; Carroll, T.J. Cross education: Possible mechanisms for the contralateral effects of unilateral resistance training. *Sports Med.* **2007**, *37*, 1–14. [CrossRef] [PubMed]
30. Kidgell, D.J.; Frazer, A.K.; Daly, R.M.; Rantalainen, T.; Ruotsalainen, I.; Ahtiainen, J.; Avela, J.; Howatson, G. Increased cross-education of muscle strength and reduced corticospinal inhibition following eccentric strength training. *Neuroscience* **2015**, *300*, 566–575. [CrossRef]
31. Sharma, N.; Classen, J.; Cohen, L.G. Neural plasticity and its contribution to functional recovery. *Handb. Clin. Neurol.* **2013**, *110*, 3–12.
32. Lecce, E.; Conti, A.; Del Vecchio, A.; Felici, F.; Scotto di Palumbo, A.; Sacchetti, M.; Bazzucchi, I. Cross-education: Motor unit adaptations mediate the strength increase in non-trained muscles following 8 weeks of unilateral resistance training. *Front. Physiol.* **2025**, *15*, 1512309. [CrossRef]
33. Manca, A.; Hortobagyi, T.; Carroll, T.J.; Enoka, R.M.; Farthing, J.P.; Gandevia, S.C.; Kidgell, D.J.; Taylor, J.L.; Deriu, F. Contralateral Effects of Unilateral Strength and Skill Training: Modified Delphi Consensus to Establish Key Aspects of Cross-Education. *Sports Med.* **2021**, *51*, 11–20. [CrossRef]
34. Hendy, A.M.; Spittle, M.; Kidgell, D.J. Cross education and immobilisation: Mechanisms and implications for injury rehabilitation. *J. Sci. Med. Sport.* **2012**, *15*, 94–101. [CrossRef] [PubMed]
35. Ehrensberger, M.; Simpson, D.; Broderick, P.; Monaghan, K. Cross-education of strength has a positive impact on post-stroke rehabilitation: A systematic literature review. *Top. Stroke Rehabil.* **2016**, *23*, 126–135. [CrossRef]
36. Fimland, M.S.; Helgerud, J.; Solstad, G.M.; Iversen, V.M.; Leivseth, G.; Hoff, J. Neural adaptations underlying cross-education after unilateral strength training. *Eur. J. Appl. Physiol.* **2009**, *107*, 723–730. [CrossRef]
37. Tricco, A.C.; Lillie, E.; Zarin, W.; O'Brien, K.K.; Colquhoun, H.; Levac, D.; Moher, D.; Peters, M.D.J.; Horsley, T.; Weeks, L.; et al. PRISMA Extension for Scoping Reviews (PRISMA-ScR): Checklist and Explanation. *Ann. Intern. Med.* **2018**, *169*, 467–473. [CrossRef]
38. Peters, M.D.J.; Godfrey, C.; McInerney, P.; Munn, Z.; Tricco, A.C.; Khalil, H. Scoping Reviews. Available online: https://synthesismanual.jbi.global (accessed on 19 July 2025).
39. Rethlefsen, M.L.; Kirtley, S.; Waffenschmidt, S.; Ayala, A.P.; Moher, D.; Page, M.J.; Koffel, J.B.; Group, P.-S. PRISMA-S: An extension to the PRISMA Statement for Reporting Literature Searches in Systematic Reviews. *Syst. Rev.* **2021**, *10*, 39. [CrossRef]
40. Bonilla, D.A.; Cardozo, L.A.; Velez-Gutierrez, J.M.; Arevalo-Rodriguez, A.; Vargas-Molina, S.; Stout, J.R.; Kreider, R.B.; Petro, J.L. Exercise Selection and Common Injuries in Fitness Centers: A Systematic Integrative Review and Practical Recommendations. *Int. J. Environ. Res. Public. Health* **2022**, *19*, 12710. [CrossRef] [PubMed]
41. Giraldo-Vallejo, J.E.; Cardona-Guzman, M.A.; Rodriguez-Alcivar, E.J.; Koci, J.; Petro, J.L.; Kreider, R.B.; Cannataro, R.; Bonilla, D.A. Nutritional Strategies in the Rehabilitation of Musculoskeletal Injuries in Athletes: A Systematic Integrative Review. *Nutrients* **2023**, *15*, 819. [CrossRef] [PubMed]
42. Montazeri, A.; Mohammadi, S.; Hesari, P.M.; Ghaemi, M.; Riazi, H.; Sheikhi-Mobarakeh, Z. Preliminary guideline for reporting bibliometric reviews of the biomedical literature (BIBLIO): A minimum requirements. *Syst. Rev.* **2023**, *12*, 239. [CrossRef]
43. Payandeh, M.; Daneshmandi, H. Investigating the effect of a cross-training program on the intensity and quality of activity in selected muscles of athletes with anterior cruciate ligament (ACL) injury. *J. Bodyw. Mov. Ther.* **2025**, *42*, 265–273. [CrossRef]
44. Lepley, L.K.; Grooms, D.R.; Burland, J.P.; Davi, S.M.; Mosher, J.L.; Cormier, M.L.; Lepley, A.S. Eccentric cross-exercise after anterior cruciate ligament reconstruction: Novel case series to enhance neuroplasticity. *Phys. Ther. Sport.* **2018**, *34*, 55–65. [CrossRef]
45. Oliveira, M.; Junior, P.L.; Imoto, A.M.; Santos, T.; Borges, J.H.S.; Nunes, P.; Barin, F.R.; Damando, M.; Peccin, M.S. Unilateral Versus Bilateral Resistance Exercise in Postoperative Rehabilitation After ACL Reconstruction With Bone-Patellar Tendon-Bone Graft: A Randomized Controlled Trial. *Orthop. J. Sports Med.* **2022**, *10*, 23259671221088830. [CrossRef]
46. Papandreou, M.G.; Billis, E.V.; Antonogiannakis, E.M.; Papaioannou, N.A. Effect of cross exercise on quadriceps acceleration reaction time and subjective scores (Lysholm questionnaire) following anterior cruciate ligament reconstruction. *J. Orthop. Surg. Res.* **2009**, *4*, 2. [CrossRef]
47. Papandreou, M.; Billis, E.; Papathanasiou, G.; Spyropoulos, P.; Papaioannou, N. Cross-exercise on quadriceps deficit after ACL reconstruction. *J. Knee Surg.* **2013**, *26*, 51–58. [CrossRef]

48. Karimijashni, M.; Sarvestani, F.K.; Yoosefinejad, A.K. The Effect of Contralateral Knee Neuromuscular Exercises on Static and Dynamic Balance, Knee Function, and Pain in Athletes Who Underwent Anterior Cruciate Ligament Reconstruction: A Single Blind Randomized Controlled Trial. *J. Sport. Rehabil.* **2023**, *32*, 524–539. [CrossRef] [PubMed]
49. Zult, T.; Gokeler, A.; van Raay, J.; Brouwer, R.W.; Zijdewind, I.; Farthing, J.P.; Hortobagyi, T. Cross-education does not improve early and late-phase rehabilitation outcomes after ACL reconstruction: A randomized controlled clinical trial. *Knee Surg. Sports Traumatol. Arthrosc.* **2019**, *27*, 478–490. [CrossRef] [PubMed]
50. Papandreou, M.; Nikos, P.; Antonogiannakis, E.; Hlias, Z. The effect of cross exercise on quadriceps strength in different knee angles after the anterior cruciate ligament reconstruction. *Braz. J. Biomotricity* **2007**, *1*, 123–138.
51. Liu, C.; Li, S.; Li, J.; Zhang, H.; Li, G.; Jiang, X. The effects of contralateral limb cross-education training on post-surgical rehabilitation outcomes in patients with anterior cruciate ligament reconstruction: A randomized controlled trial. *J. Orthop. Surg. Res.* **2025**, *20*, 118. [CrossRef]
52. Hsieh, C.-P.; Wei, T.-S.; Wu, C.-C. The early effects of isokinetic muscle training on knee joint muscle strength after modified double-bundle anterior cruciate ligament reconstruction. *Int. J. Clin. Exp. Med.* **2016**, *9*, 14461–14470.
53. Irrgang, J.J.; Anderson, A.F.; Boland, A.L.; Harner, C.D.; Kurosaka, M.; Neyret, P.; Richmond, J.C.; Shelborne, K.D. Development and validation of the international knee documentation committee subjective knee form. *Am. J. Sports Med.* **2001**, *29*, 600–613. [CrossRef]
54. Roos, E.M.; Roos, H.P.; Lohmander, L.S.; Ekdahl, C.; Beynnon, B.D. Knee Injury and Osteoarthritis Outcome Score (KOOS)—Development of a self-administered outcome measure. *J. Orthop. Sports Phys. Ther.* **1998**, *28*, 88–96. [CrossRef] [PubMed]
55. Hooper, D.M.; Morrissey, M.C.; Drechsler, W.I.; McDermott, M.; McAuliffe, T.B. Validation of the Hughston Clinic subjective knee questionnaire using gait analysis. *Med. Sci. Sports Exerc.* **2001**, *33*, 1456–1462. [CrossRef]
56. Hendy, A.M.; Teo, W.-P.; Kidgell, D.J. Anodal tDCS prolongs the cross-education of strength and corticomotor plasticity. *Med. Sci. Sports Exerc.* **2015**, *47*, 1788–1797. [CrossRef]
57. Ruddy, K.L.; Leemans, A.; Woolley, D.G.; Wenderoth, N.; Carson, R.G. Structural and functional cortical connectivity mediating cross education of motor function. *J. Neurosci.* **2017**, *37*, 2555–2564. [CrossRef]
58. Stöckel, T.; Carroll, T.J.; Summers, J.J.; Hinder, M.R. Motor learning and cross-limb transfer rely upon distinct neural adaptation processes. *J. Neurophysiol.* **2016**, *116*, 575–586. [CrossRef]
59. Rojas-Valverde, D.; Herrera-Gonzalez, E.; Bonilla, D.A. Sports injuries as reversible involution: A novel approach to rehabilitation and readaptation. *Front. Sports Act. Living* **2025**, *7*, 1519404. [CrossRef] [PubMed]
60. Arumugam, A.; Bjorklund, M.; Mikko, S.; Hager, C.K. Effects of neuromuscular training on knee proprioception in individuals with anterior cruciate ligament injury: A systematic review and GRADE evidence synthesis. *BMJ Open* **2021**, *11*, e049226. [CrossRef]
61. Humes, C.; Aguero, S.; Chahla, J.; Foad, A. Blood Flow Restriction and Its Function in Post-Operative Anterior Cruciate Ligament Reconstruction Therapy: Expert Opinion. *Arch. Bone Jt. Surg.* **2020**, *8*, 570–574. [CrossRef] [PubMed]
62. Charles, D.; White, R.; Reyes, C.; Palmer, D. A Systematic Review of the Effects of Blood Flow Restriction Training on Quadriceps Muscle Atrophy and Circumference Post Acl Reconstruction. *Int. J. Sports Phys. Ther.* **2020**, *15*, 882–891. [CrossRef]
63. Smith, M.J.; Hoffman, N.J.; Jose, A.J.S.; Burke, L.M.; Opar, D.A. Nutritional Interventions to Attenuate Quadriceps Muscle Deficits following Anterior Cruciate Ligament Injury and Reconstruction. *Sports Med.* **2025**, *55*, 569–596. [CrossRef] [PubMed]
64. Greif, D.N.; Emerson, C.P.; Allegra, P.; Arizpe, A.; Mansour, K.L.; Cade, W.H., 2nd; Baraga, M.G. Supplement Use in Patients Undergoing Anterior Cruciate Ligament Reconstruction: A Systematic Review. *Arthroscopy* **2020**, *36*, 2537–2549. [CrossRef] [PubMed]

applied sciences

MDPI

Systematic Review

Aerobic Training on Mental Health in Children and Adolescents: A Systematic Review with Meta-Analysis

María del Carmen Carcelén-Fraile [1], Agustín Aibar-Almazán [2,*] and Fidel Hita-Contreras [2]

[1] Department of Educational Sciences, Faculty of Social Sciences, University of Atlántico Medio, 35017 Las Palmas de Gran Canaria, Spain

[2] Department of Health Sciences, Faculty of Health Sciences, University of Jaén, 23071 Jaen, Spain

* Correspondence: aaibar@ujaen.es

Abstract: Introduction: Mental health problems in children and adolescents are a growing public health concern. Aerobic physical activity has been identified as a promising, accessible, and low-cost intervention to promote psychological well-being. Methods: This systematic review and meta-analysis was conducted following PRISMA 2020 guidelines and a pre-registered PROSPERO protocol. We searched PubMed, Scopus, Web of Science, and CINAHL for randomized controlled trials published until December 2024, with no language restrictions. Eligibility criteria included interventions based on aerobic physical activity among children and adolescents (5–18 years) reporting outcomes on mental health (e.g., depression, anxiety, self-esteem, mood, resilience). Methodological quality was assessed with the PEDro scale. Results: From 828 initial records, 21 studies met the inclusion criteria and 19 were included in the meta-analysis. The trials, conducted in diverse countries and settings, applied structured aerobic interventions (e.g., running, dancing, cycling, active games) with durations of 6–30 weeks. The pooled results showed significant improvements in depression, anxiety, self-esteem, and stress, with the largest effects observed in stress reduction and self-esteem enhancement. Conclusions: Aerobic physical activity appears to be an effective and feasible strategy to improve mental health in children and adolescents. Its integration into school and community programs is recommended as a preventive and complementary approach to traditional treatments.

Keywords: aerobic training; children; adolescents; self-esteem; anxiety; depression; systematic review; meta-analysis

1. Introduction

Mental health during childhood and adolescence has become a growing public health priority worldwide. It is estimated that around one in seven adolescents, approximately 166 million young people between 10 and 19 years old, suffer from a mental disorder, with emotional problems, such as anxiety and depression, being the most common [1]. Specifically, anxiety affects approximately 4.4% of children between 10 and 14 years old and 5.5% of adolescents between 15 and 19 years old, while depression has a prevalence of 1.4% and 3.2%, respectively, in those same age groups [2]. These internalizing disorders are associated with a variety of negative consequences, including poor academic performance, social isolation, emotional dysfunction, and, in the most severe cases, suicidal ideation and self-harm [3].

In the Spanish context, the data reflect an equally alarming situation. According to a recent report by UNICEF Spain [4], 41% of adolescents report having had or believed they

https://doi.org/10.3390/app15179572

had a mental health problem in the past year. Similarly, a study coordinated by seven Spanish universities indicates that more than 1.5 million adolescents between the ages of 12 and 18 present significant emotional symptoms, and more than half are undergoing psychological or pharmacological treatment. These data coincide with hospital records showing a sustained increase in pediatric psychiatric admissions over the past two decades [5]. This growing incidence underscores the urgent need to establish effective prevention, early diagnosis, and intervention strategies to adequately address mental health problems in this population.

Among the various strategies proposed to address mental health problems in children and adolescents, physical activity (PA) has gained increasing relevance as a preventative, therapeutic, accessible, and low-cost alternative [6]. PA is defined as "any bodily movement produced by skeletal muscles that results in energy expenditure greater than resting levels" [7]. This definition encompasses a wide range of activities that can be classified based on their intensity, duration, context, and purpose. Unlike organized physical exercise, which is planned, repetitive, and aims to improve one or more components of physical fitness, PA encompasses both organized activities (such as team sports or supervised classes) and informal activities (spontaneous games, active movements such as walking or cycling, and recreational activities such as dancing or running in the playground) [8]. What makes PA particularly attractive in the context of promoting children's mental health is its inclusiveness, versatility, and adaptability to different educational, family, and community settings [9]. Unlike other more expensive or clinically structured interventions, PA can be relatively easily integrated into children's daily routines, without the need for advanced technological resources or specialized supervision. This makes it a particularly suitable strategy for implementation in schools, where it can be part of the physical education curriculum, extracurricular programs, or specific school health interventions [10]. Several studies have shown that PA has the ability to modulate physiological, emotional, and social responses that directly impact children's psychological well-being. Regular participation in physical activity is associated with a greater sense of belonging, improved body image, reduced social isolation, and higher levels of perceived competence [11]. Furthermore, it can promote the development of social skills, such as teamwork, respect for rules, and emotional self-regulation. Therefore, PA not only acts as a protective factor against emotional symptoms but can also play a crucial role in strengthening positive psychological resources, such as self-esteem, resilience, and self-efficacy [12,13]. These properties make physical activity an effective and multifaceted tool for promoting the comprehensive well-being of children and adolescents, and justify the growing interest in its evaluation within rigorous experimental frameworks such as randomized clinical trials [14].

To guide the interpretation of the evidence, this review is framed within the biopsychosocial model of mental health, which conceptualizes psychological well-being as the result of interactions between biological, psychological, and social factors. Aerobic physical activity represents a multidimensional intervention capable of simultaneously influencing these domains through physiological adaptations (e.g., regulation of neurotransmitters and neurotrophic factors), psychological mechanisms (e.g., improvements in self-esteem, emotional regulation), and social interaction (e.g., peer belonging and group cohesion). This integrative framework has been increasingly used to explain the role of physical activity in youth mental health.

Within the broad spectrum of physical activity modalities, aerobic exercise has been shown to be particularly effective in improving the mental health of children and adolescents [15]. This type of activity is characterized by sustained, cyclical movements that mobilize large muscle groups, at moderate to vigorous intensity, which promotes both physiological and psychological improvements. Activities such as running, swimming, cycling, dancing, jumping rope, or participating in active games are common examples of aerobic exercise and can be performed in school, community, or recreational settings. Their accessible, dynamic, and adaptable nature makes them an ideal tool for implementation with young populations [14].

The effectiveness of aerobic exercise on mental health is supported by various neurobiological and psychosocial mechanisms. In physiological terms, this type of exercise has been shown to stimulate the release of neurotransmitters such as serotonin, dopamine, and norepinephrine, which are involved in emotional regulation and pleasure [16]. Furthermore, it increases the secretion of endorphins, substances that act as natural analgesics and generate feelings of well-being, and activates the endocannabinoid system, linked to the reduction in stress and anxiety [17]. Another key effect is the increase in the expression of BDNF (brain-derived neurotrophic factor), a neurotrophic factor that facilitates brain plasticity and emotional regulation [18]. Furthermore, aerobic exercise can regulate the activity of the hypothalamic–pituitary–adrenal axis, reducing cortisol levels and promoting greater stress tolerance [11,19]. At the psychosocial level, group aerobic interventions promote a sense of belonging, social support among peers, and perceived self-efficacy, all of which are protective factors against negative emotional symptoms [10]. Regular aerobic exercise has therefore been associated with significant decreases in depressive symptoms, anxiety, anger, emotional fatigue, and perceived stress, and with increases in self-esteem, physical self-concept, positive mood, and emotional resilience in healthy children and adolescents or those at clinical risk [12,20]. These findings, along with their practical applicability, consolidate aerobic exercise as a highly recommended intervention strategy for promoting psychological well-being during development.

Given the alarming epidemiological trends and the limitations of conventional treatments such as pharmacological and psychotherapeutic approaches, which are often costly, stigmatized, or difficult to access for children and adolescents, there is a growing need for effective, scalable, and preventive strategies [12,15]. Aerobic exercise emerges as a promising public health intervention with high feasibility in school and community settings. Nevertheless, the available evidence remains heterogeneous in terms of study design, intervention modalities, and reported outcomes. Therefore, a rigorous synthesis of findings through systematic review and meta-analysis is required to clarify the magnitude and consistency of the effects of aerobic physical activity on youth mental health. In this context, this systematic review and meta-analysis aims to synthesize the available evidence from randomized controlled trials (RCTs) evaluating the effects of aerobic physical activity interventions on mental health in children and adolescents (5–18 years). Unlike previous meta-analyses, our study provides an updated synthesis of the literature published up to 2024, includes only RCTs, applies rigorous RoB 2 and PEDro assessments, and explores potential dose–response relationships through meta-regression. This approach allows us to clarify the specific contribution of aerobic interventions to child and adolescent mental health and to strengthen the evidence base for school- and community-based implementation.

2. Materials and Methods

This review was conducted following the 2020 PRISMA statement guidelines [21] and adhered to a pre-registered protocol in PROSPERO (registration number: CRD420251116424). In addition, the methodological framework was guided by the recommendations outlined in the Cochrane Handbook for Systematic Reviews of Interventions [22].

2.1. Sources of Information and Search Strategy

A comprehensive search strategy was used to identify relevant studies in multiple electronic databases during the month of July. The following databases were consulted:

PubMed, Scopus, Web of Science, and CINAHL. The search included articles published through December 2024 with no language or publication status restrictions.

The search strategy included the following combination of keywords and Boolean operators: ("aerobic exercise" OR "aerobic training" OR "cardiovascular exercise") AND ("mental health" OR "psychological well-being" OR "emotional health" OR "depression" OR "anxiety" OR "self-esteem" OR self-concept OR stress) AND ("children" OR "child" OR "youth" OR "minors" OR "adolescents").

2.2. Selection Criteria

To ensure the relevance and quality of the included studies, the following eligibility criteria were applied: (A) Inclusion criteria: (i) Studies evaluating the effects of aerobic exercise on the mental health of children and adolescents; (ii) Randomized controlled trials (RCTs); (iii) Studies with participants aged 5–18 years; (iv) Studies reporting quantitative data on mental health outcomes (e.g., depression, anxiety, psychological well-being); (B) Exclusion criteria: (i) Studies focusing on adult or older populations; (ii) Reviews, meta-analyses, conference proceedings, commentaries, editorials, or case reports; (iii) Studies that did not include physical activity interventions; and (iv) Studies with unclear outcomes, incomplete data, or data that could not be reliably extracted.

2.3. Data Extraction and Reliability

Data from the included studies were extracted using a standardized extraction form developed specifically for this review. The form captured relevant information such as: author(s), year of publication, country, sample size and characteristics, details of the intervention (type, frequency, duration), mental health outcomes assessed, measurement instruments used, and main findings. Two reviewers with expertise in systematic reviews and aerobic exercise extracted data to ensure consistency and reduce the risk of errors. Both had prior experience conducting meta-analyses in the fields of physical activity and mental health. Any discrepancies were resolved through discussion, and when necessary, a third senior reviewer with extensive experience in evidence synthesis was consulted to reach a consensus. To further enhance reliability, the results of the most recent psychology reviews were first summarized before selecting studies potentially relevant to the defined topics.

2.4. Assessment of Methodological Quality

The methodological rigor of the included trials was evaluated using the PEDro scale [23]. This instrument comprises 11 criteria, although only 10 contribute to the overall score since the first item (eligibility criteria) is reported separately and not counted. Each criterion is scored dichotomously ("Yes" = 1; "No" = 0), resulting in a total score ranging from 0 to 10. For interpretative purposes, studies were classified into four categories: poor (0–3), fair (4–5), good (6–8), and excellent quality (9–10). In addition, we applied the Cochrane Risk of Bias 2 (RoB 2) tool, which evaluates five domains: (i) randomization process, (ii) deviations from intended interventions, (iii) missing outcome data, (iv) measurement of the outcome, and (v) selection of the reported result. This dual approach allowed us to complement the PEDro scoring system with a more detailed risk-of-bias assessment.

2.5. Analytical Decisions for Meta-Analysis

The outcomes of the meta-analysis were summarized in forest plots, which detailed for each study the first author, year of publication, sample size, and the corresponding effect sizes (Hedges' g). Each estimate was displayed with its 95% confidence interval and associated p-value, facilitating the interpretation of both magnitude and statistical significance. To test the robustness of the synthesis, we performed sensitivity analyses by sequentially excluding studies with overlapping samples, statistical outliers, or atypical values that could bias the pooled results. The comparison between these reduced models and the full dataset confirmed the stability of the overall effects. Additionally, subgroup analyses were performed, stratifying studies according to the nature of the outcome variables. Primary outcomes were defined a priori as depressive symptoms, anxiety, and self-esteem. Secondary outcomes included mood, stress, resilience, and executive functioning. All outcomes were analyzed at the post-intervention timepoint, as reported in the original trials. The first category included indicators related to mental health, such as emotional and behavioral symptoms (e.g., anxiety, depression, emotional regulation, and aggression). The second grouping focused on aspects of cognitive functioning, particularly executive skills such as working memory, inhibitory control, and cognitive flexibility. Separate meta-analyses were conducted for each of these subgroups, allowing for a deeper understanding of the differences in effects and a clarification of the specific impact that physical activity could have on each dimension. Given the diverse nature of the included studies, a random-effects model was used. This methodological choice allowed for the incorporation of potential real-world variations between the reported effect sizes, thus providing greater generalizability of the results. Statistical heterogeneity was analyzed using Cochran's Q test and the I^2 statistic [22]. An I^2 value greater than 50% was interpreted as indicating moderate or high heterogeneity, which could indicate differences in the methods, settings, or characteristics of the study populations. Finally, publication bias was examined using funnel plot analysis, which allowed us to visualize potential biases caused by the tendency to report statistically significant results. This visual assessment was complemented with statistical tests such as the Egger test, which served to detect potential asymmetries and confirm the presence of bias in the set of studies reviewed.

3. Results

3.1. Study Selection Process

A total of 828 records were initially retrieved through systematic searches conducted across four databases: PubMed (255 records), Scopus (218), Web of Science (167), and CINAHL (188). After applying predefined filters, such as study type (randomized controlled trials and original research articles), language (English and Spanish), and target population (children and adolescents), the dataset was refined. Further screening involved the removal of duplicate entries and the assessment of titles and abstracts based on relevant keywords, resulting in 282 unique records. In the first selection stage, 154 articles were identified as potentially relevant and subjected to full-text review. Following this in-depth evaluation, 21 studies met all inclusion criteria and were included in the qualitative synthesis and meta-analysis. The remaining 97 studies were excluded primarily due to methodological limitations, irrelevant outcome variables, or lack of focus on physical activity interventions. The complete selection process is illustrated in Figure 1, according to the PRISMA 2020 guidelines.

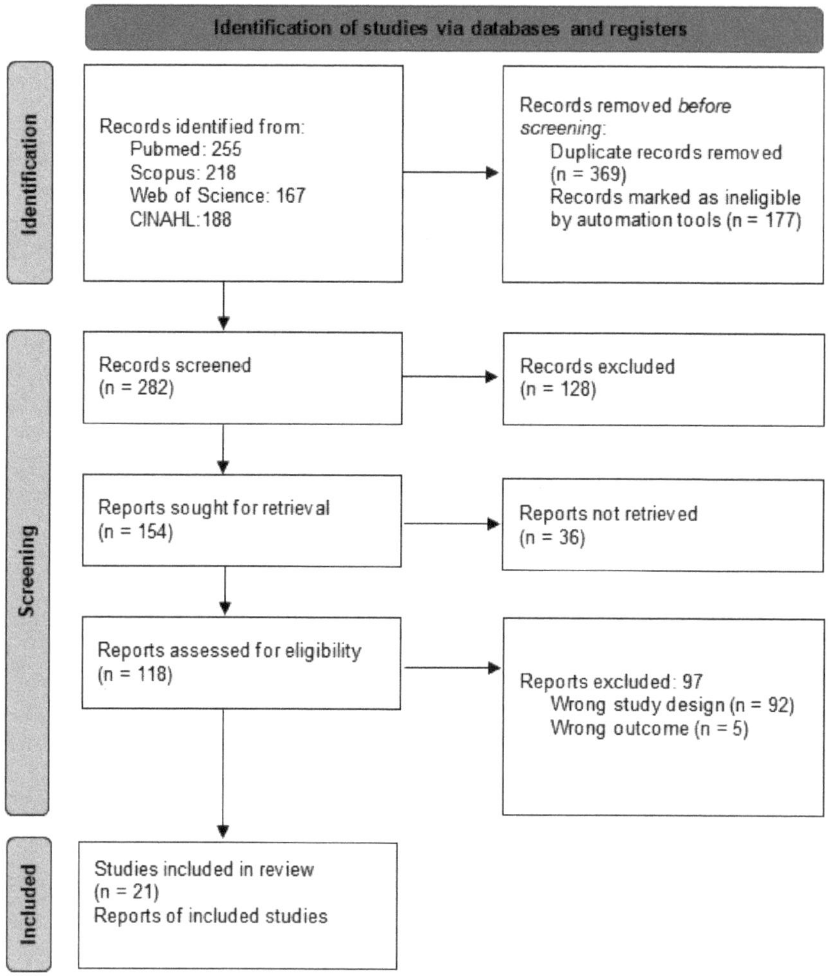

Figure 1. Study selection process flow chart.

3.2. Methodological Quality

The methodological rigor of the 21 studies reviewed in this systematic analysis was assessed using the PEDro scale, a tool designed to evaluate essential aspects of study design in randomized controlled trials. This scale produces scores from 0 to 10, excluding the eligibility criteria item, which does not contribute to the total. In this review, PEDro scores ranged from 2 to 7, indicating a generally moderate level of methodological quality. No study achieved the highest possible score, and only one study reached a score of 7, classifying it as high-quality. In contrast, three studies received scores of 3 or below, with one study obtaining the lowest score of 2. Overall, 57% of the studies scored between 5 and 6, suggesting adequate implementation of randomization, intergroup comparisons, and complete data reporting. However, 17% of the included articles scored 4 points, and 13% scored 3 or fewer, exposing significant methodological gaps. The most frequently observed shortcomings were the lack of blinding for participants, therapists, or assessors; insufficient allocation concealment; and limited application of intention-to-treat analyses. These methodological issues are consistent with previous systematic reviews in the field of pediatric exercise and mental health, which have similarly highlighted the frequent absence of blinding and inadequate reporting of randomization procedures [24,25]. Although our assessments were conducted independently by two experienced reviewers using standardized tools (PEDro, RoB 2), and disagreements were resolved by consensus with a senior researcher, it must be acknowledged that critical appraisal inherently involves a degree of subjectivity. Nevertheless, the convergence of our observations with those reported in prior reviews suggests that these shortcomings represent structural limitations of the literature, rather than an overly rigorous or idiosyncratic assessment by our review team. These deficiencies heighten the potential for bias and diminish internal validity. Thus, although many of the studies utilized broadly acceptable research methodologies, their limitations should be acknowledged when interpreting the findings. It is essential that future trials investigating physical activity interventions in youth with neurodevelopmental disorders adopt more rigorous design frameworks to enhance the credibility of their outcomes. A comprehensive summary of quality assessment results can be found in Table 1.

Table 1. Methodological quality of the included articles.

Study	1	2	3	4	5	6	7	8	9	10	11	Total Score
Annesi et al. [26]	1	0	0	0	0	0	0	0	0	1	1	2/10
Bazzano et al. [27]	1	1	1	0	1	0	0	0	0	1	1	5/10
Bohnert et al. [28]	1	1	1	0	0	1	0	0	1	1	1	6/10
Burgess et al. [29]	1	0	0	0	1	0	0	0	0	1	1	3/10
Cowley et al. [30]	1	1	1	0	1	0	0	0	0	1	1	5/10
Crews et al. [31]	1	0	1	0	0	0	0	0	0	1	1	3/10
Eather et al. [32]	1	1	1	0	1	0	0	1	1	0	1	7/10
Gaurav et al. [33]	1	1	0	0	0	1	0	0	0	1	1	4/10
Jelanian et al. [34]	1	1	0	0	0	0	0	0	1	1	1	4/10
Khalsa et al. [35]	1	1	1	0	1	0	0	1	1	1	1	7/10
Lobo et al. [36]	1	1	1	0	0	1	1	1	0	1	1	7/10
Melnyk et al. [37]	1	1	1	0	1	0	0	0	0	1	1	5/10
Melnyk et al. [38]	1	1	1	0	0	1	0	0	0	1	1	5/10
Noggle et al. [39]	1	1	1	0	1	0	0	0	0	1	1	5/10
Olive et al. [40]	1	0	1	0	1	0	1	0	1	1	1	6/10
Sharma et al. [41]	1	0	1	0	0	0	0	0	0	1	1	3/10
Smith et al. [42]	1	1	1	0	1	0	0	0	0	1	1	5/10
Telles et al. [43]	1	1	1	0	1	0	0	1	1	1	1	7/10
Wassenaar et al. [44]	1	0	1	0	1	1	0	0	0	1	1	5/10
Ho et al. [45]	1	1	1	1	1	0	0	0	1	1	1	7/10
Wang et al. [46]	1	1	1	0	1	0	0	0	1	1	1	6/10

Scoring criteria: (1) eligibility specified; (2) random allocation; (3) concealed allocation; (4) comparable groups at baseline; (5) participant blinding; (6) therapist blinding; (7) assessor blinding; (8) adequate follow-up; (9) intention-to-treat analysis; (10) between-group statistical comparisons. Each item is scored as Yes = 1 or No = 0.

The overall risk of bias assessment using the Cochrane RoB 2 tool revealed a heterogeneous methodological quality across the included trials. High-quality studies, such as those conducted by Wassenaar et al. [44], Wang et al. [46] and Smith et al. [42], demonstrated a low risk of bias, largely due to robust randomization procedures, prespecified protocols, appropriate handling of missing data, and the use of validated outcome measures. Conversely, earlier trials [26–28,31], frequently exhibited "some concerns", primarily related to insufficient reporting of allocation concealment, lack of preregistration, and reliance on self-reported outcomes without assessor blinding. Studies involving yoga- or mindfulness-based interventions [35,39,43,45] also tended to fall into the "some concerns" category, given the challenges in participant blinding and the absence of detailed analytic protocols, despite showing adequate adherence and acceptable data completeness. Importantly, none of the included trials was judged to be at high risk of bias, and sensitivity analyses confirmed that excluding studies with greater methodological concerns did not substantially alter the pooled estimates. Overall, the evidence base can be considered moderately robust, with stronger confidence derived from more recent, preregistered school-based RCTs that reported outcomes in line with their protocols (Figure 2).

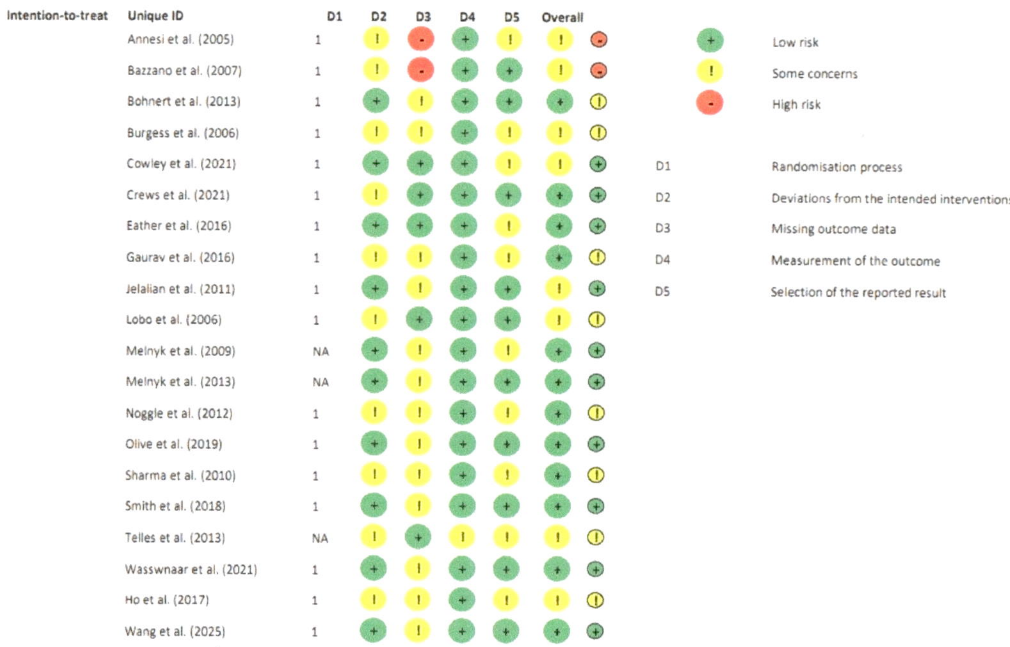

Figure 2. Risk of bias of the included articles [26–34,36–46]. D1: bias arising from the randomization process; D2: bias due to deviations from the intended intervention; D3: bias due to missing outcome data; D4: bias in the measurement of the outcome; D5: bias in the selection of the reported result.

3.3. Characteristics of the Studies

This systematic review included a total of 21 randomized controlled trials (RCTs) that investigated the effects of aerobic or movement-based physical activity interventions on the psychological and emotional well-being of children and adolescents. The studies were conducted between 2004 and 2025, and took place in a range of countries including the United States, United Kingdom, Australia, India, Mexico, China, and Iran, reflecting a geographically and culturally diverse research base. Participants across studies were predominantly school-aged children and adolescents, with mean ages ranging from 5.2 to 17.2 years. The sample sizes varied considerably, with studies enrolling between 19 and 627 participants. Most studies included comparable numbers of participants in the intervention and control groups, though some presented modest imbalances. The control conditions generally comprised standard school-based physical education classes, health education programs, or wait-list groups, and in some cases, no active intervention. Intervention characteristics showed considerable variability. The type of physical activity was predominantly aerobic in nature but implemented through diverse formats, including traditional aerobic exercise sessions, yoga-based interventions, circuit training, or sport-related group activities. The frequency of sessions ranged from 1 to 6 times per week, while the duration of each session varied from 15 to 90 min. The total length of the interventions spanned from 6 to 30 weeks, with the majority of programs lasting between 8 and 12 weeks. The intensity of exercise was commonly classified as moderate to vigorous, although several interventions also included low-intensity components. Regarding the outcomes assessed, the studies employed a wide array of validated psychological and behavioral instruments. The most frequently measured variables included depressive symptoms, anxiety, self-esteem and self-concept, and mood states. Additional outcomes included stress, executive function, resilience, social skills, and quality of life, measured with tools such as the PSS, Stroop Test, SSRS, PedsQL, and others. Overall, the studies included in this review demonstrate considerable heterogeneity in terms of intervention design, duration, intensity, and outcome assessment. However, they share a common focus on evaluating the mental and emotional effects of physical activity in youth populations, and collectively provide a comprehensive view of its potential benefits in diverse educational and clinical settings (Table S1).

3.4. Study Results

This review included 21 randomized controlled trials that examined the impact of 268 structured aerobic or movement-based physical activity programs on the emotional and psychosocial health of children and adolescents. Overall, the interventions tended to reduce depressive and anxiety symptoms and promote self-esteem, self-concept, mood regulation, stress management, and resilience, although effect sizes and consistency varied according to intervention features and outcome measures Depressive symptoms were evaluated in 11 of the included studies [26,27,29,31,33,35,37,38,40,44,46]. Of these, eight studies [27,30,32,37,38,40,45,46] reported statistically significant reductions in depression levels in the intervention groups compared to controls. These improvements were more evident in programs of moderate to vigorous intensity, conducted at least three times per week, and lasting 8 to 12 weeks. Interventions that incorporated not only physical exertion but also supportive or skill-building environments appeared particularly beneficial in reducing depressive symptomatology.

Anxiety was measured in 8 studies [26,27,31,35,37,38,41,46], using various standardized instruments including the STAIC, SCARED, and subscales of the BASC-2. Among these, five studies [27,35,41,45,46] reported significant reductions in anxiety levels among participants in the physical activity groups. These studies generally used structured, rhythmic, or mindfulness-based movement, such as yoga or aerobic exercise combined with relaxation or cognitive-behavioral components. Improvements were more likely when the interventions were delivered consistently (4–5 times per week) and included elements targeting emotional regulation.

Self-esteem and self-concept outcomes were assessed in 13 studies [26,28,30–32,34–36,42–46], using tools such as the Self-Perception Profile for Children (SPPC), CY-PSPP, and SDQ-I. Six of these studies [28,32,36,45,46] found significant improvements in self-esteem or domain-specific self-concepts, particularly in physical self-perception, general self-worth, and social competence. Programs that promoted motor skill development, mastery experiences, and social interaction appeared to have the greatest impact on perceived competence and global self-esteem.

Mood and emotional state were examined in 7 studies [26,28,29,39,42] using tools such as the Profile of Mood States-Short Form (POMS-SF) and other emotion-related subscales.

Four of these studies [28,31,39,42] reported reductions in negative affect (e.g., tension, anger, fatigue) and/or increases in positive mood following the intervention. These findings were particularly associated with programs that combined movement with mindfulness, relaxation, or social-emotional learning components.

Stress was directly evaluated in 5 studies [33,35,36,41], and three of them [33,35,41] found significant reductions in perceived stress post-intervention. Notably, both of these studies incorporated yoga-based practices and cognitive-behavioral strategies, suggesting an additive benefit when physical activity is combined with psychological techniques targeting stress regulation.

Resilience was specifically assessed in 2 studies [37,38], both of which reported significant improvements in post-intervention scores. These findings support the hypothesis that regular physical activity may enhance adaptive emotional functioning, particularly when programs include components that foster coping skills and emotional awareness.

In general, studies with higher frequency and intensity, session durations of at least 45 min, and programs lasting more than eight weeks were more likely to produce significant psychosocial benefits. The heterogeneity of intervention formats, delivery settings, and outcome measures limits direct comparisons, but overall, the results suggest that structured physical activity interventions can positively influence emotional well-being and self-perceptions in youth.

3.5. *Meta-Analysis*

Although 21 randomized controlled trials were included in the qualitative synthesis, only 19 were retained in the quantitative meta-analysis. Two studies were excluded from the pooled analyses because they did not provide sufficient data to calculate standardized effect sizes, despite fulfilling all other eligibility criteria The heterogeneity analysis yielded a Q value of 694.334 with 38 degrees of freedom, indicating substantial between-study variability. This is supported by the I^2 statistic of 94.53%, which suggests that a large proportion of the variability in effect sizes is due to true heterogeneity rather than chance. In addition, the Tau^2 was 0.593 and the Tau value was 0.770, indicating a considerable dispersion among the estimated true effects. Given this level of heterogeneity, a random-effects model was applied to obtain more accurate pooled estimates. The overall effect size was 0.081, with a 95% confidence interval ranging from 0.024 to 0.138. Figure 3 displays a forest plot summarizing these results, highlighting a general trend toward positive effects of aerobic physical activity on mental health outcomes, although with some variability across individual studies.

3.5.1. Subgroup Analysis

A subgroup analysis was performed considering the five variables included in the study. The results showed statistically significant effects in the domains of depression ($g = 0.132$; $p = 0.004$) and anxiety ($g = 0.314$; $p < 0.001$), with small to moderate effect sizes in the expected direction. However, in the subgroups corresponding to mood and self-esteem, the effects did not reach statistical significance. Even so, the effect sizes show a consistent trend across the different domains, which provide support for the robustness of the main findings.

Depression

The results of the meta-analysis showed an overall standardized effect size (g) of 0.132, indicating a small to moderate positive effect of aerobic interventions on depression levels in children and adolescents. This effect was statistically significant ($p = 0.004$), and the 95% confidence interval [0.042, 0.222] does not include zero, supporting the presence of a real effect (Figure 4). To assess between-study heterogeneity, the Q test was applied, yielding a value of 188.146 with 10 degrees of freedom and a $p < 0.001$, indicating significant heterogeneity among the included studies. This conclusion is reinforced by the I^2 statistic of 94.69%, suggesting that approximately 95% of the observed variability among effect sizes is due to real differences between studies, rather than chance. The values of $Tau^2 = 0.482$ and $Tau = 0.694$ confirm a high degree of dispersion among the estimated true effects. Given this high level of heterogeneity, a random-effects model was used for the analysis, which is methodologically appropriate considering the variability between studies in terms of design, sample characteristics, and intervention applied. Although the Egger test showed $p = 0.054$, suggesting some possibility of publication bias, this evidence should be interpreted with caution due to the relatively small number of included studies and the high level of heterogeneity (Figure 5).

Study name	Subgroup within study	Statistics for each study						
		Std diff in means	Standard error	Variance	Lower limit	Upper limit	Z-Value	p-Value
Annesi et al. [24]	Anxiety	0,367	0,211	0,045	-0,047	0,780	1,737	0,082
Annesi et al. [24]	Depression	0,803	0,218	0,047	0,377	1,230	3,693	0,000
Annesi et al. [24]	Mood and emotional state	-0,314	0,211	0,044	-0,727	0,098	-1,492	0,136
Annesi et al. [24]	Self-esteem	0,314	0,211	0,044	-0,099	0,727	1,491	0,136
Bazzano et al. [25]	Anxiety	-0,243	0,217	0,047	-0,668	0,181	-1,124	0,261
Bazzano et al. [25]	Depression	-0,497	0,219	0,048	-0,926	-0,067	-2,268	0,023
Bohnert et al. [26]	Self-esteem	0,049	0,247	0,061	-0,435	0,532	0,197	0,844
Burgess et al. [27]	Depression	0,262	0,201	0,040	-0,132	0,656	1,304	0,192
Burgess et al. [27]	Mood and emotional state	0,235	0,201	0,040	-0,158	0,629	1,173	0,241
Cowley et al. [28]	Self-esteem	0,450	0,313	0,098	-0,163	1,063	1,438	0,150
Crews et al. [29]	Anxiety	-0,436	0,249	0,062	-0,924	0,053	-1,748	0,081
Crews et al. [29]	Depression	-0,897	0,258	0,067	-1,404	-0,391	-3,472	0,001
Crews et al. [29]	Self-esteem	-6,752	0,637	0,406	-8,001	-5,503	-10,596	0,000
Eather et al. [30]	Self-esteem	0,270	0,205	0,042	-0,132	0,673	1,315	0,188
Gaurav et al. [31]	Anxiety	-1,984	0,262	0,069	-2,498	-1,471	-7,575	0,000
Gaurav et al. [31]	Depression	-2,285	0,276	0,076	-2,825	-1,745	-8,289	0,000
Gaurav et al. [31]	Stress	-2,284	0,276	0,076	-2,824	-1,744	-8,287	0,000
Ho et al. [43]	Self-esteem	0,222	0,257	0,066	-0,282	0,725	0,863	0,388
Jelanian et al. [32]	Self-esteem	0,037	0,258	0,067	-0,469	0,543	0,145	0,885
Khalsa et al. [33]	Anxiety	-0,009	0,199	0,040	-0,398	0,381	-0,044	0,965
Khalsa et al. [33]	Depression	0,016	0,199	0,040	-0,374	0,406	0,080	0,936
Khalsa et al. [33]	Self-esteem	-0,103	0,199	0,040	-0,493	0,287	-0,518	0,605
Khalsa et al. [33]	Stress	0,075	0,199	0,040	-0,314	0,465	0,379	0,705
Lobo et al. [34]	Self-esteem	0,587	0,323	0,105	-0,046	1,221	1,817	0,069
Lobo et al. [34]	Stress	0,415	0,320	0,102	-0,213	1,042	1,296	0,195
Melnyk et al. [35]	Anxiety	0,695	0,489	0,239	-0,263	1,653	1,422	0,155
Melnyk et al. [35]	Depression	-0,127	0,476	0,227	-1,060	0,806	-0,266	0,790
Melnyk et al. [36]	Anxiety	0,912	0,084	0,007	0,747	1,077	10,827	0,000
Melnyk et al. [36]	Depression	0,760	0,083	0,007	0,597	0,922	9,151	0,000
Noggle et al. [37]	Mood and emotional state	-0,716	0,315	0,099	-1,334	-0,098	-2,271	0,023
Noggle et al. [37]	Stress	-0,284	0,309	0,095	-0,889	0,321	-0,921	0,357
Olive et al. [38]	Depression	-0,035	0,078	0,006	-0,188	0,118	-0,450	0,653
Sharma et al. [39]	Stress	-2,391	0,239	0,057	-2,860	-1,922	-10,001	0,000
Smith et al. [40]	Self-esteem	0,163	0,090	0,008	-0,014	0,340	1,809	0,070
Telles et al.[41]	Self-esteem	0,047	0,202	0,041	-0,349	0,443	0,232	0,816
Wang et al. [44]	Anxiety	-0,960	0,244	0,059	-1,438	-0,482	-3,936	0,000
Wang et al. [44]	Depression	-0,791	0,240	0,058	-1,261	-0,321	-3,299	0,001
Wassenaar et al. [42]	Depression	0,111	0,196	0,039	-0,274	0,496	0,565	0,572
Wassenaar et al. [42]	Self-esteem	-0,070	0,196	0,038	-0,454	0,315	-0,357	0,721
		0,081	0,029	0,001	0,024	0,138	2,801	0,005

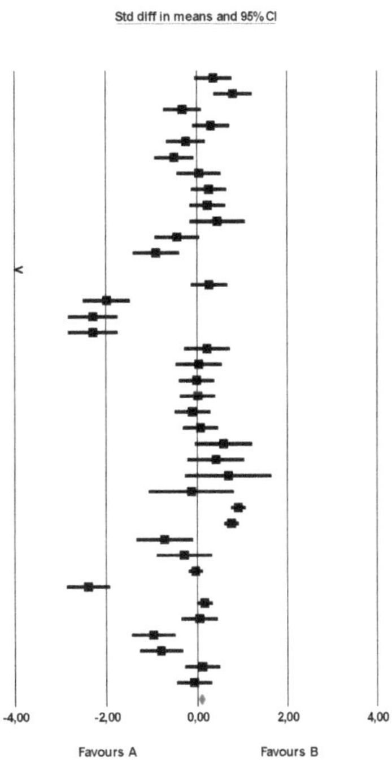

Figure 3. Forest plot of the meta-analysis on the effects of aerobics interventions on mental health in children and adolescents.

Study name	Subgroup within study	Statistics for each study						
		Std diff in means	Standard error	Variance	Lower limit	Upper limit	Z-Value	p-Value
Annesi et al. [24]	Depression	0,803	0,218	0,047	0,377	1,230	3,693	0,000
Bazzano et al. [25]	Depression	-0,497	0,219	0,048	-0,926	-0,067	-2,268	0,023
Burgess et al. [27]	Depression	0,262	0,201	0,040	-0,132	0,656	1,304	0,192
Crews et al. [29]	Depression	-0,897	0,258	0,067	-1,404	-0,391	-3,472	0,001
Gaurav et al. [31]	Depression	-2,285	0,276	0,076	-2,825	-1,745	-8,289	0,000
Khalsa et al. [33]	Depression	0,016	0,199	0,040	-0,374	0,406	0,080	0,936
Melnyk et al. [35]	Depression	-0,127	0,476	0,227	-1,060	0,806	-0,266	0,790
Melnyk et al. [36]	Depression	0,760	0,083	0,007	0,597	0,922	9,151	0,000
Olive et al. [38]	Depression	-0,035	0,078	0,006	-0,188	0,118	-0,450	0,653
Wang et al. [44]	Depression	-0,791	0,240	0,058	-1,261	-0,321	-3,299	0,001
Wassenaar et al. [42]	Depression	0,111	0,196	0,039	-0,274	0,496	0,565	0,572
		0,132	0,046	0,002	0,042	0,222	2,885	0,004

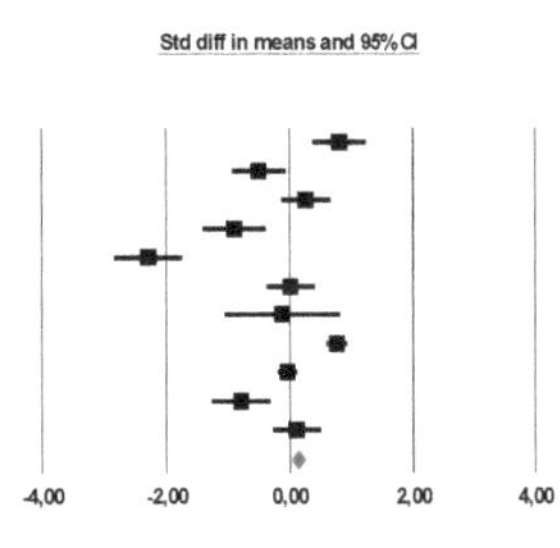

Figure 4. Forest plot of the meta-analysis on the effects of aerobics interventions on depression in children and adolescents.

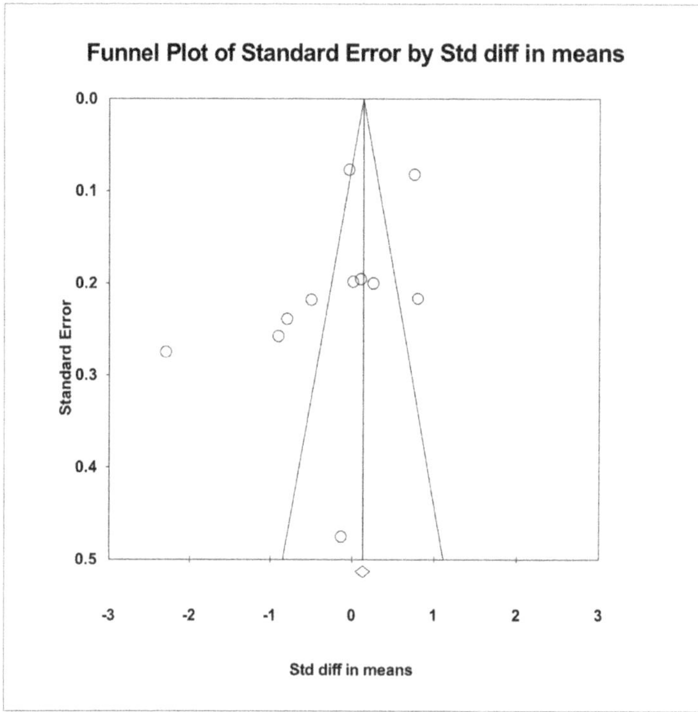

Figure 5. Funnel plot for depression.

Anxiety

The results of the meta-analysis indicated an overall standardized effect size (Hedges' g) of 0.314, representing a small to moderate positive effect of aerobic interventions on anxiety levels in children and adolescents. This effect was statistically significant ($p < 0.001$), and the 95% confidence interval [0.193, 0.435] excluded zero, supporting the presence of a real and reliable effect (Figure 6). To evaluate the degree of heterogeneity across studies, the Q test yielded a value of 173.626 with 7 degrees of freedom ($p < 0.001$), indicating significant between-study variability. This was further supported by an I^2 statistic of 95.97%, suggesting that nearly all the variability in observed effect sizes is due to actual differences between studies rather than random error. Moreover, the Tau^2 value of 0.943 and Tau value of 0.971 confirmed a high level of dispersion in the true effect estimates. Given the substantial heterogeneity, a random-effects model was employed, which the appropriate analytical approach is considering potential differences across studies in terms of sample demographics, study design, and implementation of interventions. Although Egger's test indicated potential publication bias ($p = 0.019$), this result should be interpreted with caution due to the small number of studies included in the analysis and the high degree of heterogeneity (Figure 7).

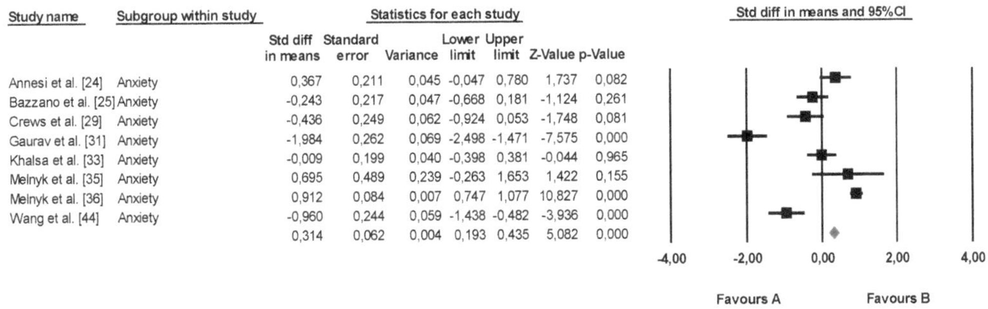

Figure 6. Forest plot of the meta-analysis on the effects of aerobics interventions on anxiety in children and adolescents.

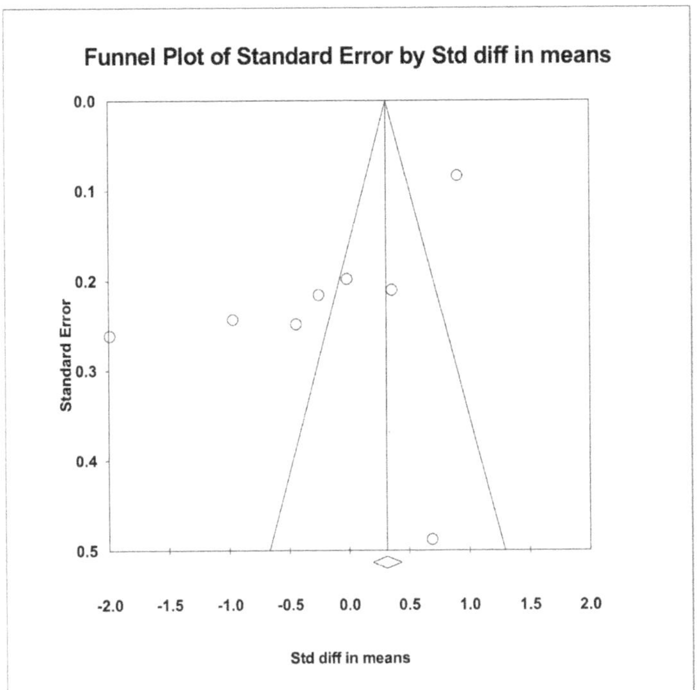

Figure 7. Funnel plot for anxiety.

Mood and Emotional State

The results of the meta-analysis indicated an overall standardized effect size (Hedges' g) of −0.147, suggesting a small negative and non-significant effect of aerobic interventions on mood and emotional state in children and adolescents. This effect did not reach statistical significance (p = 0.265), and the 95% confidence interval [−0.406, 0.112] includes zero, indicating uncertainty about the presence of a true effect (Figure 8). To assess heterogeneity among the included studies, the Q test yielded a value of 7.517 with 2 degrees of freedom (p = 0.023), indicating significant variability between studies. This finding is supported by an I^2 statistic of 73.39%, which shows that a considerable portion of the observed variance in effect sizes can be attributed to real differences rather than random error. Additionally, the Tau^2 value of 0.153 and Tau value of 0.391 reflect a moderate level of dispersion in the true effects. Given this level of heterogeneity, the use of a random-effects model was justified, as it appropriately accounts for differences in study populations, designs, and intervention characteristics. Although Egger's test ($p = 0.208$) did not indicate clear evidence of publication bias, the limited number of studies and observed heterogeneity suggest that this result should be interpreted with caution (Figure 9).

Study name	Subgroup within study	Std diff in means	Standard error	Variance	Lower limit	Upper limit	Z-Value	p-Value
Annesi et al. [24]	Mood and emotional state	-0,314	0,211	0,044	-0,727	0,098	-1,492	0,136
Burgess et al. [27]	Mood and emotional state	0,235	0,201	0,040	-0,158	0,629	1,173	0,241
Noggle et al. [37]	Mood and emotional state	-0,716	0,315	0,099	-1,334	-0,098	-2,271	0,023
		-0,147	0,132	0,017	-0,406	0,112	-1,114	0,265

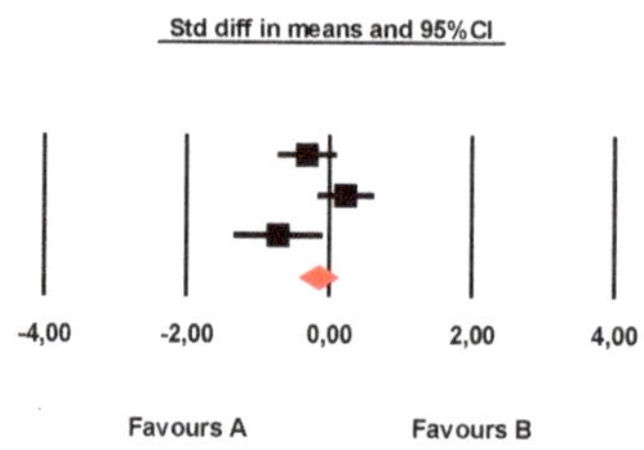

Figure 8. Forest plot of the meta-analysis on the effects of aerobics interventions on mood in children and adolescents.

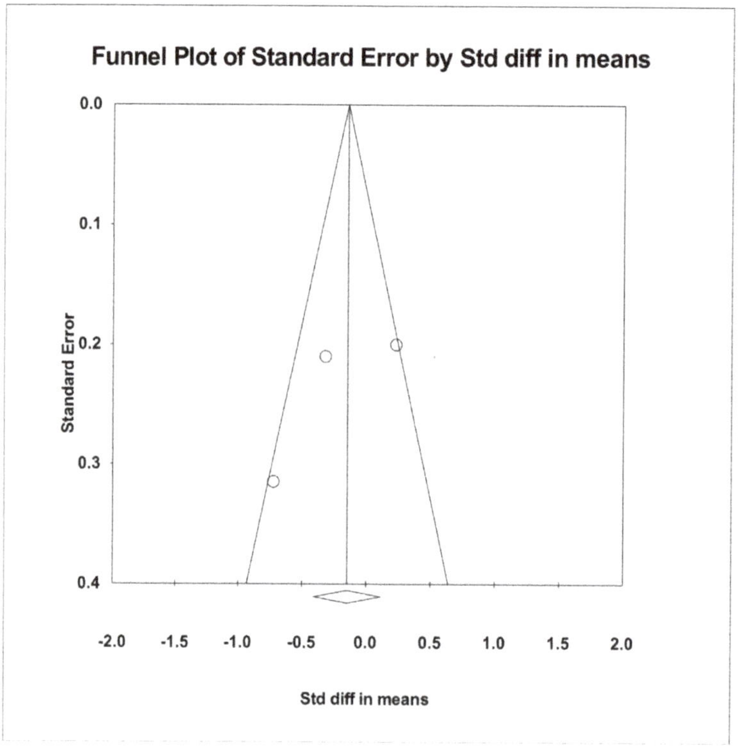

Figure 9. Funnel plot for mood.

Self-Esteem

The results of the meta-analysis indicated an overall standardized effect size (Hedges' g) of 0.091, suggesting a small and non-significant positive effect of aerobic interventions on self-esteem levels in children and adolescents. This effect did not reach statistical significance ($p = 0.107$), and the 95% confidence interval [−0.020, 0.202] includes zero, reflecting uncertainty regarding the presence of a true underlying effect (Figure 10). To assess the degree of heterogeneity among the studies, the Q test produced a value of 123.528 with 11 degrees of freedom ($p < 0.001$), indicating substantial between-study variability. This was further corroborated by an I^2 statistic of 91.10%, which suggests that a large proportion of the observed variation in effect sizes is due to real differences between studies rather than chance. The Tau^2 value of 0.444 and Tau value of 0.667 also indicate a moderate level of dispersion in the true effect estimates. Given the considerable heterogeneity, the application of a random-effects model was appropriate to account for variations in study populations, intervention protocols, and methodological approaches. Although Egger's test yielded a *p*-value of 0.077, which does not provide strong evidence for publication bias, this finding should be interpreted with caution due to the relatively small number of included studies and the high heterogeneity detected (Figure 11).

Study name	Subgroup within study	Std diff in means	Standard error	Variance	Lower limit	Upper limit	Z-Value	p-Value
Annesi et al. [24]	Self-esteem	0,314	0,211	0,044	-0,099	0,727	1,491	0,136
Bohnert et al. [26]	Self-esteem	0,049	0,247	0,061	-0,435	0,532	0,197	0,844
Cowley et al. [28]	Self-esteem	0,450	0,313	0,098	-0,163	1,063	1,438	0,150
Crews et al. [29]	Self-esteem	-6,752	0,637	0,406	-8,001	-5,503	-10,596	0,000
Eather et al. [30]	Self-esteem	0,270	0,205	0,042	-0,132	0,673	1,315	0,188
Ho et al. [43]	Self-esteem	0,222	0,257	0,066	-0,282	0,725	0,863	0,388
Jelanian et al. [32]	Self-esteem	0,037	0,258	0,067	-0,469	0,543	0,145	0,885
Khalsa et al. [33]	Self-esteem	-0,103	0,199	0,040	-0,493	0,287	-0,518	0,605
Lobo et al. [34]	Self-esteem	0,587	0,323	0,105	-0,046	1,221	1,817	0,069
Smith et al. [40]	Self-esteem	0,163	0,090	0,008	-0,014	0,340	1,809	0,070
Telles et al.[41]	Self-esteem	0,047	0,202	0,041	-0,349	0,443	0,232	0,816
Wassenaar et al. [42]	Self-esteem	-0,070	0,196	0,038	-0,454	0,315	-0,357	0,721
		0,091	0,056	0,003	-0,020	0,202	1,614	0,107

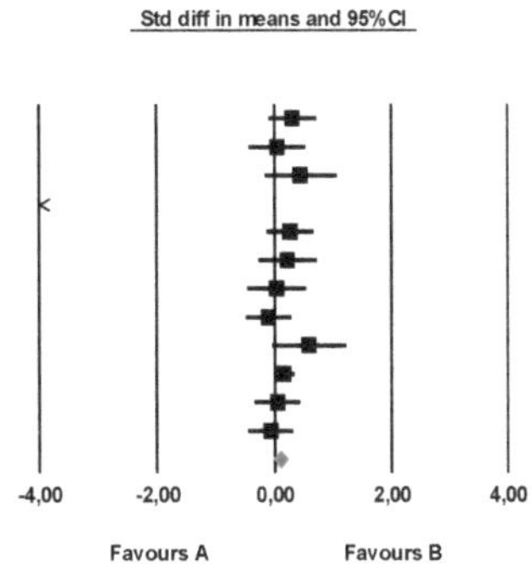

Figure 10. Forest plot of the meta-analysis on the effects of aerobics interventions on self-esteem in children and adolescents.

Figure 11. Funnel plot for self-esteem.

Stress

The results of the meta-analysis indicated an overall standardized effect size (Hedges' g) of −0.904, reflecting a large and statistically significant reduction in stress levels among children and adolescents following aerobic interventions. This effect was statistically significant ($p < 0.001$), and the 95% confidence interval [−1.129, −0.680] did not include zero, confirming the presence of a true and robust effect (Figure 12). To evaluate heterogeneity among the included studies, the Q test yielded a value of 109.020 with 11 degrees of freedom ($p < 0.001$), indicating substantial variability between studies. This conclusion is further supported by an I^2 statistic of 96.33%, showing that the vast majority of observed variance across studies is due to real differences rather than sampling error. Additionally, $Tau^2 = 1.785$ and $Tau = 1.336$ reflect a high degree of dispersion in the true effect sizes. Given the magnitude of heterogeneity, the use of a random-effects model was appropriate, accounting for differences in study design, intervention types, and sample characteristics. Although Egger's test yielded a p-value of 0.475, suggesting no strong evidence of publication bias, this result should still be interpreted cautiously due to the small number of studies and the high heterogeneity observed (Figure 13).

Study name	Subgroup within study	Std diff in means	Standard error	Variance	Lower limit	Upper limit	Z-Value	p-Value
Gaurav et al. [31]	Stress	-2,284	0,276	0,076	-2,824	-1,744	-8,287	0,000
Khalsa et al. [33]	Stress	0,075	0,199	0,040	-0,314	0,465	0,379	0,705
Lobo et al. [34]	Stress	0,415	0,320	0,102	-0,213	1,042	1,296	0,195
Noggle et al. [37]	Stress	-0,284	0,309	0,095	-0,889	0,321	-0,921	0,357
Sharma et al. [39]	Stress	-2,391	0,239	0,057	-2,860	-1,922	-10,001	0,000
		-0,904	0,115	0,013	-1,129	-0,680	-7,896	0,000

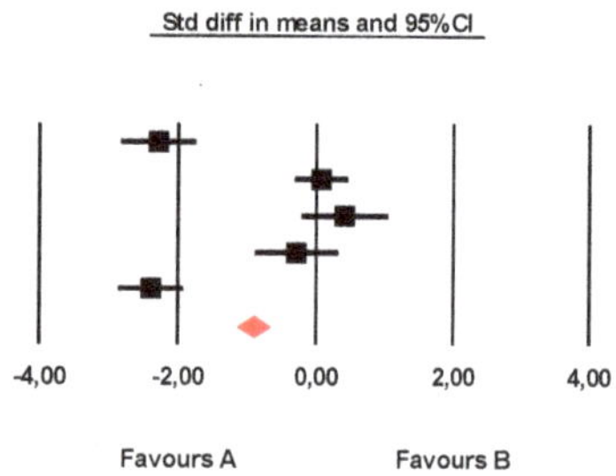

Figure 12. Forest plot of the meta-analysis on the effects of aerobics interventions on stress in children and adolescents.

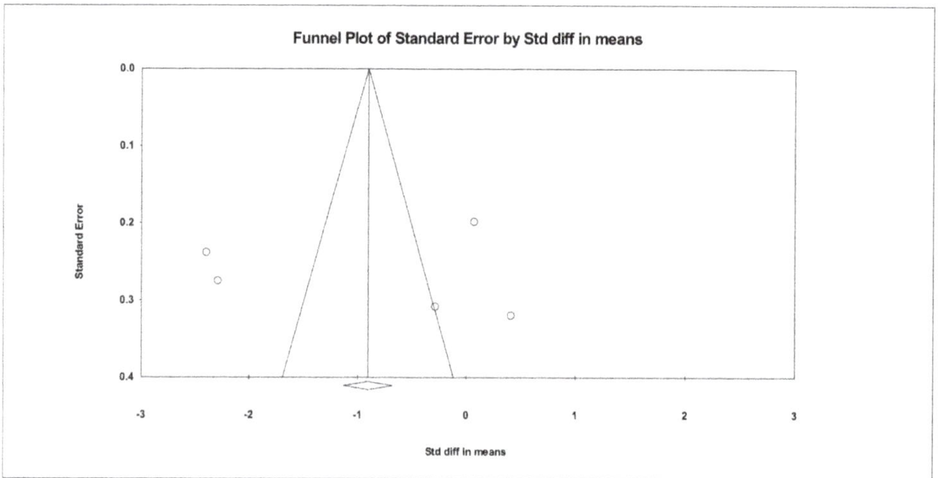

Figure 13. Funnel plot for stress.

3.5.2. Sensitivity Analyses

To assess the robustness of the findings, several sensitivity analyses were conducted. First, studies with duplicate or overlapping samples were excluded to avoid redundancy in effect estimates. Second, statistical outliers presenting extreme effect sizes were removed to evaluate their influence on the pooled results. Third, interventions primarily based on yoga or mindfulness were excluded, since these could be considered conceptually distinct from conventional aerobic training.

The exclusion of duplicate or overlapping studies did not substantially modify the overall effect sizes across outcomes. Similarly, the removal of outliers slightly reduced heterogeneity values but did not change the direction or statistical significance of the results. Regarding the exclusion of yoga/mindfulness-based interventions, the pooled effect on anxiety decreased modestly (from Hedges' g = 0.31 to g = 0.26) but remained statistically significant, supporting the robustness of the association between aerobic activity and improved mental health outcomes. Overall, these analyses confirmed that the main conclusions of the meta-analysis were not driven by any single study or subgroup of interventions.

3.5.3. Meta-Regression

To explore the high level of heterogeneity observed in the main analyses ($I^2 > 90\%$), meta-regressions were performed to explore potential sources of heterogeneity. Variables included were intervention duration (weeks), frequency of weekly sessions, mean age of participants, and intervention type. The duration of the interventions (weeks) did not show a significant association with the outcomes (coeff. = −0.0208; 95% CI: −0.0829 to 0.0413; $p = 0.511$), explaining zero proportion of the variance between studies ($R^2 = 0.0\%$). Similarly, the mean age of participants was not found to be a significant moderator (coeff. = 0.0105; 95% CI: −0.0755 to 0.0965; $p = 0.811$), with no reduction in heterogeneity ($R^2 = 0.0\%$). Regarding the type of intervention (aerobic, school-based, combined), the comparisons did not reach overall statistical significance ($Q = 1.75$; $p = 0.781$), indicating that the exercise modality did not consistently explain the variability in the effect sizes. In contrast, the number of sessions per week emerged as a significant moderator. The analysis showed that a higher number of sessions was associated with a reduced effect size (coefficient = −0.3499; 95% CI: −0.5068 to −0.1930; $p < 0.001$), explaining approximately 24% of the variance between studies ($R^2 = 0.24$). Overall, the analyses suggest that the weekly frequency of sessions may be a determining factor in the magnitude of the observed effects, while duration, participant age, and intervention type did not appear to significantly influence the effect size.

4. Discussion

The primary objective of this systematic review and meta-analysis was to evaluate the effects of aerobic physical activity-based interventions on multiple dimensions of mental health in children and adolescents, including depressive symptoms, anxiety, self-esteem, mood/emotional state, and resilience. The systematic review was designed to provide a comprehensive synthesis of the available randomized controlled trials, mapping the breadth and quality of the evidence. The meta-analysis aimed to quantitatively pool the results in order to estimate the overall magnitude of the effects, explore sources of heterogeneity, and identify potential dose–response patterns. Conducting both approaches in parallel allowed us not only to summarize the existing literature but also to generate robust effect estimates and examine moderators that cannot be addressed through qualitative synthesis alone. The main findings indicate that these interventions produce small to moderate benefits on most of the variables analyzed, with the most consistent effects being improved self-esteem and stress reduction. Although the magnitude of the effects varies across studies, the results support the integration of regular aerobic programs as a preventive and therapeutic tool for the psychological well-being of children and adolescents.

Although significant improvements were observed in domains such as anxiety and depressive symptoms, the pooled effects were small in magnitude ($g \approx 0.08$) and accompanied by substantial heterogeneity ($I^2 = 94.5\%$). These findings highlight the need for cautious interpretation. In particular, potential dose–response signals (≥45 min, ≥3 sessions/week, ≥8 weeks) require further confirmation, and methodological shortcomings such as limited allocation concealment, lack of blinding, and modest PEDro scores (mostly 5–6/10) limit the internal validity. Therefore, while aerobic interventions appear promising, the clinical impact of the effects observed in this review should be considered modest and highly context-dependent. Depression in children and adolescents represents a critical public health problem, associated with low self-esteem, academic difficulties, social isolation, and risk of self-harm [47]. The present meta-analysis observed that aerobic exercise-based interventions generated a positive effect on depressive symptoms in this population, with small to moderate effect sizes. These findings are consistent with the results obtained by Korczak et al. [48], who in their systematic review concluded that aerobic physical activity programs produce significant improvements in mood in young people with mild to moderate depressive symptoms. Similarly, in a recent meta-analysis, Bailey et al. [8] highlighted that regular physical exercise is associated with a significant reduction in depression levels, particularly when interventions are implemented in structured and frequently supervised settings. Compared to psychotherapeutic treatments such as cognitive-behavioral therapy, aerobic exercise does not require specialized clinical infrastructure, facilitating its implementation in school or community settings. Notably, research such as that by Kvam et al. [49] showed that, although CBT remains an effective option, physical activity programs offer similar benefits, with fewer barriers to access. Studies such as that by Tan et al. [50], conducted in multicultural school settings, also corroborate that structured exercise improves not only mood, but also academic engagement and social interaction, acting as an integral psychosocial resource. Therefore, these findings suggest that aerobic exercise is an accessible, low-cost, and effective tool for the prevention and complementary treatment of depressive symptoms in children and adolescents.

Anxiety is one of the most common psychological conditions in adolescence, significantly affecting academic, social, and emotional functioning [51]. This stage of development is characterized by numerous neurobiological and psychosocial changes that increase the vulnerability to experiencing anxiety symptoms, both in their transient (state) and more stable and persistent (trait) forms [52]. In the present meta-analysis, the effects of aerobic interventions on anxiety levels were small and did not reach statistical significance. These findings align with those reported by Pascoe et al. [53], who indicated that the benefits of physical activity on anxiety in adolescents are modest, especially when the intervention does not include a specific emotion regulation component. The variability in results could be due, in part, to differences in the intensity, duration, and format of the interventions. Structured group programs that promote social cohesion and the perception of emotional safety often generate better outcomes compared to individual or unsupervised programs [54]. Furthermore, the lack of robust effects may be related to the type of anxiety assessed, given that measurement tools do not always accurately discriminate between trait and state anxiety [55]. Compared with treatments specifically designed to reduce anxiety, such as cognitive-behavioral therapy or mindfulness programs, aerobic physical activity may offer a lesser effect, although its inclusion within multicomponent interventions has been shown to be effective in enhancing overall mental health outcomes [56]. Therefore, although aerobic PA is not a specific treatment for anxiety, it does represent a preventive and complementary alternative that can benefit school and clinical populations when implemented appropriately.

Self-esteem is one of the central pillars of healthy psychological development during adolescence, as it influences emotional regulation, social behavior, and the ability to cope with stressful situations. Positive self-esteem has been associated with a lower risk of developing disorders such as depression, anxiety, and eating disorders [57]. In the present meta-analysis, aerobic interventions showed a moderate positive effect on self-esteem in children and adolescents, with notable consistency across the included studies. This finding is consistent with the results of Ekeland et al. [58], who in their Cochrane review emphasized that physical activity programs have a beneficial impact on general self-concept and self-esteem, especially when implemented in educational settings. Furthermore, recent studies such as that of Gentile et al. [59] have identified that improvements in the perception of physical competence, one of the key dimensions of self-concept, mediate the relationship between physical activity and self-esteem. This improvement is enhanced when activities are perceived as enjoyable, challenging, and socially integrative. Interventions based on dance, functional training, or adapted sports have also demonstrated positive effects on body perception and self-image, as evidenced by the work of Fernández-Bustos et al. [60] and Saavedra et al. [61]. These findings suggest that, beyond the specific modality, regular participation in structured and supervised aerobic physical activity represents an effective strategy for strengthening self-esteem during adolescence, especially in vulnerable populations.

Resilience, defined as the ability to effectively cope with adverse situations and recover from them with positive adaptation, is considered an essential psychosocial competency in adolescent development. This skill acts as a protective factor against the onset of emotional disorders and facilitates better school and social integration [62]. Although it has received less attention than other psychological variables, the studies included in this review that measured resilience as an outcome showed positive effects derived from participation in aerobic interventions. These results are consistent with those proposed by Paulus et al. [3], who point out that physical activity promotes coping mechanisms, emotional strengthening, and perseverance, all of which are fundamental components in building resilience. Furthermore, research such as that of McMahon et al. [63] has proposed that physical exercise promotes resilience through neurobiological processes such as neurogenesis and regulation of the HPA (hypothalamic–pituitary–adrenal) axis, as well as by improving emotional regulation and perceived self-efficacy. Furthermore, regular structured physical activity, especially in group and supervised settings, has been shown to foster autonomy, decision-making, and problem-solving, all of which are key elements in resilient response [64]. While these findings are promising, it is necessary to increase the number and methodological quality of studies that explicitly assess this dimension as a primary variable, as well as to employ validated and standardized psychometric instruments in adolescent populations. This will consolidate the evidence on the role of aerobic physical exercise as an effective strategy for fostering resilience in young people.

Meta-regression analyses allowed for the exploration of dose–response signals. Weekly session frequency emerged as a significant moderator, indicating that interventions with $\geq$3 sessions per week produced more consistent effects. In contrast, total duration in weeks, participant age, and exercise type did not show significant associations. These preliminary findings suggest that program regularity, rather than duration alone, may be a key factor.

Despite the methodological soundness applied in this systematic review with meta-analysis, there are several limitations that should be considered when interpreting the results. First, the high heterogeneity observed in several analyses ($I^2 > 90\%$) suggests significant variability among the included studies, possibly attributable to differences in methodological designs, intervention characteristics, program durations, population types, and assessment tools used. Although random-effects models and subgroup analyses were applied, this heterogeneity could affect the generalizability of the results. Second, the limited number of studies per specific variable, especially in the case of resilience and emotional well-being, reduces the statistical power of the analyses and limits the ability to draw definitive conclusions about these dimensions. Furthermore, some studies lacked long-term follow-up, making it impossible to determine whether the observed positive effects are sustainable over time. Another relevant limitation is the lack of homogeneity in the measurement of psychological variables. Multiple instruments were used to assess similar constructs, such as self-esteem or mood, complicating direct comparison of results across studies. Furthermore, not all included trials clearly reported exercise intensity, program adherence, or level of supervision, which affects intervention fidelity and replicability. Unlike previous reviews, the present study focused exclusively on randomized controlled trials in populations aged 5 to 18 years and included recent literature published since 2020. These criteria allow for a more up-to-date, rigorous, and targeted synthesis for mental health promotion in school settings.

5. Conclusions

This meta-analysis provides relevant evidence on the impact of aerobic physical activity interventions on the mental health of children and adolescents. The results suggest that these interventions have a moderate positive effect on self-esteem and depressive symptoms, and more modest effects on mood, anxiety, and resilience. These conclusions support the value of physical activity as a non-pharmacological tool for promoting psychological well-being in children and adolescents. Given its accessibility, low cost, and potential for implementation in school and community settings, aerobic physical activity represents a promising public health strategy. However, more studies with rigorous designs, larger samples, and longitudinal follow-up are needed, especially focusing on less-explored variables such as resilience and emotional well-being. Therefore, promoting regular aerobic exercise could constitute an effective intervention for strengthening mental health during key developmental stages such as childhood and adolescence. This strategy should be considered an integral part of educational and preventive programs aimed at promoting psychological well-being at early ages.

Supplementary Materials: The following supporting information can be downloaded at: https://www.mdpi.com/article/10.3390/app15179572/s1, Table S1: Characteristics of the included studies.

Author Contributions: Conceptualization, M.d.C.C.-F.; methodology, F.H.-C.; formal analysis, A.A.-A.; writing—original draft preparation, F.H.-C.; writing—review and editing, M.d.C.C.-F.; supervision, A.A.-A. All authors have read and agreed to the published version of the manuscript.

Funding: This research received no external funding.

Institutional Review Board Statement: Not applicable.

Informed Consent Statement: Not applicable.

Data Availability Statement: Not applicable.

Conflicts of Interest: The authors declare no conflicts of interest.

Abbreviations

The following abbreviations are used in this manuscript:

PA	physical activity
BDNF	Brain-derived neurotrophic factor
RCTs	Randomized controlled trials
PHQA	Patient Health Questionnaire–A
SCARED	The Screen for Child Anxiety and Related Emotional Disorders
SDQ	The Self-Description Questionnaire
SSRS	Social Skills Rating System
CY-PSPP	Children and Youth Physical Self-Perception Profile
PSDQ	Physical Self-Description Questionnaire
SAS–A	the Social Anxiety Scale for Adolescents
SPPA	Self-Perception Profile for Adolescents
BASC-2	Behavior Assessment System for Children–Second Edition

RS	Resilience Scale
PSS	The Perceived Stress Scale
BYI-II	The Beck Depression Inventory and The Beck anxiety Inventory
STAIC	The State–Trait Anxiety Inventory for Children
SDQ-I	Self-Description Questionnaire I
POMS-SF	Profile of Mood States–Short Form
CDI-2	Children's Depression Inventory-2
ASS	the Academic Stress Scale
CDI	The Children's Depression Inventory
ISEQ	the Indian adaptation of Battle's self-esteem questionnaire
SPPC	Self-Perception Profile for Children
SF-12v2	Short Form-12 Health Survey version 2
RSES	Rosenberg Self-Esteem Scale
SEAS	Student Examination Anxiety Scale
ADSS	Anxiety, Depression and Stress Scale

References

1. UNICEF. *The State of the World's Children 2021: On My Mind—Promoting, Protecting and Caring for Children's Mental Health*; UNICEF: New York, NY, USA, 2021. Available online: https://data.unicef.org/topic/child-health/mental-health (accessed on 13 July 2025).
2. World Health Organization (WHO). *Adolescent Mental Health*; WHO: Geneva, Switzerland, 2021. Available online: https://www.who.int/news-room/fact-sheets/detail/adolescent-mental-health (accessed on 13 July 2025).
3. Paulus, F.W.; Ohmann, S.; Möhler, E.; Plener, P.; Popow, C. Emotional Dysregulation in Children and Adolescents With Psychiatric Disorders: A Narrative Review. *Front. Psychiatry* **2021**, *12*, 628252. [CrossRef]
4. UNICEF España. *La Salud Mental de los Adolescentes en España: Situación Actual y Retos Futuros*; UNICEF España: Madrid, Spain, 2023. Available online: https://www.unicef.es/educa/salud-mental-infanto-juvenil (accessed on 13 July 2025).
5. Soriano, V.; Ramos, J.M.; López-Ibor, M.I.; Chiclana-Actis, C.; Faraco, M.; González-Cabrera, J.; González-Fraile, E.; Mestre-Bach, G.; Pinargote, H.; Corpas, M.; et al. Hospital admissions in adolescents with mental disorders in Spain over the last two decades: A mental health crisis? *Eur. Child Adolesc. Psychiatry* **2023**, *34*, 1125–1134.
6. Alhusseini, N.; Lin, T.K.; Werner, K.; Lin, G.; Altwaijri, Y.; Baattaiah, B.A.; Bruckner, T.; AlAhmed, R.; Alkabbani, A.; Alsukait, R.F.; et al. Cost-effectiveness of physical activity-oriented interventions for improving mental health: A systematic review. *BMC Public Health* **2025**, *25*, 1766. [CrossRef] [PubMed]
7. Caspersen, C.J.; Powell, K.E.; Christenson, G.M. Physical activity, exercise, and physical fitness: Definitions and distinctions for health-related research. *Public Health Rep.* **1985**, *100*, 126–131. [PubMed]
8. Bailey, R.; Vasícková, J.; Vlček, P.; Demidoff, A.R.; Pühse, U.; Heck, S.; Scheuer, C. *An International Review of the Contributions of School-Based Physical Activity, Physical Education, and School Sport to the Promotion of Health-Enhancing Physical Activity*; Health and Physically Active Schools in Europe: Bratislava, Slovakia, 2022.
9. Janssen, I.; LeBlanc, A.G. Systematic review of the health benefits of physical activity and fitness in school-aged children and youth. *Int. J. Behav. Nutr. Phys. Act.* **2010**, *7*, 40. [CrossRef] [PubMed]
10. Biddle, S.J.H.; Ciaccioni, S.; Thomas, G.; Vergeer, I. Physical activity and mental health in children and adolescents: An updated review of reviews and an analysis of causality. *Psychol. Sport Exerc.* **2019**, *42*, 146–155. [CrossRef]
11. Lubans, D.R.; Richards, J.; Hillman, C.H.; Faulkner, G.; Beauchamp, M.R.; Nilsson, M.; Kelly, P.; Smith, J.J.; Raine, L.B.; Biddle, S.J.H. Physical activity for cognitive and mental health in youth: A systematic review of mechanisms. *Pediatrics* **2016**, *138*, e20161642. [CrossRef] [PubMed]
12. Rodríguez, L.M.; González-Cutre, D.; Cervelló, E.; Sicilia, Á. Effects of school-based physical activity programs on psychological well-being: A meta-analysis. *BMC Public Health* **2024**, *24*, 526. [CrossRef]
13. Cadenas-Sánchez, C.; Mora-González, J.; Migueles, J.H.; Esteban-Cornejo, I.; Ortega, F.B. El efecto de la actividad física sobre la salud mental de los adolescentes: Una revisión sistemática y metaanálisis. *Psicol. Educ.* **2023**, *29*, 45–55.
14. Neil-Sztramko, S.E.; Caldwell, H.; Dobbins, M. School-based physical activity programs for promoting physical activity and fitness in children and adolescents aged 6 to 18. *Cochrane Database Syst. Rev.* **2021**, *9*, CD007651. [CrossRef]
15. Carney, R.; Firth, J. Exercise interventions in child and adolescent mental health care: An overview of the evidence and recommendations for implementation. *JCPP Adv.* **2021**, *1*, e12031. [CrossRef]
16. Dishman, R.K.; Heath, G.W.; Lee, I.M. *Physical Activity Epidemiology*, 3rd ed.; Human Kinetics: Champaign, IL, USA, 2021.

17. Heyman, E.; Gamelin, F.X.; Goekint, M.; Piscitelli , F.; Roelands , B.; Leclair, E.; Di Marzo, V.; Meeusen , R. Intense exercise increases circulating endocannabinoid and BDNF levels in humans--possible implications for reward and depression. *Psychoneuroendocrinology* 2012, 37, 844–851. [CrossRef]
18. Stillman, C.M.; Cohen, J.; Lehman, M.E.; Erickson, K.I. Mediators of physical activity on neurocognitive function: A review at multiple levels of analysis. *Prog. Neurobiol.* **2020**, *193*, 101938. [CrossRef]
19. Kandola, A.; Ashdown-Franks, G.; Hendrikse, J.; Sabiston, C.M.; Stubbs, B. Physical activity and depression: Towards understanding the antidepressant mechanisms of physical activity. *Neurosci. Biobehav. Rev.* **2019**, *107*, 525–539. [CrossRef] [PubMed]
20. Huang, W.; Wong, T.L. Exercise Prescriptions for Young People's Emotional Wellbeing: A Systematic Review of Physical Activity Intensity, Duration, and Modality. *Front. Psychol.* **2025**, *16*, 1552531. [CrossRef] [PubMed]
21. Page, M.J.; McKenzie, J.E.; Bossuyt, P.M.; Boutron, I.; Hoffmann, T.C.; Mulrow, C.D.; Shamseer, L.; Tetzlaff, J.M.; Akl, E.A.; Brennan, S.E.; et al. The PRISMA 2020 statement: An updated guideline for reporting systematic reviews. *BMJ* **2021**, *372*, n71. [CrossRef]
22. Higgins, J.P.T.; Thomas, J.; Chandler, J.; Cumpston, M.; Li, T.; Page, M.J.; Welch, V.A. *Cochrane Handbook for Systematic Reviews of Interventions*; Version 6.4; Cochrane: London, UK, 2023. Available online: https://www.cochrane.org/authors/handbooks-and-manuals/handbook (accessed on 13 July 2025).
23. Cashin, A.G.; McAuley, J.H. Clinimetrics: Physiotherapy Evidence Database (PEDro) Scale. *J. Physiother.* **2020**, *66*, 59. [CrossRef] [PubMed]
24. Carter, T.; Morres, I.; Meade, O.; Callaghan, P. The Effect of Exercise on Depressive Symptoms in Adolescents: A Systematic Review and Meta-Analysis. *J. Am. Acad. Child Adolesc. Psychiatry* **2016**, *55*, 580–590. [CrossRef]
25. Rodríguez-Ayllón, M.; Cadenas-Sánchez, C.; Estévez-López, F.; Muñoz, N.E.; Mora-González, J.; Migueles, J.H.; Molina-García, P.; Henriksson, H.; Mena-Molina, A.; Martínez-Vizcaíno, V.; et al. Role of Physical Activity and Sedentary Behavior in the Mental Health of Preschoolers, Children and Adolescents: A Systematic Review and Meta-Analysis. *Sports Med.* **2019**, *49*, 1383–1410. [CrossRef] [PubMed]
26. Annesi, J.J. Improvements in self-concept associated with reductions in negative mood in preadolescents enrolled in an after-school physical activity program. *Psychol. Rep.* **2005**, *97*, 400–404. [CrossRef]
27. Bazzano, A.N.; Sun, Y.; Chavez-Gray, V.; Akintimehin, T.; Gustat, J.; Barrera, D.; Roi, C. Effect of yoga and mindfulness intervention on symptoms of anxiety and depression in young adolescents attending middle school: A pragmatic community-based cluster randomized controlled trial in a racially diverse urban setting. *Int. J. Environ. Res. Public Health* **2022**, *19*, 12076. [CrossRef] [PubMed]
28. Bohnert, A.M.; Ward, A.K. Making a difference: Evaluating the Girls in the Game (GIG) after-school program. *J. Early Adolesc.* **2013**, *33*, 104–130. [CrossRef]
29. Burgess, G.; Grogan, S.; Burwitz, L. Effects of a 6-week aerobic dance intervention on body image and physical self-perceptions in adolescent girls. *Body Image* **2006**, *3*, 57–66. [CrossRef]
30. Cowley, E.S.; Watson, P.M.; Foweather, L.; Belton, S.; Mansfield, C.; Whitcomb-Khan, G.; Cacciatore, I.; Thompson, A.; Thijssen, D.; Wagenmakers, A.J. Formative evaluation of a home-based physical activity intervention for adolescent girls—The HERizon Project: A randomised controlled trial. *Children* **2021**, *8*, 76. [CrossRef] [PubMed]
31. Crews, D.J.; Lochbaum, M.R.; Landers, D.M. Aerobic physical activity effects on psychological well-being in low-income Hispanic children. *Percept. Mot. Ski.* **2004**, *98*, 319–324. [CrossRef]
32. Eather, N.; Morgan, P.J.; Lubans, D.R. Effects of exercise on mental health outcomes in adolescents: Findings from the CrossFit™ Teens randomized controlled trial. *Psychol. Sport Exerc.* **2016**, *26*, 14–23. [CrossRef]
33. Gaurav, P.; Bera, T.K.; Uddhav, S. Yoga for controlling examination anxiety, depression and academic stress among students appearing for Indian board examination. *Int. J. Recent Sci. Res.* **2013**, *4*, 1216–1219.
34. Jelalian, E.; Sato, A.; Hart, C.N. The effect of group-based weight-control intervention on adolescent psychosocial outcomes: Perceived peer rejection, social anxiety, and self-concept. *Child. Health Care* **2011**, *40*, 197–211. [CrossRef] [PubMed]
35. Khalsa, S.B.; Hickey-Schultz, L.; Cohen, D.; Steiner, N.; Cope, S. Evaluation of the mental health benefits of yoga in a secondary school: A preliminary randomized controlled trial. *J. Behav. Health Serv. Res.* **2012**, *39*, 80–90. [CrossRef] [PubMed]
36. Lobo, Y.B.; Winsler, A. The effects of a creative dance and movement program on the social competence of Head Start preschoolers. *Soc. Dev.* **2006**, *15*, 501–519. [CrossRef]
37. Melnyk, B.M.; Jacobson, D.; Kelly, S.; O'HAver, J.; Small, L.; Mays, M.Z. Improving the mental health, healthy lifestyle choices, and physical health of Hispanic adolescents: A randomized controlled pilot study. *J. Sch. Health* **2009**, *79*, 575–584. [CrossRef]
38. Melnyk, B.M.; Jacobson, D.; Kelly, S.; Belyea, M.; Shaibi, G.; Small, L.; O'hAver, J.; Marsiglia, F.F. Promoting healthy lifestyles in high school adolescents: A randomized controlled trial. *Am. J. Prev. Med.* **2013**, *45*, 407–415. [CrossRef]
39. Noggle, J.J.; Steiner, N.J.; Minami, T.; Khalsa, S.B.S. Benefits of yoga for psychosocial well-being in a US high school curriculum: A preliminary randomized controlled trial. *J. Dev. Behav. Pediatr.* **2012**, *33*, 193–201. [CrossRef]

40. Olive, L.S.; Byrne, D.; Cunningham, R.B.; Telford, R.M.; Telford, R.D. Can physical education improve the mental health of children? The LOOK study cluster-randomized controlled trial. *J. Educ. Psychol.* **2019**, *111*, 1331. [CrossRef]
41. Sharma, V.; Shrivastava, S.; MalhSharma, U.K.; Shrivastava, S.; Malhotra, S.; Singh, R.; Singh, T.B. Yoga and cognitive behavior techniques for academic stress and mental wellbeing among school students. *Delhi Psychiatry J.* **2010**, *13*, 12–19.
42. Smith, J.J.; Beauchamp, M.R.; Faulkner, G.; Morgan, P.J.; Kennedy, S.G.; Lubans, D.R. Intervention effects and mediators of well-being in a school-based physical activity program for adolescents: The 'Resistance Training for Teens' cluster RCT. *Ment. Health Phys. Act.* **2018**, *15*, 88–94. [CrossRef]
43. Telles, S.; Singh, N.; Bhardwaj, A.; Kumar, A.; Balkrishna, A. Effect of yoga or physical exercise on physical, cognitive and emotional measures in children: A randomized controlled trial. *Child Adolesc. Psychiatry Ment. Health* **2013**, *7*, 37. [CrossRef] [PubMed]
44. Wassenaar, T.M.; Wheatley, C.M.; Beale, N.; Nichols, T.; Salvan, P.; Meaney, A.; Atherton, K.; Diaz-Ordaz, K.; Dawes, H.; Johansen-Berg, H. The effect of a one-year vigorous physical activity intervention on fitness, cognitive performance and mental health in young adolescents: The Fit to Study cluster randomised controlled trial. *Int. J. Behav. Nutr. Phys. Act.* **2021**, *18*, 47. [CrossRef] [PubMed]
45. Ho, F.K.; Louie, L.H.T.; Wong, W.H.S.; Chan, K.L.; Tiwari, A.; Chow, C.B.; Ip, P. Effect of a mind–body exercise on children with autism spectrum disorders: A randomized controlled trial. *Pediatr. Res.* **2017**, *81*, 581–589. [CrossRef]
46. Wang, Y.; Chan, C.H.Y.; Li, W.H.C.; Lau, Y.; Ho, K.Y.; Lee, I.K.M.; Chung, J.O.K. Effects of a school-based physical activity intervention on mental health in Chinese adolescents: A randomized controlled trial. *Int. J. Environ. Res. Public Health* **2025**, *22*, 263. [CrossRef]
47. Almeida, I.L.L.; Rego, J.F.; Teixeira, A.C.G.; Moreira, M.R. Social isolation and its impact on child and adolescent development: A systematic review. *Rev. Paul. Pediatr.* **2021**, *40*, e2020385. [CrossRef] [PubMed]
48. Korczak, D.J.; Madigan, S.; Colasanto, M. Children's Physical Activity and Depression: A Meta-analysis. *Pediatrics* **2017**, *139*, e20162266. [CrossRef] [PubMed]
49. Kvam, S.H.; Kleppe, C.L.; Nordhus, I.H.; Hovland, A. Exercise as a treatment for depression: A meta-analysis. *J. Affect. Disord.* **2016**, *202*, 67–86. [CrossRef]
50. Vaquero-Solís, M.; Gallego, D.I.; Tapia-Serrano, M.Á.; Pulido, J.J.; Sánchez-Miguel, P.A. School-Based Physical Activity Interventions in Children and Adolescents: A Systematic Review. *Int. J. Environ. Res. Public Health* **2020**, *17*, 999. [CrossRef]
51. Beesdo, K.; Knappe, S.; Pine, D.S. Anxiety and anxiety disorders in children and adolescents: Developmental issues and implications for DSM-V. *Psychiatr. Clin. N. Am.* **2009**, *32*, 483–524. [CrossRef]
52. Caouette, J.D.; Guyer, A.E. Gaining insight into adolescent vulnerability for social anxiety from developmental cognitive neuroscience. *Dev. Cogn. Neurosci.* **2014**, *8*, 65–76. [CrossRef]
53. Pascoe, M.C.; Bailey, A.P.; Craike, M.; Carter, T.; Patten, R.; Stepto, N.K.; Parker, A.G. Physical activity and exercise in youth mental health promotion: A scoping review. *BMJ Open Sport Exerc. Med.* **2020**, *6*, e000677. [CrossRef]
54. Hoare, E.; Milton, K.; Foster, C.; Allender, S. The associations between sedentary behaviour and mental health among adolescents: A systematic review. *Int. J. Behav. Nutr. Phys. Act.* **2016**, *13*, 108. [CrossRef]
55. Hegberg, N.J.; Tone, E.B. Physical activity and stress resilience: Considering those at risk for developing mental health problems. *Ment. Health Phys. Act.* **2015**, *8*, 1–7. [CrossRef]
56. Nemoto, M.; Nemoto, K.; Sasai, H.; Ota, M.; Haneda, M.; Sekine, A.; Arai, T. Emotional Response to Various Exercise Types in Patients With Mental Disorders. *Cureus* **2024**, *16*, e75371. [CrossRef] [PubMed]
57. Henriksen, I.O.; Ranøyen, I.; Indredavik, M.S.; Stenseng, F. The role of self-esteem in the development of psychiatric problems: A three-year prospective study in a clinical sample of adolescents. *Child Adolesc. Psychiatry Ment. Health* **2017**, *11*, 68. [CrossRef] [PubMed]
58. Ekeland, E.; Heian, F.; Hagen, K.B.; Abbott, J.; Nordheim, L. Exercise to improve self-esteem in children and young people. *Cochrane Database Syst. Rev.* **2019**, *1*, CD003683. [CrossRef]
59. Gentile, A.; Boca, S.; Giammusso, I.; Rosa, S.; Agostini, T. Self-concept and perceived physical competence in adolescents: Mediating role in the relationship between physical activity and psychological well-being. *Int. J. Environ. Res. Public Health* **2022**, *19*, 2395. [CrossRef]
60. Fernández-Bustos, J.G.; González-Martí, I.; Contreras, O.R.; Cuevas, R. Physical self-concept and physical activity levels in adolescents: A systematic review. *Children* **2021**, *8*, 672. [CrossRef]
61. Saavedra, J.M.; Domínguez, A.M.; García-Hermoso, A.; Escalante, Y. Effects of creative dance on self-concept in adolescents: A meta-analysis. *Int. J. Environ. Res. Public Health* **2020**, *17*, 4755. [CrossRef]
62. Sisto, A.; Vicinanza, F.; Campanozzi, L.L.; Ricci, G.; Tartaglini, D.; Tambone, V. Towards a Transversal Definition of Psychological Resilience: A Literature Review. *Medicina* **2019**, *55*, 745. [CrossRef]

63. McMahon, E.M.; Corcoran, P.; O'Regan, G.; Keeley, H.; Cannon, M.; Carli, V.; Wasserman, C.; Hadlaczky, G.; Sarchiapone, M.; Apter, A.; et al. Physical activity in European adolescents and associations with anxiety, depression and well-being. *Eur. Child Adolesc. Psychiatry* **2017**, *26*, 111–122. [CrossRef]
64. Benttä, G.; Grape, F.; Holmström, T.; Hassmén, N. Physical activity as a pathway to resilience: A scoping review. *Int. J. Environ. Res. Public Health* **2020**, *17*, 8584. [CrossRef]

MDPI AG
Grosspeteranlage 5
4052 Basel
Switzerland
Tel.: +41 61 683 77 34

Applied Sciences Editorial Office
E-mail: applsci@mdpi.com
www.mdpi.com/journal/applsci

www.ingramcontent.com/pod-product-compliance
Lightning Source LLC
Chambersburg PA
CBHW040750250926
51850CB00007BB/1462
* 9 7 8 3 7 2 5 8 7 9 6 5 6 *